MATHEMATICS RESEARCH DEVELOPMENTS

PARTIAL DIFFERENTIAL EQUATIONS: CLASSIFICATION, PROPERTIES AND APPLICATIONS

MATHEMATICS RESEARCH DEVELOPMENTS

Additional books in this series can be found on Nova's website
under the Series tab.

Additional e-books in this series can be found on Nova's website
under the e-book tab.

MATHEMATICS RESEARCH DEVELOPMENTS

PARTIAL DIFFERENTIAL EQUATIONS: CLASSIFICATION, PROPERTIES AND APPLICATIONS

DEBORAH E. RICHARDS
EDITOR

nova publishers
New York

Library of Congress Cataloging-in-Publication Data

Partial differential equations : classification, properties and applications / Deborah E. Richards.
pages cm. -- (Mathematics research developments)
Includes index.
ISBN 978-1-63482-643-3 (hardcover)
1. Differential equations, Partial. I. Richards, Deborah E.
QA377.P2976 2015
515'.353--dc23

2015013339

Published by Nova Science Publishers, Inc. †New York

CONTENTS

PREFACE

This book includes research on the Lax-Milgram theorem, which can be used to prove existence and uniqueness of weak solutions to partial differential equations and several examples of its application to relevant boundary value problems are presented. The authors also investigate nonlinear control problems for couple partial differential equations arising from climate and circulation dynamics in the equatorial zone; the integration of partial differential equations (PDE) with the help of non-commutative analysis over octonions and Cayley-Dickson algebras; and the existence and properties of solutions, applications in sequential optimal control with pointwise in time state constraints.

The Lax-Milgram theorem can be used to prove existence and uniqueness of weak solutions to partial differential equations, in particular, elliptic boundary value problems. After introducing some pertinent information required for the development of the theorem, it is stated and two proofs are given in detail. Several examples of its application to relevant boundary value problems are presented in Chapter 1. Some consequences of these results and their relevance to computational aspects of the subject are summarized at the end.

Chapter 2 investigates nonlinear control problems for coupled partial differential equations arising from climate and circulation dynamics in equatorial zone. The phenomenon and processes that we want to modelize, by coupled and time-dependent PDEs, occur in tropical Atlantic and Pacific regions: the circulation is there characterized by steady zonal currents varying monthly or seasonally and by long waves propagating westward along the equator known as "equatorial waves", driven by the variability of atmospheric currents and superimposed to the steady mean currents. The equatorial waves can be connected with strong vertical velocities and then induce "upwellings" or "downwellings". This study corresponds to the modeling of tropical instability waves (taking into account the complex interaction between the atmosphere and the ocean circulation) which are illustrated by the "*El Niño*" phenomena and tropical cyclones. After the introduction in which we give background, motivation and fundamental principles governing circulation dynamics, we present a mathematical model of circulation mechanisms, operating in the equatorial zone, in a global climate system, and give some assumptions and notations. The equations are non-linear, time-dependent and coupled, and are of Navier-Stokes type for the velocity and pressure, and of transport-diffusion type for the temperature, with non linear boundary and interface conditions. Afterwards, we investigate the well-posedness of governing nonlinear systems. For provide the uniqueness, we prove some regularity results. Next, sensitivity and control problems are presented and analyzed. The problem is controlled by the variability of the heat interface-stress between the ocean and atmosphere. The observation is the vorticity of the ocean and (or) of the atmosphere. Finally, we conclude this work with some comments and possible future developments.

Chapter 3 is devoted to the integration of partial differential equations (PDE) with the help of non-commutative analysis over octonions and Cayley-Dickson algebras. For this purpose a non-commutative line integration of functions and generalized functions is studied. First, second and higher order PDE with continuous, piecewise continuous and generalized variable coefficients over Cayley-Dickson algebras are investigated. Then solutions of some types of non-linear PDE over Cayley-Dickson algebras are studied.

This technique is based on decompositions of partial differential operators (PDO) into products of lower order PDO over octonions and Cayley-Dickson algebras.

Moreover, PDE of higher order with real and complex coefficients and with variable coefficients with or without boundary conditions are considered. Applications to Sobolev's mixed type PDE are investigated as well. Local, global solutions and problems with boundary conditions are described.

Chapter 4 is devoted to the study of a sequential optimization problem for a semilinear hyperbolic divergent equation with pointwise state inequality constraint and with mixed boundary conditions. The chapter consists of two main parts.

The first part is devoted to the study of the existence and uniqueness of solutions to semilinear hyperbolic divergent equations. Also in this part we study a linear hyperbolic divergent equation with a Radon measure in the right part. We study the following questions: existence, uniqueness and stability of solutions to such equations with mixed boundary condition; special integral representations of solutions to such equations; and the stability of solutions of linear hyperbolic divergent equations.

In the second part of the paper, using the results of its first part, we have studied in detail the optimal control problem with pointwise in time state constraints. The state constraint contains a functional parameter that belongs to the class of continuous functions and occurs as an additive term. We compute the first variations of functionals on the basis of a so-called two-parameter needle variation of controls. We consider necessary conditions for minimizing sequences in an optimal control problem with a pointwise in time state constraint of inequality type and with dynamics described by a semilinear hyperbolic equation in divergence form with mixed boundary condition. For the parametric optimization problem, we also consider regularity and normality conditions stipulated by the differential properties of its value function.

In: Partial Differential Equations
Editor: Deborah E. Richards, pp. 1-13
ISBN: 978-1-63482-643-3

Chapter 1

THE LAX-MILGRAM THEOREM AND SOME APPLICATIONS TO PARTIAL DIFFERENTIAL EQUATIONS

Paul Bracken*
Department of Mathematics, University of Texas, Edinburg, TX, US

Abstract

The Lax-Milgram theorem can be used to prove existence and uniqueness of weak solutions to partial differential equations, in particular, elliptic boundary value problems. After introducing some pertinent information required for the development of the theorem, it is stated and two proofs are given in detail. Several examples of its application to relevant boundary value problems are presented. Some consequences of these results and their relevance to computational aspects of the subject are summarized at the end.

Keywords: partial differential equation, functional, weak solution, Hilbert space, bilinear form

MSc: 35A02, 35A15, 35G05

1. Introduction

Let L and S be two linear differential operators and $u(\mathbf{x})$ a function which belongs to a Hilbert or Banach space of functions on a region $\Omega \subset \mathbb{R}^n$ with boundary $\partial\Omega$ and $\mathbf{x} = (x_1, \dots, x_n) \in \Omega \subset \mathbb{R}^n$. The system which is defined by

$$L\,u(\mathbf{x}) = f(\mathbf{x}), \qquad \mathbf{x} \in \Omega, \qquad S\,u(\mathbf{x}) = g(\mathbf{x}), \qquad x \in \partial\Omega \tag{1.1}$$

where $f(\mathbf{x})$ and $g(\mathbf{x})$ are functions of $\mathbf{x}$, is referred to as a linear boundary value problem for $u(\mathbf{x})$ when L is a linear operator on Ω. In problem (1.1), $u(\mathbf{x})$ is an unknown function referred to as the solution. It is usually the objective to determine $u(\mathbf{x})$ in some way, or at

*E-mail address: bracken@utpa.edu

least establish its existence and uniqueness. When f in (1.1) is also a function of $u(\mathbf{x})$ and perhaps its derivatives as well, then (1.1) is called a nonlinear boundary value problem. It is remarkable that problems of the form (1.1) can be expressed in the form of an abstract variational problem which can be expressed in the following way: on Hilbert space H, find a $u(\mathbf{x}) \in H$ such that

$$B(u, v) = h(v), \tag{1.2}$$

for all $v(\mathbf{x}) \in H$. In (1.2), B is a bilinear form and $h \in H^*$, the dual of H **[1-4]**. It may be stated further that boundary value problems of the form (1.1) can also be put in the form of an operator equation

$$Au = h(v), \tag{1.3}$$

where A is a linear or nonlinear operator on a Hilbert or Banach space into another. The Lax-Milgram theorem is a result which provides information about the existence and uniqueness of solutions of the form (1.1)-(1.3) under certain conditions **[5]**. It is the intention here to give a brief introduction to this theorem and to see how it may be proved. It is also possible to present some applications to problems where the theorem is applicable. The proofs of the theorem make use of several important results from functional analysis, the Riesz representation theorem in particular. It is not hard to find other examples from current research in which the topics of this work come up **[6]**.

Let us begin by presenting a more concrete version of a specific problem which has the form (1.1). Given measurable functions a^{ij}, b^i, $c : \Omega \to \mathbb{R}$ with a^{ij}, b^i, $c \in L^\infty(\Omega)$, let L denote the linear, second-order differential operator defined by

$$Lu = -\sum_{i,j=1}^{n} (a^{ij}(\mathbf{x})u_{x_i})_{x_j} + \sum_{i=1}^{n} (b^i(\mathbf{x})u)_{x_i} + c(\mathbf{x})u. \tag{1.4}$$

A classical solution of the boundary value problem is a function $u \in C^2(\Omega)$ which satisfies the equation and the boundary conditions at every point. Due to lack of regularity of the coefficients in the equation, the boundary value problem may not have classical solutions, hence a weaker concept of solution is required.

A weak solution of problem (1.1) is a function u in some Sobolov space, for example $H_0^1(\Omega)$, in which $g(\mathbf{x}) = 0$ on $\partial\Omega$ and for all $v \in H_0^1(\Omega)$, the following holds

$$\int_\Omega [\sum_{i,j=1}^{n} a^{ij}(\mathbf{x})u_{x_i}v_{x_j} - \sum_{i=1}^{n} b^i(\mathbf{x})uv_{x_i} + c(\mathbf{x})uv]\,dx = \int_\Omega f(\mathbf{x})v\,dx, \tag{1.5}$$

Integration by parts is applied to obtain (1.5). Thus a useful way in which to reformulate the idea of a weak solution that relates to the Lax-Milgram theorem is based on the possibility of defining a bilinear form, such as that on the left in (1.5), With regard to the problem formulated here, it is defined to be

$$B(u, v) = \int_\Omega [\sum_{i,j=1}^{n} a^{ij}(\mathbf{x})u_{x_i}v_{x_j} - \sum_{i=1}^{n} b^i(\mathbf{x})uv_{x_i} + c(\mathbf{x})uv]\,dx. \tag{1.6}$$

The function $u \in H_0^1(\Omega)$ is a weak solution of (1.5) provided that

$$B(u, v) = \langle f, v \rangle_{L^2}, \tag{1.7}$$

for all $v \in H_0^1(\Omega)$. In (1.7), we have used

$$\langle f, g \rangle_{L^2} = \int_\Omega fg\,dx \tag{1.8}$$

for the inner product on $L_2(\Omega)$, while the $H^1(\Omega)$ inner product is denoted by

$$\langle f, g \rangle_{H^1} = \int_\Omega fg\,dx + \int_\Omega \sum_{i=1}^{n} f_{x_i} g_{x_i}\,dx. \tag{1.9}$$

The only integration measure used is abbreviated to be $\int_\Omega f\,dx = \int \cdots \int_\Omega f(x_1, x_2, \dots, x_n)\,dx_1 \dots dx_n$.

2. The Lax-Milgram Theorem

To prepare the way for the Lax-Milgram theorem, it will be useful to introduce some preliminary definitions and required results.

Definition 2.1. Let H be a Hilbert space. A mapping $B(\cdot,\cdot) : H \times H \to \mathbb{C}$ is called a sesquilinear functional provided the following properties apply,

$$\begin{aligned} B(x_1 + x_2, y) &= B(x_1, y) + B(x_2, y), \\ B(\lambda x, y) &= \lambda B(x, y), \\ B(x, y_1 + y_2) &= B(x, y_1) + B(x, y_2), \\ B(x, \lambda y) &= \bar{\lambda} B(x, y). \end{aligned} \tag{2.1}$$

A functional which is linear in the second variable is called a bilinear form, so when H is real, the two coincide. The applications discussed here will be over real Hilbert spaces. The real inner product is an example of a bilinear form.

Definition 2.2. The form B is called a symmetric bilinear form if $B(y, x) = B(x, y)$ for all $(x, y) \in H \times H$. It is positive definite if $B(x, x) \geq 0$ for all $x \in H$ and $B(x, x) = 0$ implies $x = 0$. Defining $F(x) = B(x, x)$, then F is called a quadratic form.

Both of the following concepts are important with regard to the theorem in question and arise frequently.

Definition 2.3. (1) B is called bounded or continuous if there exists a constant $M > 0$ such that

$$|B(x, y)| \leq M \cdot ||x|| \cdot ||y||. \tag{2.2}$$

(2) B is said to be coercive if there exists a constant $\vartheta > 0$ such that

$$B(x, x) \geq \vartheta\, ||x||^2, \tag{2.3}$$

for all $x \in H$.

Theorem 2.1. Let T be a bounded linear operator on a Hilbert space H. The complex-valued function $B(\cdot,\cdot)$ on $H \times H$ which is defined by

$$B(x,y) = \langle x, Ty \rangle \tag{2.4}$$

is a bounded bilinear form on H and $||B|| = ||T||$. Conversely, let B be a bounded bilinear form on the Hilbert space H. Then there exists a unique bounded linear operator T on H such that $B(x,y) = \langle x, Ty \rangle$ for all $(x,y) \in H \times H$.

Proof. Let T be a bounded linear operator on H. Since $\langle , \rangle$ is linear, $B(x,y) = \langle x, Ty \rangle$ satisfies the linearity property

$$\begin{gathered} B(x_1 + x_2, y) = \langle x_1 + x_2, Ty \rangle = \langle x_1, Ty \rangle + \langle x_2, Ty \rangle = B(x_1, y) + B(x_2, y), \\ B(\lambda x, y) = \langle \lambda x, Ty \rangle = \lambda \langle x, Ty \rangle = \lambda B(x,y). \end{gathered} \tag{2.5}$$

and moreover, it follows that

$$|B(x,y)| = |\langle x, Ty \rangle| \leq ||T|| \cdot ||x|| \cdot ||y||. \tag{2.6}$$

This implies that $\sup_{||x||=||y||=1} |B(x,y)| \leq ||T||$, or $||B|| \leq ||T||$.

Conversely, define $h_y(x)$ to be

$$h_y(x) = B(x,y) \tag{2.7}$$

and linearity of h_y follows from that of B and as well, $|h_y(x)| = |B(x,y)| \leq ||B|| \cdot ||x|| \cdot ||y||$ so

$$||h_y|| \leq ||B|| \cdot ||y||.$$

Since h_y is a bounded linear functional on H, by the Riesz representation theorem, there exists a unique $Ty \in H$ such that

$$h_y(x) = \langle x, Ty \rangle, \tag{2.8}$$

for all $x \in H$ and the following bound holds

$$||T_y|| = ||h_y|| \leq ||B|| \cdot ||y||. \tag{2.9}$$

This implies that $||T|| \leq ||B||$, hence, since $||B|| \leq ||T||$, it follows that $||B|| = ||T||$. Moreover the operator T defined by $T(y) = \langle x, Ty \rangle$ is linear since

$$\begin{gathered} \langle x, T(\lambda y) \rangle = h_{\lambda y}(x) = B(x, \lambda y) = \lambda^* B(x,y) = \\ = \lambda^* h_y(x) = \lambda^* \langle x, Ty \rangle = \langle x, \lambda Ty \rangle, \\ \langle x, T(y_1 + y_2) \rangle = h_{y_1 + y_2}(x) = \\ = B(x, y_1 + y_2) = B(x, y_1) + B(x, y_2) = \end{gathered}$$

$$= h_{y_1}(x) + h_{y_2}(x) = \langle x, Ty_1\rangle + \langle x, Ty_2\rangle$$
$$= \langle x, Ty_1 + Ty_2\rangle.$$

Combining (2.7) and (2.8), we have

$$B(x,y) = h_y(x) = \langle x, Ty\rangle,$$

where T is linear for all $(x,y) \in H \times H$. Moreover, for every fixed $y \in H$, it follows that $\langle x, Ty\rangle = \langle x, Sy\rangle$ implies that $\langle x, (T-S)y\rangle = 0$, or $(T-S)y = 0$ for all $y \in H$ and therefore $T = S$. This proves that there exists a unique, bounded linear operator T such that $B(x,y) = \langle x, Ty\rangle$. □

The following is a Corollary to Theorem 2.1 and will be used as well:

Let T be a bounded linear operator on a Hilbert space H. Then the complex-valued function $B(\cdot,\cdot)$ on $H \times H$ defined by $B(x,y) = \langle Tx, y\rangle$ is a bounded linear form on H and $||B|| = ||T||$. Conversely, if $B(\cdot,\cdot)$ is a bounded bilinear form on H, there exists a unique, bounded linear operator T on H such that $B(x,y) = \langle Tx, y\rangle$ for all $(x,y) \in H \times H$.

Theorem 2.2. (Lax-Milgram) Suppose that $B : H \times H \to \mathbb{R}$ is a bilinear mapping for which there exist constants $\alpha, \vartheta > 0$ such that B is bounded

$$|B(u,v)| \leq \alpha||u|| \cdot ||v||, \qquad u, v \in H, \tag{2.10}$$

and moreover, B is coercive as well

$$B(u,v) \geq \vartheta||u||^2, \qquad u \in H, \tag{2.11}$$

Let $f : H \to \mathbb{R}$ be a bounded linear functional on H. Then there exists a unique element $u \in H$ such that

$$B(u,v) = \langle f, v\rangle, \tag{2.12}$$

for all $v \in H$.

Proof 1. Since $B(\cdot,\cdot)$ is bounded, there exists an $\alpha > 0$ such that (2.10) holds. By Theorem 2.1 and corollary, there must exist a bounded linear operator T on H and a functional $h_u \in H^*$ such that $B(u,v) = h_u(v)$ takes the form

$$\langle Tu, v\rangle = \langle h_u, v\rangle.$$

This is equivalent to $\langle Tu - h_u, v\rangle = 0$ for all $v \in H$, which implies that $Tu = h_u$. It will be shown that $Tu = h_u$ has a unique solution by proving that the affine mapping $u \to u - \rho(Tu - h_u) \in H$ is a contraction mapping for appropriate values of the parameter $\rho > 0$. Using the properties of the inner product, it follows that

$$||u - \rho Tu||^2 = \langle u - \rho Tu, u - \rho Tu\rangle = ||u||^2 - 2\rho\langle Tu, u\rangle + \rho^2||Tu||^2.$$

Since B is assumed to be coercive, $B(u,u) = \langle Tu, u\rangle \geq \vartheta||u||^2$ holds, and T is bounded, $||Tu|| \leq \alpha||u||^2$, so the following lower bound must hold,

$$||u - \rho Tu||^2 \leq (1 - 2\rho\vartheta + \alpha^2\rho^2)||u||^2. \tag{2.13}$$

If the operator M is defined as $Mu = u - \rho(Tu - h_u)$, then (2.13) implies that

$$||Mu - Mw|| = ||(u - \rho(Tu - h_u)) - (w - \rho(Tw - h_u)|| = ||u - w - \rho T(u - w)||$$
$$\leq (1 - 2\rho\vartheta + \rho^2\alpha^2)^{1/2}\,||u - w||. \tag{2.14}$$

This means the operator M is a contraction mapping if the value of ρ is chosen appropriately. For M to be a contraction mapping, the parameter ρ must satisfy the following bounds $0 < 1 - 2\vartheta\rho + \alpha^2\rho^2 < 1$, hence $\rho \in (0, 2\vartheta/\alpha^2)$. By the Banach contraction fixed point theorem, M has a unique fixed point, which is the required unique solution, and we are done. □

Before giving a second proof, it should be pointed out that if the bilinear form B is symmetric so that $B(u, v) = B(v, u)$, $u, v \in H$, a much simpler proof can be given by realizing that

$$\langle\langle u, v\rangle\rangle = B(u, v)$$

is a new inner product on H to which the Riesz theorem directly applies. The Lax-Milgram theorem has its primary significance in the fact that it does not require symmetry of B. This is very important in applications to differential equations. Another proof of the theorem emphasizes this.

Proof 2. For each $u \in H$, the mapping $v \to B(u, v)$ is a bounded linear functional on H. The Riesz theorem asserts the existence of a unique element $w \in H$ which satisfies

$$B(u, v) = \langle w, v\rangle, \qquad v \in H. \tag{2.15}$$

By the corollary, let us write $Au = w$ whenever (2.15) holds so that

$$B(u, v) = \langle Au, v\rangle$$

for $u, v \in H$. Now $A : H \to H$ is a bounded linear operator. Indeed, if $\lambda_1, \lambda_2 \in \mathbb{R}$ and $u_1, u_2 \in H$, it can be observed that for each $v \in H$, it follows that

$$\langle A(\lambda_1 u_1 + \lambda_2 u_2), v\rangle = B(\lambda_1 u_1 + \lambda_2 u_2, v) = \lambda_1 B(u_1, v) + \lambda_2 B(u_2, v)$$
$$= \lambda_1\langle Au_1, v\rangle + \lambda_2\langle Au_2, v\rangle = \langle\lambda_1 Au_1 + \lambda_2 Au_2, v\rangle.$$

This equality holds for each $v \in H$ and so A is linear and moreover,

$$||Au||^2 = \langle Au, Au\rangle = B(u, Au) \leq \alpha||u|| \cdot ||Au||. \tag{2.16}$$

Consequently, from (2.16), it follows that $||Au|| \leq \alpha||u||$ for all $u \in H$ and so the operator A is bounded.

It is asserted that A is one-to-one and the range of A is closed in H. This follows from (2.3) and from the observation

$$\vartheta||u||^2 \leq B(u, u) = \langle Au, u\rangle \leq ||Au|| \cdot ||u||. \tag{2.17}$$

This implies that $\vartheta||u|| \leq ||Au||$.

The range of A is H, that is $R(A) = H$. Since the range $R(A)$ is closed, if not there would exist a nonzero $w \in H$ such that $w \in R(A)^{\perp}$. However, this would imply that $\vartheta||w||^2 \leq B(w,w) = \langle Aw, w\rangle = 0$, which is a contradiction. Again from the Riesz theorem, it may be concluded that

$$\langle f, v\rangle = \langle w, v\rangle$$

for all $v \in H$ and some element $w \in H$. Since $R(A) = H$, a $u \in H$ is found which satisfies $Au = w$ so that

$$B(u,v) = \langle Au, v\rangle = \langle w, v\rangle = \langle f, v\rangle.$$

This is the claim (2.12). Finally, let us show that there is at most one element $u \in H$ which satisfies (2.12). If both $B(u,v) = \langle f, v\rangle$ and $B(\bar{u}, v) = \langle f, v\rangle$, then it follows by linearity that $B(u - \bar{u}, v) = 0$ for all $v \in H$. Now substitute $v = u - \bar{u}$ back in this to obtain the bound $\vartheta||u - \bar{u}||^2 \leq B(u - \bar{u}, u - \bar{u}) = 0$. □

3. Some Applications of the Theorem to Partial Differential Equations

1. The first example will examine the elliptic boundary value problem defined by,

$$Lu(\mathbf{x}) = f(\mathbf{x}), \qquad \mathbf{x} \in \Omega,$$
$$u(\mathbf{x}) = 0, \qquad x \in \partial\Omega. \tag{3.1}$$

It is assumed that operator L contains only second-order terms and takes the form,

$$Lu(\mathbf{x}) = -\sum_{i,j=1}^{n} (a^{ij}(\mathbf{x})u_{x_i})_{x_j}. \tag{3.2}$$

For the operator L defined by (3.2), a weak solution of (3.1) is an element $u \in H_0^1(\Omega)$ such that

$$B(u,v) = \langle f, v\rangle_{L^2(\Omega)} = \int_{\Omega} f(\mathbf{x})v(\mathbf{x})\,dx. \qquad v \in H_0^1(\Omega). \tag{3.3}$$

Then $B : H_0^1(\Omega) \times H_0^1(\Omega) \to \mathbb{R}$ is the continuous bilinear form

$$B(u,v) = \int_{\Omega} \sum_{i,j=1} a^{ij}(\mathbf{x})u_{x_i}u_{x_j}\,dx. \tag{3.4}$$

There exists a unique solution to the elliptic boundary value problem (3.1).

Theorem 3.1. Let $\Omega \subset \mathbb{R}^n$ be a bounded open set and suppose the operator L in (3.2) is coercive and $a^{ij}(\Omega) \in L^{\infty}(\Omega)$. For every $f(\mathbf{x}) \in L^2(\Omega)$, the problem (3.1) has a unique weak solution $u \in H_0^1(\Omega)$. The solution operator which is denoted by $L^{-1} : f \to u$ is a compact, linear operator from $L^2(\Omega)$ into $H_0^1(\Omega)$.

Proof. To prove the existence and uniqueness of a weak solution, it suffices to check that the bilinear form B here satisfies all the hypotheses of Theorem 2.2. Boundedness of B follows from the inequality

$$|B(u,v)| \leq \sum_{i,j=1}^{n} \int_{\Omega} |a^{ij} u_{x_i} u_{x_j}| \, dx \leq$$

$$\leq \sum_{i,j=1}^{n} ||a^{ij}||_{L^{\infty}} ||u_{x_i}||_{L^2} ||v_{x_j}||_{L^2} \leq C||u||_{H^1} \cdot ||v||_{H^1}. \tag{3.5}$$

Let us show that B is strictly positive definite. Since $u|_{\partial\Omega} = 0$, and Ω is bounded, the Poincaré-Friedrich's inequality implies there exists a constant $K > 0$ such that

$$||u||^2_{L^2(\Omega)} \leq K \int_{\Omega} |\nabla u|^2 \, dx, \qquad u \in H^1_0(\Omega). \tag{3.6}$$

since B is also coercive, there exists a constant $\vartheta > 0$ such that

$$B(u,u) = \int_{\Omega} \sum_{i,j=1}^{n} a^{ij}(\mathbf{x}) \, u_{x_i} u_{x_j} \, dx \geq \vartheta \int_{\Omega} \sum_{j=1}^{n} u^2_{x_j} \, dx = \vartheta \int_{\Omega} |\nabla u|^2 \, dx. \tag{3.7}$$

The inequalities (3.6) and (3.7) provide the following bound

$$||u||^2_{H^1} = ||u||^2_{L^2(\Omega)} + ||\nabla u||^2_{L^2(\Omega)} \leq \frac{K+1}{\vartheta} B(u,u),$$

which implies that

$$B(u,u) \geq \frac{\vartheta}{K+1} ||u||^2_{H^1}. \tag{3.8}$$

Hence, by Theorem 2.2, it may be concluded that for every $\bar{f} \in H^1_0(\Omega)$, there exists a unique element $u \in H^1_0(\Omega)$ such that

$$B(u,v) = \langle \bar{f}, v \rangle_{H^1}, \qquad v \in H^1_0(\Omega). \tag{3.9}$$

It follows then that the map $S : \bar{f} \to u$ is continuous since

$$||u||^2_{H^1} \leq \frac{K+1}{\vartheta} ||\bar{f}||_{H^1}.$$

Since the canonical embedding $j : H^1_0(\Omega) \to L^2(\Omega)$ is compact by the Rellich-Kondrachow theorem. Hence, its dual operator j^* is also compact. Since H^1_0 and L^2 are Hilbert spaces, they can be identified with their duals. Choosing $\bar{f} = j^* f \in H^1_0(\Omega)$ and from (3.9) it follows that

$$B(u,v) = \langle j^* f, v \rangle_{H^1} = \langle f, v \rangle_{L^2(\Omega)}, \tag{3.10}$$

for all $v \in H^1_0(\Omega)$. By definition, u is a weak solution. Since S is continuous and the linear operator j^* is compact, the solution operator $L^{-1} : f \to u$, which is the composition $L^{-1} = S \circ j$, is compact.

2.

Theorem 3.2. The Robin boundary value problem which is defined on $\Omega \subset \mathbb{R}^n$ as

$$-\Delta u + u = f(\mathbf{x}), \qquad x \in \Omega, \quad f \in L^2(\Omega), \tag{3.11}$$

$$\frac{\partial u}{\partial \nu} + \alpha\, u = 0, \qquad x \in \partial\Omega, \tag{3.12}$$

where $\partial/\partial\nu$ is the normal derivative on $\partial\Omega$ and $\alpha > 0$, has a unique solution.

To obtain the weak solution formulation of (3.11), (3.12), each equation in the pair is multiplied by a function $v \in H_0^1(\Omega)$ and integrated over Ω. In the case of the first equation, Green's formula for integration by parts is used to obtain

$$\int_\Omega \nabla u \cdot \nabla v\, dx = \int_{\partial\Omega} \frac{\partial u}{\partial \nu} v\, d\Gamma + \int_\Omega uv dx = \int_\Omega fv\, dx. \tag{3.13}$$

In (3.13), $d\Gamma$ is the integration measure on $\partial\Omega$. From the integral of the second equation, replace the term with the normal derivative in (3.13) to obtain

$$\int_\Omega \nabla u \cdot \nabla v\, dx + \alpha \int_{\partial\Omega} uv\, d\Gamma + \int_\Omega uv\, dx = \int_\Omega fv\, dx. \tag{3.14}$$

Consequently, the bilinear form is made up of the terms on the left-hand side of (3.14), and defined to be

$$B(u,v) = \sum_{i=1}^n \int_\Omega \frac{\partial u}{\partial x_i} \frac{\partial v}{\partial x_i}\, dx + \alpha \int_{\partial\Omega} uv d\Gamma + \int_\Omega uv\, dx. \tag{3.15}$$

Defining the functional $\mathcal{L}$ to be

$$\mathcal{L}(v) = \int_\Omega fv\, dx, \tag{3.16}$$

then equation (3.14) assumes the form

$$B(u,v) = \mathcal{L}(v). \tag{3.17}$$

Expression (3.15) is clearly symmetric as $B(u,v) = B(v,u)$. By the Poincaré-Friedrich's inequality, the following lower bound on B can be established,

$$\begin{aligned} B(u,u) &= \sum_{i=1}^n \int_\Omega \frac{\partial u}{\partial x_i} \frac{\partial u}{\partial x_i}\, dx + \alpha \int_{\partial\Omega} u^2 d\Gamma + \int_\Omega u^2\, dx \\ &\geq \sum_{i=1}^n \int_\Omega \frac{\partial u}{\partial x_i} \frac{\partial u}{\partial x_i}\, dx + \int_{\partial\Omega} u^2\, d\Gamma + \alpha \int_{\partial\Omega} u^2\, d\Gamma \\ &\geq ||u||_{H^1}^2 + \alpha \int_\Gamma u^2\, d\Gamma \geq c||u||_{H^1}^2. \end{aligned} \tag{3.18}$$

Therefore, it follows that B given by (3.15) is coercive. Moreover, the functional $\mathcal{L}$ is linear since given $v_1, v_2 \in H_0^1(\Omega)$ and scalars λ_1, λ_2, it follows that

$$\mathcal{L}(\lambda_1 v_1 + \lambda_2 v_2) = \int_\Omega f(\lambda_1 v_1 + \lambda_2 v_2)\, dx$$

$$= \lambda_1 \int_\Omega fv_1\, dx + \lambda_2 \int_\Omega fv_2\, dx = \lambda_1 \mathcal{L}(v_1) + \lambda_2 \mathcal{L}(v_2).$$

Using the Cauchy-Schwarz inequality, (3.16) satsifies

$$|\int_{\Omega} fv\, dx| \leq (\int_{\Omega} |f|^2\, dx)^{1/2} (\int_{\Omega} |v|^2\, dx)^{1/2} = ||f||_{L^2(\Omega)} \cdot ||v||_{L^2(\Omega)}. \tag{3.19}$$

Since $f \in L^2(\Omega)$, there exists a $K > 0$ such that $||f||_{L^2(\Omega)} \leq K$ and $v \in H^1(\Omega)$ implies that $||v||_{L^2(\Omega)} \leq ||v||_{H^1(\Omega)}$. Therefore, it follows that $\mathcal{L}$ satisfies the bound,

$$|\mathcal{L}(v)| \leq K||v||_{H^1}.$$

Consequently, by Theorem 2.2, it follows that (3.14) or (3.17) has a unique solution in $H_0^1(\Omega)$.

3. The Stokes equations are considered as a final example. These equations constitute a more complicated type of system and are given by

$$-\mu\Delta u + \nabla p - f = 0, \quad \operatorname{div} u = 0, \quad x \in \Omega, \quad u = 0 \quad x \in \partial\Omega \tag{3.20}$$

In system (3.20), $u = (u_1, u_2, u_3)$ is the velocity vector of a fluid with viscosity μ, p is the physical pressure and $f = (f_1, f_2, f_3) \in L_2(\Omega)$ is a force vector. Although $n = 3$ is the physically relevant case, system (3.20) can be studied for arbitrary n, as done here. The following spaces will be important,

$$D = \{u \in C_0^{\infty}(\Omega) | \operatorname{div} u = 0\},$$
$$H = \{u \in H_0^1(\Omega) | \operatorname{div} u = 0\},$$
$$N = \{p \in L_2(\Omega) | \int_{\Omega} p dx = 0\}. \tag{3.21}$$

The space H can be given the inner product

$$(u, v)_H = \sum_{i=1}^{n} \int_{\Omega} \nabla u_i \cdot \nabla v_i\, dx. \tag{3.22}$$

To obtain the weak formulation of (3.20), the first equation is multiplied using the Euclidean scalar product by an n-component test function $v \in D$ and then integrated. Green's formula for integration by parts will be required and presented in the following form,

$$\int_{\Omega} \sum_{i=1}^{n} v_i \Delta u_i\, dx + \int_{\Omega} \sum_{i=1}^{n} \nabla u_i \cdot \nabla v_i\, dx = \int_{\partial\Omega} \sum_{i=1}^{n} v_i \frac{\partial u_i}{\partial \nu}\, d\Gamma \tag{3.23}$$

Green's formula is applied to the result of multiplying the first equation and integrating. Scalar multiply then the first equation in (3.20) by v, to obtain,

$$-\int_{\Omega} \mu \sum_{i=1}^{n} \Delta u_i v_i\, dx + \int_{\Omega} \sum_{i=1}^{n} (\nabla p)_i v_i\, dx - \int_{\Omega} \sum_{i=1}^{n} f_i v_i\, dx = 0. \tag{3.24}$$

Using Green's formula (3.23), (3.24) becomes

$$\mu \int_{\Omega} \sum_{i=1}^{n} \nabla u_i \nabla v_i \, dx - \mu \int_{\partial\Omega} \sum_{i=1}^{n} v_i \frac{\partial u_i}{\partial \nu} \, d\Gamma$$

$$= - \int_{\Omega} \sum_{i=1}^{n} (\nabla p)_i v_i \, dx + \int_{\Omega} \sum_{i=1}^{n} f_i v_i \, dx. \tag{3.25}$$

For any $v \in D$, it is the case that $v = 0$ on $\partial\Omega$, and since

$$-\langle v, \nabla p \rangle = \langle \operatorname{div} v, p \rangle, \qquad p \in N, \tag{3.26}$$

so $\operatorname{div} v = 0$ implies that $\langle v, \nabla p \rangle = 0$. This transforms the problem into the following form,

$$B(u, v) = \mu \sum_{i=1}^{n} \int_{\Omega} \nabla u_i \nabla v_i \, dx = \sum_{i=1}^{n} \int_{\Omega} f_i v_i \, dx. \tag{3.27}$$

The weak solution of (3.27) is the classical solution of (3.20) since $\langle v, -\mu\Delta u - f \rangle_{L^2(\Omega)} = 0$ does not imply that $-\mu\Delta u - f = 0$ since v satisfies the constraint $\operatorname{div} v = 0$ for $v \in H$. Instead, this implies that $-\mu\Delta u - f = -\nabla p$ due to (3.26).

The bilinear form B satisfies all the conditions of the Lax-Milgram theorem. The form B is symmetric, $B(u, v) = B(v, u)$, and it is bounded since the Cauchy-Schwarz inequality implies that

$$|B(u, v)| = |\sum_{i=1}^{n} \mu \int_{\Omega} \nabla u_i \nabla v_i \, dx|$$

$$\leq \mu (\sum_{i=1}^{n} \int_{\Omega} |\nabla u_i|^2 \, dx)^{1/2} \cdot (\sum_{i=1}^{n} \int_{\Omega} |\nabla v_i|^2 \, dx)^{1/2} \tag{3.28}$$

$$= \mu ||u||_H ||v||_H.$$

It is clear that B is coercive since

$$|B(u, v)| = \mu \sum_{i=1}^{n} \int_{\Omega} \nabla v_i \nabla v_i \, dx = \mu ||v||_H^2 \geq \vartheta ||v||_H^2,$$

for all $\mu \geq \vartheta$. Therefore, by the Lax-Milgram Theorem 2.2, the problem (3.27) has a unique solution.

4. Consequences for Applications and Summary

The Lax-Milgram theorem can be very useful in the study of partial differential equations or variational problems that can be put in the form (1.2) or perhaps (1.3). Some remarks which pertain to the computational aspects of the subject will be presented. This leads, for example, to an application such as the Rayleigh-Ritz-Galerkin method which deals with the approximate solution of an operator equation in the form of a finite series such as

$$u_M = \sum_{j=1}^{M} c_j \phi_j + \phi_0, \tag{4.1}$$

and its weak formulation $B(u,v) = F(v)$. The coefficients c_j in (4.1) are called the Rayleigh-Ritz-Galerkin coefficients. These coefficients are chosen such that the abstract variational formulation $B(u,v) = F(v)$ holds for $v = \phi_i$,

$$B(\phi_i, \sum_{j=1}^{M} c_j\phi_j + \phi_0) = F(\phi_i), \qquad i = 1, 2, \dots, M. \tag{4.2}$$

Moreover, let $B(u,v)$ be a symmetric coercive and bounded bilinear form defined on H associated with the operator equation $T : H \subset L^2(\Omega) \to H$. The problem of finding a number λ and a nonzero function $u \in H$ such that

$$B(u,v) = \lambda\langle u, v\rangle_{L^2(\Omega)} \tag{4.3}$$

is called the eigenvalue problem of the bilinear form $B(u,v)$ **[1]**. The weak problem associated with the operator equation $Tu - \lambda u = y$ consists of finding a $u \in H$ such that $B(u,v) - \lambda\langle u, v\rangle_{L^2(\Omega)} = \langle y, v\rangle_{L^2(\Omega)}$.

A further comment can be made in the case that B is coercive. Suppose this is not the case. By choosing a constant $\gamma > 0$ sufficiently large, a new operator L_γ can be defined for (1.1)

$$L_\gamma u = Lu + \gamma u. \tag{4.4}$$

which is strictly positive definite. Then the corresponding bilinear form $B_\gamma(u,v)$ satisfies

$$B_\gamma(u,u) \geq \vartheta||u||_{H^1}^2, \tag{4.5}$$

for all $u \in H_0^1(\Omega)$ and some constant $\vartheta > 0$. The Lax-Milgram theorem can then be applied to conclude that $L_\gamma u = f$ has a weak solution $u \in H_0^1(\Omega)$ and $f \to u = L_\gamma^{-1} f$ is a linear compact operator from $L^2(\Omega)$ into $H_0^1(\Omega) \subset L^2(\Omega)$. The original problem now takes the form

$$Lu = L_\gamma u - \gamma u = f. \tag{4.6}$$

Since γL_γ^{-1} is a compact operator from $L^2(\Omega)$ into itself, so Fredholm's theory can be applied to it.

Finally, the following theorem can be mentioned with regard to Example 1. Since $H_0^1(\Omega) \subset L^2(\Omega)$, the solution operator can also be considered a compact operator from $L^2(\Omega)$ into $L^2(\Omega)$. Moreover, the operator L^{-1} is self-adjoint in the case $a^{ij} = a^{ji}$. Hence, by the Hilbert-Schmidt theorem, it admits a representation in terms of a countable basis of eigenfunctions φ_k of L^{-1}, and there exists the representation

$$L^{-1} f = \sum_{k=1}^{\infty} \lambda_k \langle f, \varphi_k\rangle_{L^2} \varphi_k.$$

The corresponding eigenvalues approach zero as $k \to \infty$.

References

[1] A. H. Siddiqi, *Applied Functional Analysis*, Marcel Dekker Inc., New York, Basel, (2004).

[2] A. Bressan, Lecture Notes on Functional Analysis with Applications to Linear Partial Differential Equations, AMS, *Graduate Studies in Mathematics*, vol. 43, Providence, RI, (2013).

[3] E. Kreyszig, *Introductory Functional Analysis with Applications*, J. Wiley and Sons, New York, (1978).

[4] A. W. Naylor and G. R. Sell, *Linear Operator Theory in Engineering and Science*, Springer-Verlag, New York, Heidelberg, (1982).

[5] P. D. Lax, A. N. Milgram, Parabolic equations, Contributions to the theory of partial differential equations, *Ann. Math. Studies*, 33, 167-190, (1954).

[6] A. Moiola and E. Spence, *SIAM Review*, 56, 274-312, (2014).

In: Partial Differential Equations
Editor: Deborah E. Richards, pp. 15-70
ISBN: 978-1-63482-643-3

Chapter 2

Coupled PDEs and Control Systems Arising in Climate Dynamics: Ocean-Atmosphere Interactions and Tropical Instability Waves

Aziz Belmiloudi*
Institut de Recherche MAthématique de Rennes (IRMAR),
Rennes, France

Abstract

This chapter investigates nonlinear control problems for coupled partial differential equations arising from climate and circulation dynamics in equatorial zone. The phenomenon and processes that we want to modelize, by coupled and time-dependent PDEs, occur in tropical Atlantic and Pacific regions: the circulation is there characterized by steady zonal currents varying monthly or seasonally and by long waves propagating westward along the equator known as "equatorial waves", driven by the variability of atmospheric currents and superimposed to the steady mean currents. The equatorial waves can be connected with strong vertical velocities and then induce "upwellings" or "downwellings". This study corresponds to the modeling of tropical instability waves (taking into account the complex interaction between the atmosphere and the ocean circulation) which are illustrated by the "*El Niño*" phenomena and tropical cyclones. After the introduction in which we give background, motivation and fundamental principles governing circulation dynamics, we present a mathematical model of circulation mechanisms, operating in the equatorial zone, in a global climate system, and give some assumptions and notations. The equations are non-linear, time-dependent and coupled, and are of Navier-Stokes type for the velocity and pressure, and of transport-diffusion type for the temperature, with non linear boundary and interface conditions. Afterwords, we investigate the well-posedness of governing nonlinear systems. For provide the uniqueness, we prove some regularity results. Next, sensitivity and control problems are presented and analyzed. The problem is controlled by

*E-mail addresses: Aziz.Belmiloudi@insa-rennes.fr; Aziz.belmiloudi@math.cnrs.fr; Address: IRMAR-INSA de Rennes, 20 avenue des Buttes de Coësmes, CS 14315,35043 Rennes Cédex, France

the variability of the heat interface-stress between the ocean and atmosphere. The observation is the vorticity of the ocean and (or) of the atmosphere. Finally, we conclude this work with some comments and possible future developments.

Keywords: Coupled PDEs, Nonlinear optimal control, coupled atmosphere-ocean systems, tropical instability waves, equatorial waves, minimizing sequences, penalization, oceanography, climatology

1. Introduction and Mathematical Setting of the Problem

1.1. Motivation

There are two types of currents whose main causes are solar radiation (the warming of the surface by downwelling radiation coming from the atmosphere), wind and gravity. The Earth unevenly receives solar energy, which is not the same in the polar region as in the equatorial region. Consequently, the intertropical zone receives so much more energy than the rest of the planet, which creates a thermal imbalance. This imbalance is the origin of the two types of oceanic currents (surface currents and deep currents), and, indeed, puts in motion the atmosphere and oceanic currents which trying to thermally readjust the whole system. These movements are influenced by a force caused by the rotation of the Earth, called the Coriolis force. Moreover, it generates winds, which are the main causes of surface currents (which affect approximately 10% of the water of the oceans and are generally limited to the top 400 meters of the oceans). The location of these surface currents changes significantly with the seasons; this phenomenon is particularly sensitive for the equatorial currents. The thermal imbalance leads also to differences in temperature depending on latitude[1] (between the various layers of the ocean). This temperature difference leads to a difference in salinity and thus in density, creating deep oceanic currents. Consequently, the surface currents and deep currents are interconnected. Thanks to the heat capacity of water (its thermal inertia is much larger than that of the air), the ocean tempers the seasonal temperature changes of the air masses, which otherwise would be much more important. Thus currents of the warm surface layers can warm a region's climate. In contrast, the rise of cold water from the sea depths to the surface moderates water temperatures in the equatorial region.[2] Therefore, the surface temperature variation plays an important role in oceanic current movements, which play a key role in the regulation of the climate, and ensure that a heat transport from the equator to the poles is as important as the atmosphere. So, We can not neglect the fluctuation caused by this variation (even if the variation is small) and consequently the analysis of its influence, is very important.

The phenomenon that we want to model occurs in the tropical Atlantic and Pacific oceans: the circulation there is characterized by steady zonal currents and by long waves propagating westward along the equator, driven by the variability of the atmospheric currents and superposed on the mean currents. The equatorial waves are connected with strong vertical currents, called convective currents, which are very sensitive to small changes in

[1]The deep waters sink in the ocean basins located at high latitudes, where temperatures are low enough so that the density increases.

[2]But this movement is still poorly understood because it is difficult to measure directly.

temperature and induce "upwellings" or "downwellings". These phenomena modify the properties of the sea water near the surface: decreasing or increasing the temperature, supply of plankton etc. and are therefore of great importance for climate, fishing activities etc. This climate variability can be illustrated, for example, by the *El Niño* phenomenon, which occurs every two to seven years. *El Niño* warming begins with a modest temperature elevation (2 or $3°C$) of surface waters in the Pacific ocean along the equatorial coast of South America. The *El Niño* current then cuts the cold and deep upwelling (it results in a reduction of food, and, consequently, causes fish populations to decline). This phenomenon is accompanied with a sharp increase in rainfall and flooding. In contrast, around Australia and Asia, the high pressures and cold water temperature decrease the precipitation, and then cause severe drought, which often lead to a series of fires. These equatorial waves have been evidenced from "in situ" observations, and from altimetric measurements[3] and data assimilation in connection with control theory, see Belmiloudi (1999, 2000a,b, 2001a, 2002a,b); Belmiloudi et al. (1997a), in which the authors analyze mathematically the process of these tropical instability waves. The equations of the large-scale systems of the ocean are derived from Navier-Stokes equations but take into account oceanographic assumptions such as Boussinesq approximation, i.e., the density variations ρ are neglected in the system except in the buoyancy term (on the other hand, density is assumed to be constant and equal to a mean value ρ_{av} in the equations describing the horizontal motion) and in the equation of state. We will afterwards take $\rho_{av} = 1$. Of course, there is a vast literature concerning the control theory and inverse problems in connection with fluid mechanics problems. The reader is referred, for example, to Belmiloudi (2008); Marchuk (1995); S. Park and L. Xu (eds.); Sritharan (ed.); Wunsch (1996) and the references therein.

To understand this phenomenon, we have introduced a model of "perturbation" based on the coupled atmosphere-ocean model given in Lions et al. (1995), and the references therein. The equations of the atmosphere and the large scale equations of the ocean are derived from Navier-Stokes equations but take into account oceanographic and atmospheric assumptions such as Hydrostatic approximation with vertical viscosity (see the equation (3)) and Boussinesq approximation, Recently, oceanographic and atmospheric problems have been the object of numerous studies either from a physical, from a theoretical or from a numerical point of view (see, e.g., Belmiloudi (1999, 2000a,b, 2001a, 2002a,b, 2008); Belmiloudi et al. (1997a,b, 1999); Chelton et al. (2001); Cummings et al. (2013); Gejadze et al. (2006a,b); Parmuzin et al. (2007); Lions et al. (1992a, 1995); Majda (2003); Jin et al. (2008); Wen et al. (2012) and the references therein). In this chapter, we modelize the physical phenomenon of long waves occurring in the tropical domain known as "tropical instabilities". These waves are supposed to be at the origin of important climatic changes in Pacific region.

1.2. Primitive Equations and Study Domain

We consider an oceanic (respectively an atmospheric) domain Ω_o (respectively Ω_a) extending on both sides of the equator ($10°S - 10°N$), and of constant depth H (respectively L). The curvature of the earth is neglected. The vertical extension of Ω_o (respectively Ω_a) ($-H \leq z \leq 0$) (respectively $0 \leq z \leq L$) corresponds to a part of the physical domain. We

[3]Altimetric measurements give the distance between the satellite and the sea surface.

assume that for depths greater than H (respectively than L), the variability is negligible. The perturbation of the current is made of zonal propagating waves. So, we can choose the zonal extension of Ω_o (respectively Ω_a) as the greatest wavelength, and impose periodic conditions on the eastern and western boundaries. The prognostic variable are the velocity $U^{(o)} = (\mathbf{u}^{(o)}, w^{(o)})$, the pressure $p^{(o)}$ and the temperature $\mathcal{T}^{(o)}$ defined in Ω_o, and the velocity $U^{(a)} = (\mathbf{u}^{(a)}, w^{(a)})$, the geopotential $p^{(a)}$ and the temperature $\mathcal{T}^{(a)}$ defined in Ω_a; they are the unknowns of our model.

The total equations satisfied by the coupled atmosphere-ocean are:

- in the ocean

$$
\begin{aligned}
&\frac{\partial U^{(o)}}{\partial t} + (U^{(o)}.\nabla_3)U^{(o)} + \mathbf{F}\wedge U^{(o)} - \mathrm{div}_3(\nu_1^{(o)}\,\nabla_3 U^{(o)}) + \nabla_3 p^{(o)} = \delta_{\mathcal{T}^{(o)}}\mathcal{T}^{(o)}\mathbf{G},\\
&\frac{\partial \mathcal{T}^{(o)}}{\partial t} + (U^{(o)}.\nabla_3)\mathcal{T}^{(o)} - \mathrm{div}_3(\nu_2^{(o)}\,\nabla_3 \mathcal{T}^{(o)}) = 0,\\
&\mathrm{div}_3(U^{(o)}) = 0,
\end{aligned}
\tag{1}
$$

- in the atmosphere

$$
\begin{aligned}
&\frac{\partial U^{(a)}}{\partial t} + (U^{(a)}.\nabla_3)U^{(a)} + \mathbf{F}\wedge U^{(a)} - \mathrm{div}_3(\nu_1^{(a)}\,\nabla_3 U^{(a)}) + \nabla_3 p^{(a)}\\
&\qquad\qquad = -\delta_{\mathcal{T}^{(a)}}\mathcal{T}^{(a)}\mathbf{G} + \mathbf{f}_1^{(a)},\\
&\frac{\partial \mathcal{T}^{(a)}}{\partial t} + (U^{(a)}.\nabla_3)\mathcal{T}^{(a)} - \mathrm{div}_3(\nu_2^{(a)}\,\nabla_3 \mathcal{T}^{(a)}) + \delta_{\mathcal{T}^{(a)}}\,w^{(a)} = g^{(a)},\\
&\mathrm{div}_3(U^{(a)}) = 0,
\end{aligned}
\tag{2}
$$

where $\mathbf{G} = (0,0,-1)$, $\mathbf{f}_1^{(a)} = (\mathbf{f}^{(a)},0)$ and $\mathbf{f}^{(a)} = (f_1^a, f_2^a)$; (x,y,z) are the cartesian coordinates: x and y are measured in the horizontal plane of the undisturbed sea-surface (x towards the east and y towards the north), and z is vertically ascendant; $\mathbf{F} = (0,0,2\omega\sin\phi)$ is the Coriolis force with ω the rotation rate of earth and ϕ the latitude; $(\mathbf{f}^{(a)}, g^{(a)})$ is the external forcing term; $\nabla_3 = (\frac{\partial}{\partial x}, \frac{\partial}{\partial y}, \frac{\partial}{\partial z})$; $\mathrm{div}_3(\mathbf{v}) = \nabla_3.\mathbf{v}$; $\nabla = (\frac{\partial}{\partial x}, \frac{\partial}{\partial y})$ and $\mathrm{div}(\mathbf{u}) = \nabla.\mathbf{u}$.

The sensitivity of ocean general circulation models to the parametrization of the vertical turbulent diffusion has been known for a long time and has been proved by Bryan (1987). The turbulent flux are usually modelized by the dissipative terms and linked to large-scale oceans by the use of mixing coefficients: $\nu_{1h}^{(o)}$ and $\nu_{2h}^{(o)}$ (respectively $\nu_{1v}^{(o)}$ and $\nu_{2v}^{(o)}$) denote the coefficients of horizontal (respectively vertical) eddy viscosity and diffusivity. The choice of a constant for these coefficients is not a good physical representation of the ocean, because the turbulent mixing is high above the thermocline and low in stratified regions. According to the parametrization described in Philander et al. (1981), $\nu_{1v}^{(o)}$ and $\nu_{2v}^{(o)}$ are variable, positive and bounded functions. They are deduced from the given mean circulation. $\nu_{1h}^{(o)}$ and $\nu_{2h}^{(o)}$ are constant and positive. For the coefficients of eddy viscosity and diffusivity of the air, we suppose the same hypothesis as in the ocean.

We are now going to simplify equations (1)-(2). It is known that the differences of the density are significant only in the term of buoyancy and in the equation of state. On the other hand, density will be assumed to be constant in the equations describing the horizontal motion. This assumption is known as Boussinesq approximation. The horizontal scale is much bigger than the vertical one. Moreover the viscosity will play an important role in the dynamics of the ocean and the atmosphere. Therefore, we can replace the equation describing the vertical motion by the following equation called Hydrostatic approximation with vertical viscosity:

- in the ocean

$$-\nu_{1h}^{(o)}\Delta w^{(o)} - \frac{\partial}{\partial z}(\nu_{1v}^{(o)}\frac{\partial w^{(o)}}{\partial z}) + \frac{\partial p^{(o)}}{\partial z} + \delta_{\mathcal{T}^{(o)}}\mathcal{T}^{(o)} = 0, \tag{3}$$

- in the atmosphere

$$-\nu_{1h}^{(a)}\Delta w^{(a)} - \frac{\partial}{\partial z}(\nu_{1v}^{(a)}\frac{\partial w^{(a)}}{\partial z}) + \frac{\partial p^{(a)}}{\partial z} - \delta_{\mathcal{T}^{(a)}}\mathcal{T}^{(a)} = 0. \tag{4}$$

Applying these physical assumptions to (1)-(2), we obtain the following equations:

- in the ocean

$$\frac{\partial \mathbf{u}^{(o)}}{\partial t} + (\mathbf{u}^{(o)}.\nabla)\mathbf{u}^{(o)} + w^{(o)}\frac{\partial \mathbf{u}^{(o)}}{\partial z} + \mathbf{F}\wedge_2 \mathbf{u}^{(o)} - \nu_{1h}^{(o)}\Delta\mathbf{u}^{(o)}$$
$$-\frac{\partial}{\partial z}(\nu_{1v}^{(o)}\frac{\partial \mathbf{u}^{(o)}}{\partial z}) + \nabla p^{(o)} = 0,$$

$$-\nu_{1h}^{(o)}\Delta w^{(o)} - \frac{\partial}{\partial z}(\nu_{1v}^{(o)}\frac{\partial w^{(o)}}{\partial z}) + \frac{\partial p^{(o)}}{\partial z} + \delta_{\mathcal{T}^{(o)}}\mathcal{T}^{(o)} = 0, \tag{5}$$

$$\frac{\partial \mathcal{T}^{(o)}}{\partial t} + (\mathbf{u}^{(o)}.\nabla)\mathcal{T}^{(o)} + w^{(o)}\frac{\partial \mathcal{T}^{(o)}}{\partial z} - \nu_{2h}^{(o)}\Delta\mathcal{T}^{(o)} - \frac{\partial}{\partial z}(\nu_{2v}^{(o)}\frac{\partial \mathcal{T}^{(o)}}{\partial z}) = 0,$$

$$\mathrm{div}(\mathbf{u}^{(o)}) + \frac{\partial w^{(o)}}{\partial z} = 0,$$

• in the atmosphere

$$\begin{aligned}
&\frac{\partial \mathbf{u}^{(a)}}{\partial t} + (\mathbf{u}^{(a)}.\nabla)\mathbf{u}^{(a)} + w^{(a)}\frac{\partial \mathbf{u}^{(a)}}{\partial z} + \mathbf{F}\wedge_2 \mathbf{u}^{(a)} - \nu_{1h}^{(a)}\Delta \mathbf{u}^{(a)} \\
&\qquad\qquad - \frac{\partial}{\partial z}(\nu_{1v}^{(a)}\frac{\partial \mathbf{u}^{(a)}}{\partial z}) + \nabla p^{(a)} = \mathbf{f}^{(a)}, \\
&-\nu_{1h}^{(a)}\Delta w^{(a)} - \frac{\partial}{\partial z}(\nu_{1v}^{(a)}\frac{\partial w^{(a)}}{\partial z}) + \frac{\partial p^{(a)}}{\partial z} - \delta_{\mathcal{T}^{(a)}}\mathcal{T}^{(a)} = 0, \\
&\frac{\partial \mathcal{T}^{(a)}}{\partial t} + (\mathbf{u}^{(a)}.\nabla)\mathcal{T}^{(a)} + w^{(a)}\frac{\partial \mathcal{T}^{(a)}}{\partial z} - \nu_{2h}^{(a)}\Delta \mathcal{T}^{(a)} \\
&\qquad\qquad - \frac{\partial}{\partial z}(\nu_{2v}^{(a)}\frac{\partial \mathcal{T}^{(a)}}{\partial z}) + \delta_{\mathcal{T}^{(a)}}\, w^{(a)} = g^{(a)}, \\
&\mathrm{div}(\mathbf{u}^{(a)}) + \frac{\partial w^{(a)}}{\partial z} = 0,
\end{aligned} \tag{6}$$

with $\mathbf{F}\wedge_2 \mathbf{u}^{(s)} = \mathbf{F}\wedge(\mathbf{u}^{(s)},0)$, $s = o, a$.

Remark 1. *If we assume that the influence of the viscosity is negligible compared to the ratios between the vertical and horizontal scales of the ocean-atmosphere domain. The Hydrostatic approximation with vertical viscosity can be replaced by the classical Hydrostatic pressure approximation as follows:*

• *in the ocean*

$$\frac{\partial p^{(o)}}{\partial z} + \delta_{\mathcal{T}^{(o)}}\mathcal{T}^{(o)} = 0, \tag{7}$$

• *in the atmosphere*

$$\frac{\partial p^{(a)}}{\partial z} - \delta_{\mathcal{T}^{(a)}}\mathcal{T}^{(a)} = 0. \tag{8}$$

Moreover, the arguments and developments presented in this work extent directly to the present case without further estimates. □

We now split up the circulation $(\mathbf{u}^{(o)}, w^{(o)}, \mathcal{T}^{(o)}, p^{(o)}, \mathbf{0}, 0)$ (respectively the circulation $(\mathbf{u}^{(a)}, w^{(a)}, \mathcal{T}^{(a)}, p^{(a)}, \mathbf{f}^{(a)}, g^{(a)})$ into a given mean value $(\mathbf{u}_0^{(o)}, w_0^{(o)}, \mathcal{T}_0^{(o)}, p_0^{(o)}, \mathbf{0}, 0)$ (respectively $(\mathbf{u}_0^{(a)}, w_0^{(a)}, \mathcal{T}_0^{(a)}, p_0^{(a)}, \mathbf{f}_0^{(a)}, g_0^{(a)})$ and variability (i.e. the total circulation is the sum of the mean circulation and a variability)

$$\begin{aligned}
&(\tilde{\mathbf{u}^{(o)}}, \tilde{w^{(o)}}, \tilde{\mathcal{T}^{(o)}}, \tilde{p^{(o)}}, \mathbf{0}, 0) \\
&\quad = (\mathbf{u}^{(o)} - \mathbf{u}_0^{(o)}, w^{(o)} - w_0^{(o)}, \mathcal{T}^{(o)} - \mathcal{T}_0^{(o)}, p^{(o)} - p_0^{(o)}, \mathbf{0}, 0), \\
&(\tilde{\mathbf{u}^{(a)}}, \tilde{w^{(a)}}, \tilde{\mathcal{T}^{(a)}}, \tilde{p^{(a)}}, \tilde{\mathbf{f}^{(a)}}, \tilde{g^{(a)}}) \\
&\quad = (\mathbf{u}^{(a)} - \mathbf{u}_0^{(a)}, w^{(a)} - w_0^{(a)}, \mathcal{T}^{(a)} - \mathcal{T}_0^{(a)}, p^{(a)} - p_0^{(a)}, \mathbf{f}^{(a)} - \mathbf{f}_0^{(a)}, g^{(a)} - g_0^{(a)}),
\end{aligned} \tag{9}$$

that will be computed by the model. This expansion is justified in tropical region: the steady mean circulation $(\mathbf{u}_0^{(o)}, w_0^{(o)}, \mathcal{T}_0^{(o)}, p_0^{(o)})$ (respectively $(\mathbf{u}_0^{(a)}, w_0^{(a)}, \mathcal{T}_0^{(a)}, p_0^{(a)})$) is known for each tropical season; the variability $(\mathbf{u}^{\widetilde{(o)}}, w^{\widetilde{(o)}}, \mathcal{T}^{\widetilde{(o)}}, p^{\widetilde{(o)}})$ (respectively $(\mathbf{u}^{\widetilde{(a)}}, w^{\widetilde{(a)}}, \mathcal{T}^{\widetilde{(a)}}, p^{\widetilde{(a)}})$) is made of westward propagating waves. The circulation $(\mathbf{u}_0^{(o)}, w_0^{(o)}, \mathcal{T}_0^{(o)}, p_0^{(o)})$ (respectively $(\mathbf{u}_0^{(a)}, w_0^{(a)}, \mathcal{T}_0^{(a)}, p_0^{(a)})$) has to satisfy steady-state system (5) (respectively (6)).

Nota bene. notation "$\tilde{\ }$" used for the perturbation of the mean circulation will now be omitted.

The equations satisfied by the perturbation of the total circulation can be deduced from (5)-(6) and then can be written as follows:

- <u>in the ocean</u>

$$
\begin{aligned}
&\frac{\partial \mathbf{u}^{(o)}}{\partial t} + (\mathbf{u}^{(o)}.\nabla)\mathbf{u}^{(o)} + w^{(o)}\frac{\partial \mathbf{u}^{(o)}}{\partial z} + (\mathbf{u}_0^{(o)}.\nabla)\mathbf{u}^{(o)} + w_0^{(o)}\frac{\partial \mathbf{u}^{(o)}}{\partial z} + (\mathbf{u}^{(o)}.\nabla)\mathbf{u}_0^{(o)} \\
&\qquad + w^{(o)}\frac{\partial \mathbf{u}_0^{(o)}}{\partial z} + \mathbf{F}\wedge_2 \mathbf{u}^{(o)} - \nu_{1h}^{(o)}\Delta \mathbf{u}^{(o)} - \frac{\partial}{\partial z}(\nu_{1v}^{(o)}\frac{\partial \mathbf{u}^{(o)}}{\partial z}) + \nabla p^{(o)} = 0, \\
&-\nu_{1h}^{(o)}\Delta w^{(o)} - \frac{\partial}{\partial z}(\nu_{1v}^{(o)}\frac{\partial w^{(o)}}{\partial z}) + \frac{\partial p^{(o)}}{\partial z} + \alpha^{(o)}\mathcal{T}^{(o)} = 0, \\
&\frac{\partial \mathcal{T}^{(o)}}{\partial t} + (\mathbf{u}^{(o)}.\nabla)\mathcal{T}^{(o)} + w^{(o)}\frac{\partial \mathcal{T}^{(o)}}{\partial z} + (\mathbf{u}_0^{(o)}.\nabla)\mathcal{T}^{(o)} + w_0^{(o)}\frac{\partial \mathcal{T}^{(o)}}{\partial z} + (\mathbf{u}^{(o)}.\nabla)\mathcal{T}_0^{(o)} \\
&\qquad + w^{(o)}\frac{\partial \mathcal{T}_0^{(o)}}{\partial z} - \nu_{2h}^{(o)}\Delta \mathcal{T}^{(o)} - \frac{\partial}{\partial z}(\nu_{2v}^{(o)}\frac{\partial \mathcal{T}^{(o)}}{\partial z}) = 0, \\
&\mathrm{div}(\mathbf{u}^{(o)}) + \frac{\partial w^{(o)}}{\partial z} = 0,
\end{aligned}
\tag{10}
$$

- <u>in the atmosphere</u>

$$
\begin{aligned}
&\frac{\partial \mathbf{u}^{(a)}}{\partial t} + (\mathbf{u}^{(a)}.\nabla)\mathbf{u}^{(a)} + w^{(a)}\frac{\partial \mathbf{u}^{(a)}}{\partial z} + (\mathbf{u}_0^{(a)}.\nabla)\mathbf{u}^{(a)} + w_0^{(a)}\frac{\partial \mathbf{u}^{(a)}}{\partial z} + (\mathbf{u}^{(a)}.\nabla)\mathbf{u}_0^{(a)} \\
&\qquad + w^{(a)}\frac{\partial \mathbf{u}_0^{(a)}}{\partial z} + \mathbf{F}\wedge_2 \mathbf{u}^{(a)} - \nu_{1h}^{(a)}\Delta \mathbf{u}^{(a)} - \frac{\partial}{\partial z}(\nu_{1v}^{(a)}\frac{\partial \mathbf{u}^{(a)}}{\partial z}) + \nabla p^{(a)} = \mathbf{f}^{(a)}, \\
&-\nu_{1h}^{(o)}\Delta w^{(a)} - \frac{\partial}{\partial z}(\nu_{1v}^{(a)}\frac{\partial w^{(a)}}{\partial z}) + \frac{\partial p^{(a)}}{\partial z} - \alpha^{(a)}\mathcal{T}^{(a)} = 0, \\
&\frac{\partial \mathcal{T}^{(a)}}{\partial t} + (\mathbf{u}^{(a)}.\nabla)\mathcal{T}^{(a)} + w^{(a)}\frac{\partial \mathcal{T}^{(a)}}{\partial z} + (\mathbf{u}_0^{(a)}.\nabla)\mathcal{T}^{(a)} + w_0^{(a)}\frac{\partial \mathcal{T}^{(a)}}{\partial z} + (\mathbf{u}^{(a)}.\nabla)\mathcal{T}_0^{(a)} \\
&\qquad + w^{(a)}\frac{\partial \mathcal{T}_0^{(a)}}{\partial z} - \nu_{2h}^{(a)}\Delta \mathcal{T}^{(a)} - \frac{\partial}{\partial z}(\nu_{2v}^{(a)}\frac{\partial \mathcal{T}^{(a)}}{\partial z}) + \alpha^{(a)}w^{(a)} = g^{(a)}, \\
&\mathrm{div}(\mathbf{u}^{(a)}) + \frac{\partial w^{(a)}}{\partial z} = 0,
\end{aligned}
\tag{11}
$$

where $\alpha^{(o)} = \delta_{\mathcal{T}^{(o)}}$ and $\alpha^{(a)} = \delta_{\mathcal{T}^{(a)}}$.

The oceanic (respectively atmospheric) domain Ω_o (respectively Ω_a) can be defined as follows.

$$\Omega_o =]0,Lx[\times]-Ly,Ly[\times]-H,0[\ \text{(respectively } \Omega_a =]0,Lx[\times]-Ly,Ly[\times]0,L[).$$

We denote by Γ_I the interface between the atmosphere and the ocean $(z=0)$.
The boundary of the domain Ω_o is $\Gamma_o = \Gamma_I \cup \Gamma_{oN} \cup \Gamma_{oS} \cup \Gamma_{oE} \cup \Gamma_{oW} \cup \Gamma_{oB}$, where Γ_{oB} the bottom of the ocean$(z=-H)$, Γ_{oE} and Γ_{oW} the eastern and western boundaries of the ocean, Γ_{oN} and Γ_{oS} the northern and southern boundaries of the ocean.
The boundary of the domain Ω_a is $\Gamma_a = \Gamma_I \cup \Gamma_{aN} \cup \Gamma_{aS} \cup \Gamma_{aE} \cup \Gamma_{aW} \cup \Gamma_{aU}$, where Γ_{aU} the upper boundary of the atmosphere $(z=L)$, Γ_{aE} and Γ_{aW} the eastern and western boundaries of the atmosphere, Γ_{aN} and Γ_{aS} the northern and southern boundaries of the atmosphere.

We note that the vertical extension $-H \leq z \leq L$ of the domain doesn't correspond to the physical domain but only to the layer where the variability is computed. We assume that for depths greater than H and than L the perturbation is negligible. To take into account the phenomena we want to describe, we set mixed boundary conditions:

- The interface condition on Γ_I is

$$w^{(o)} = w^{(a)} = 0,$$

$$-\nu_{1v}^{(o)} \frac{\partial \mathbf{u}^{(o)}}{\partial z} = \nu_{1v}^{(a)} \frac{\partial \mathbf{u}^{(a)}}{\partial z} = G_1 \mid \mathbf{u}^{(a)} - \mathbf{u}^{(o)} \mid_2^r (\mathbf{u}^{(a)} - \mathbf{u}^{(o)}), \tag{12}$$

$$-\nu_{2v}^{(o)} \frac{\partial \mathcal{T}^{(o)}}{\partial z} = \nu_{2v}^{(a)} \frac{\partial \mathcal{T}^{(a)}}{\partial z}$$

$$= F + G_2(\mathcal{T}^{(a)} - \mathcal{T}^{(o)}) + G_3 \mid \mathbf{u}^{(a)} - \mathbf{u}^{(o)} \mid_2^r (\mathcal{T}^{(a)} - \mathcal{T}^{(o)}),$$

where $r \geq 0$ and $|\,.\,|_2$ is the euclidean norm. The functions $(G_i)_{i=1,3}$ are given positive, time independent, smooth, and bounded functions. The function F is given smooth function. The boundary conditions are given by

- the flow is periodic in the x-direction

$$\mathbf{u}^{(o)} |_{\Gamma_{oE}} = \mathbf{u}^{(o)} |_{\Gamma_{oW}}, w^{(o)} |_{\Gamma_{oE}} = w^{(o)} |_{\Gamma_{oW}}, \mathcal{T}^{(o)} |_{\Gamma_{oE}} = \mathcal{T}^{(o)} |_{\Gamma_{oW}},$$

$$\mathbf{u}^{(a)} |_{\Gamma_{aE}} = \mathbf{u}^{(a)} |_{\Gamma_{aW}}, w^{(a)} |_{\Gamma_{aE}} = w^{(a)} |_{\Gamma_{aW}}, \mathcal{T}^{(a)} |_{\Gamma_{aE}} = \mathcal{T}^{(a)} |_{\Gamma_{aW}},$$

- on the northern and southern boundaries we impose

$$v^{(o)} = 0, \frac{\partial u^{(o)}}{\partial y} = \frac{\partial w^{(o)}}{\partial y} = \frac{\partial \mathcal{T}^{(o)}}{\partial y} = 0 \ on \ \Gamma_{oN} \ and \ \Gamma_{oS}, \tag{13}$$

$$v^{(a)} = 0, \frac{\partial u^{(a)}}{\partial y} = \frac{\partial w^{(a)}}{\partial y} = \frac{\partial \mathcal{T}^{(a)}}{\partial y} = 0 \ on \ \Gamma_{aN} \ and \ \Gamma_{aS},$$

- the perturbation vanishes at $z=-H$ and $z=L$

$$\mathbf{u}^{(o)} = \mathbf{0}, w^{(o)} = \mathcal{T}^{(o)} = 0 \ on \ \Gamma_{oB},$$

$$\mathbf{u}^{(a)} = \mathbf{0}, w^{(a)} = \mathcal{T}^{(a)} = 0 \ on \ \Gamma_{aU},$$

with $\mathbf{u}^{(o)} = (u^{(o)}, v^{(o)})$ and $\mathbf{u}^{(a)} = (u^{(a)}, v^{(a)})$.
We assume that, at initial time $t = 0$, the mean circulation is not disturbed. Therefore the initial condition is

$$\begin{aligned} &\mathbf{u}^{(o)}(t=0) = \mathbf{u}^{(a)}(t=0) = 0, \\ &\mathcal{T}^{(o)}(t=0) = \mathcal{T}^{(a)}(t=0) = 0. \end{aligned} \tag{14}$$

The equations (10) and (11) are not of a Cauchy-Kovalevsky type with respect to all variables because of the continuity equation and of Hydrostatic equation with vertical viscosity. The method proposed by Lions et al. (1992a) is the integration of these two diagnostic equations with respect to the vertical variable, and then we obtain another formulation of the primitive equations, which is a three dimensional evolution system. Therefore we can used Faedo-Galerkin method to solve the weak formulations.
By integrating the continuity equations with respect to z and taking the boundary conditions for $w^{(o)}$ and $w^{(a)}$ into account, we obtain

$$\begin{aligned} &w^{(o)}(x,y,z;t) = W^{(o)}(\mathbf{u}^{(o)})(x,y,z;t) = -\mathrm{div}\Big(\int_{-H}^{z} \mathbf{u}^{(o)}(z')dz'\Big), \\ &w^{(a)}(x,y,z;t) = W^{(a)}(\mathbf{u}^{(a)})(x,y,z;t) = \mathrm{div}\Big(\int_{z}^{L} \mathbf{u}^{(a)}(z')dz'\Big), \\ &\mathrm{div}\Big(\int_{-H}^{0} \mathbf{u}^{(o)}(z')dz'\Big) = 0, \\ &\mathrm{div}\Big(\int_{0}^{L} \mathbf{u}^{(a)}(z')dz'\Big) = 0. \end{aligned} \tag{15}$$

By integrating Hydrostatic equations with vertical viscosity in (10)-(11), we obtain

$$\begin{aligned} &p^{(o)} = P^{(o)} + \alpha^{(o)} \int_{z}^{0} \mathcal{T}^{(o)}(z')dz' + \int_{z}^{0} L^{(o)}(\mathbf{u}^{(o)})(z')dz', \\ &p^{(a)} = P^{(a)} + \alpha^{(a)} \int_{0}^{z} \mathcal{T}^{(a)}(z')dz' - \int_{0}^{z} L^{(a)}(\mathbf{u}^{(a)})(z')dz', \end{aligned} \tag{16}$$

where

$$L^{(o)}(\mathbf{u}^{(o)}) = -\nu_{1h}^{(o)} \Delta W^{(o)}(\mathbf{u}^{(o)}) - \frac{\partial}{\partial z}\Big(\nu_{1v}^{(o)} \frac{\partial W^{(o)}(\mathbf{u}^{(o)})}{\partial z}\Big),$$

$$L^{(a)}(\mathbf{u}^{(a)}) = -\nu_{1h}^{(a)} \Delta W^{(a)}(\mathbf{u}^{(a)}) - \frac{\partial}{\partial z}\Big(\nu_{1v}^{(a)} \frac{\partial W^{(a)}(\mathbf{u}^{(a)})}{\partial z}\Big).$$

$P^{(o)}$ is the pressure of the sea water on the surface of the ocean and $P^{(a)}$ is the geopotential at the isobaric surface z=0 of the air. The unknown functions $P^{(o)}$ and $P^{(a)}$ depend only on the first two space variables x and y and they are the Lagrangian multipliers associated with the constraints of the third and fourth parts of (15). Then we obtain the following reformulation of equations (10)-(11):

• in the ocean

$$\frac{\partial \mathbf{u}^{(o)}}{\partial t} - \nu_{1h}^{(o)}\Delta\mathbf{u}^{(o)} - \frac{\partial}{\partial z}(\nu_{1v}^{(o)}\frac{\partial \mathbf{u}^{(o)}}{\partial z}) + \nabla P^{(o)} + (\mathbf{u}^{(o)}.\nabla)\mathbf{u}^{(o)} + W^{(o)}(\mathbf{u}^{(o)})\frac{\partial \mathbf{u}^{(o)}}{\partial z}$$
$$+(\mathbf{u}_0^{(o)}.\nabla)\mathbf{u}^{(o)} + w_0^{(o)}\frac{\partial \mathbf{u}^{(o)}}{\partial z} + (\mathbf{u}^{(o)}.\nabla)\mathbf{u}_0^{(o)} + W^{(o)}(\mathbf{u}^{(o)})\frac{\partial \mathbf{u}_0^{(o)}}{\partial z}$$
$$+\mathbf{F}\wedge_2 \mathbf{u}^{(o)} + \alpha^{(o)}\nabla R(\mathcal{T}^{(o)}) + \nabla\mathcal{L}(\mathbf{u}^{(o)}) = 0,$$
$$\frac{\partial \mathcal{T}^{(o)}}{\partial t} - \nu_{2h}^{(o)}\Delta\mathcal{T}^{(o)} - \frac{\partial}{\partial z}(\nu_{2v}^{(o)}\frac{\partial \mathcal{T}^{(o)}}{\partial z}) + (\mathbf{u}^{(o)}.\nabla)\mathcal{T}^{(o)} + W^{(o)}(\mathbf{u}^{(o)})\frac{\partial \mathcal{T}^{(o)}}{\partial z} \tag{17}$$
$$+(\mathbf{u}_0^{(o)}.\nabla)\mathcal{T}^{(o)} + w_0^{(o)}\frac{\partial \mathcal{T}^{(o)}}{\partial z} + (\mathbf{u}^{(o)}.\nabla)\mathcal{T}_0^{(o)} + W^{(o)}(\mathbf{u}^{(o)})\frac{\partial \mathcal{T}_0^{(o)}}{\partial z} = 0,$$
$$\text{div}(\int_{-H}^{0} \mathbf{u}^{(o)}(z')dz') = 0,$$

• in the atmosphere

$$\frac{\partial \mathbf{u}^{(a)}}{\partial t} - \nu_{1h}^{(a)}\Delta\mathbf{u}^{(a)} - \frac{\partial}{\partial z}(\nu_{1v}^{(a)}\frac{\partial \mathbf{u}^{(a)}}{\partial z}) + \nabla P^{(a)} + (\mathbf{u}^{(a)}.\nabla)\mathbf{u}^{(a)} + W^{(a)}(\mathbf{u}^{(a)})\frac{\partial \mathbf{u}^{(a)}}{\partial z}$$
$$+(\mathbf{u}_0^{(a)}.\nabla)\mathbf{u}^{(a)} + w_0^{(a)}\frac{\partial \mathbf{u}^{(a)}}{\partial z} + (\mathbf{u}^{(a)}.\nabla)\mathbf{u}_0^{(a)} + W^{(a)}(\mathbf{u}^{(a)})\frac{\partial \mathbf{u}_0^{(a)}}{\partial z}$$
$$+\mathbf{F}\wedge_2 \mathbf{u}^{(a)} + \alpha^{(a)}\nabla R(\mathcal{T}^{(a)}) + \nabla\tilde{\mathcal{L}}(\mathbf{u}^{(a)}) = \mathbf{f}^{(a)},$$
$$\frac{\partial \mathcal{T}^{(a)}}{\partial t} - \nu_{2h}^{(a)}\Delta\mathcal{T}^{(a)} - \frac{\partial}{\partial z}(\nu_{2v}^{(a)}\frac{\partial \mathcal{T}^{(a)}}{\partial z}) + (\mathbf{u}^{(a)}.\nabla)\mathcal{T}^{(a)} + W^{(a)}(\mathbf{u}^{(a)})\frac{\partial \mathcal{T}^{(a)}}{\partial z} \tag{18}$$
$$+(\mathbf{u}_0^{(a)}.\nabla)\mathcal{T}^{(a)} + w_0^{(a)}\frac{\partial \mathcal{T}^{(a)}}{\partial z} + (\mathbf{u}^{(a)}.\nabla)\mathcal{T}_0^{(a)}$$
$$+W^{(a)}(\mathbf{u}^{(a)})\frac{\partial \mathcal{T}_0^{(a)}}{\partial z} - \alpha^{(a)}W^{(a)}(\mathbf{u}^{(a)}) = g^{(a)},$$
$$\text{div}(\int_{0}^{L} \mathbf{u}^{(a)}(z')dz') = 0,$$

where

$$R(\rho) = \int_z^0 \rho(z')dz' \text{ if } z \leq 0 \text{ and } R(\rho) = \int_0^z \rho(z')dz' \text{ if } z \geq 0, \tag{19}$$
$$\mathcal{L}(\mathbf{u}^{(o)}) = \int_z^0 L^{(o)}(\mathbf{u}^{(o)})(z')dz' \text{ and } \tilde{\mathcal{L}}(\mathbf{u}^{(a)}) = \int_0^z L^{(a)}(\mathbf{u}^{(a)})(z')dz'.$$

1.3. Outline of the Chapter

In section 2, we study the properties of the different forms appearing in the weak equations and we prove the existence of the perturbation of the mean flow by using Faedo-Galerkin procedure. For provide the uniqueness, we prove some regularity results for the perturbation. We give sufficient conditions on the perturbation of the external forcing $(\mathbf{f}^{(a)}, g^{(a)})$, on the variability of the interface-stress F and on the mean circulation$(\mathbf{u}_0^{(o)}, w_0^{(o)}, \mathcal{T}_0^{(o)}, \mathbf{u}_0^{(a)}, w_0^{(a)}, \mathcal{T}_0^{(a)})$ in order to prove the regularity results. The regularity being satisfied, we then obtain the uniqueness of the solution. As the purpose of our model is to modelize a physical phenomenon occurring in tropical domain and especially the long waves known as "tropical instabilities", the formulation differs from the model in Lions et al. (1995), and the references therein, by the advection terms related to the mean current, and by specific boundary conditions. Consequently different operators and then difficulties are appearing. In section 3, we study the optimal control problem for the linearized primitive equations with the nonlinear boundary condition. The problem is controlled by the variability of the interface-stress F. So the control is the Neumann type and acts on part of the boundary during a time T. The observation is the vorticity of the ocean and (or) of the atmosphere. We prove the existence of an optimal control by means of minimizing sequences. To characterize the control, we introduce a family of penalized control problems corresponding to a linearization of the direct problem. Then the optimal penalized control is characterized by a set of linear equations: direct and adjoints problems, linked by an inequality. The optimality conditions for the initial problem is obtained as the limit of the penalization. This type of method was introduced by Lions (1969) to control singular distributed systems. The specificity of this work derives from the following features: the time-dependent and coupled atmosphere-ocean type problem with nonlinear boundary conditions, the nature of the equations and especially the non-local constraints: the third part of (17) and of (18), the control acts as Neumann condition on the interface between the atmosphere and the ocean. The assumptions of the model correspond to the physical features of equatorial waves. It is then easy to deduce from the characteristics of the mean circulation the characteristics of the excited waves.

1.4. Assumptions and Notations

Let $L^2(\Omega_o)$ and $L^2(\Omega_a)$ be the L^2-spaces of functions on the respective domains with all the norms and inner products denoted by the same notations $|\,.\,|$ and $(.,.)$.
The usual norm and scalar product in the Sobolev spaces $H^s(\Omega_o)$ and $H^s(\Omega_a)$ $(s \in \mathbb{R})$ are respectively denoted by $\|\,.\,\|_s$ and $((.,.))_s$. Of particular interest are $L^s(\Gamma_I) - norm$ $\|\,.\,\|_{L^s(\Gamma_I)}$ $(s \in \mathbb{R}\,)$ and $L^2(\Gamma_I) - scalar\ product$ $(.)_{\Gamma_I}$. For any Banach space X, we define in the usual way the spaces $L^p(0,T,X)$ $(1 \leq p < \infty)$, $L^\infty(0,T,X)$ and $C(0,T,X)$ of X-valued function $u(t)$ of the time variable t. In order to study Problem (17)-(18), we introduce the following functional spaces:

$$\mathcal{V}_{o0} = \{(\mathbf{v}^{(o)}, w^{(o)}) \in (H^1(\Omega_o))^3 / \ \mathrm{div}(\mathbf{v}^{(o)}) + \frac{\partial w^{(o)}}{\partial z} = 0, \ \ (\mathbf{v}^{(o)}, w^{(o)}).\mathbf{n} = 0 \ on\ \Gamma_o\},$$

$$\mathcal{V}_{a0} = \{(\mathbf{v}^{(a)}, w^{(a)}) \in (H^1(\Omega_a))^3 / \ \mathrm{div}(\mathbf{v}^{(a)}) + \frac{\partial w^{(a)}}{\partial z} = 0, \ \ (\mathbf{v}^{(a)}, w^{(a)}).\mathbf{n} = 0 \ on\ \Gamma_a\},$$

$$\mathcal{H}_{o1} = \{\mathbf{v}^{(o)} \in (L^2(\Omega_o))^2/\ (\mathbf{v}^{(o)},0).\mathbf{n} = 0\ on\ \Gamma_I \cup \Gamma_{oN} \cup \Gamma_{oS} \cup \Gamma_{oB},\ (\mathbf{v}^{(o)},0).\mathbf{n}|_{\Gamma_{oE}} = -(\mathbf{v}^{(o)},0).\mathbf{n}|_{\Gamma_{oW}}\},$$

$$\mathcal{H}_{a1} = \{\mathbf{v}^{(a)} \in (L^2(\Omega_a))^2/\ (\mathbf{v}^{(a)},0).\mathbf{n} = 0\ on\ \Gamma_I \cup \Gamma_{aN} \cup \Gamma_{aS} \cup \Gamma_{aL},\ (\mathbf{v}^{(a)},0).\mathbf{n}|_{\Gamma_{aE}} = -(\mathbf{v}^{(a)},0).\mathbf{n}|_{\Gamma_{aW}}\},$$

$$\mathcal{W}_{o1} = \{\mathbf{v}^{(o)} \in (H^1(\Omega_o))^2/\ W^{(o)}(\mathbf{v}^{(o)}) = -\int_{-H}^{z} \mathrm{div}(\mathbf{v}^{(o)})dz' \in H^1(\Omega_o),\ \mathbf{v}^{(o)} = \mathbf{0}\ on\ \Gamma_{oB},\ (\mathbf{v}^{(o)},0).\mathbf{n} = 0\ on\ \Gamma_I \cup \Gamma_{oN} \cup \Gamma_{oS},\ \mathbf{v}^{(o)}|_{\Gamma_{oE}} = \mathbf{v}^{(o)}|_{\Gamma_{oW}}\},$$

$$\mathcal{W}_{a1} = \{\mathbf{v}^{(a)} \in (H^1(\Omega_a))^2/\ W^{(a)}(\mathbf{v}^{(a)}) = \int_{z}^{L} \mathrm{div}(\mathbf{v}^{(a)})dz' \in H^1(\Omega_a),\ \mathbf{v}^{(a)} = \mathbf{0}\ on\ \Gamma_{aL},\ (\mathbf{v}^{(a)},0).\mathbf{n} = 0\ on\ \Gamma_I \cup \Gamma_{aN} \cup \Gamma_{aS},\ \mathbf{v}^{(a)}|_{\Gamma_{aE}} = \mathbf{v}^{(a)}|_{\Gamma_{aW}}\},$$

$$\mathcal{V}_{o1} = \{\mathbf{v}^{(o)} \in \mathcal{W}_{o1}/\ \int_{-H}^{0} \mathrm{div}(\mathbf{v}^{(o)})dz = 0\},\ \mathcal{V}_{a1} = \{\mathbf{v}^{(a)} \in \mathcal{W}_{a1}/\ \int_{0}^{L} \mathrm{div}(\mathbf{v}^{(a)})dz = 0\},$$

$$\mathcal{H}_{o2} = L^2(\Omega_o),\ \mathcal{H}_{a2} = L^2(\Omega_a),$$

$$\mathcal{V}_{o2} = \{\phi^{(o)} \in H^1(\Omega_o)/\ \phi^{(o)} = 0\ on\ \Gamma_{oN} \cup \Gamma_{oS} \cup \Gamma_{oB}, \phi^{(o)}|_{\Gamma_{oE}} = \phi^{(o)}|_{\Gamma_{oW}}\},$$

$$\mathcal{V}_{a2} = \{\phi^{(a)} \in H^1(\Omega_a)/\ \phi^{(a)} = 0\ on\ \Gamma_{aN} \cup \Gamma_{aS} \cup \Gamma_{aL}, \phi^{(a)}|_{\Gamma_{aE}} = \phi^{(a)}|_{\Gamma_{aW}}\},$$

$$\mathcal{H}_o = \mathcal{H}_{o1} \times \mathcal{H}_{o2},\ \mathcal{H}_a = \mathcal{H}_{a1} \times \mathcal{H}_{a2},\ \mathcal{V}_o = \mathcal{V}_{o1} \times \mathcal{V}_{o2},\ \mathcal{V}_a = \mathcal{V}_{a1} \times \mathcal{V}_{a2},$$

$$\mathcal{H}_1 = \mathcal{H}_{o1} \times \mathcal{H}_{a1},\ \mathcal{V}_1 = \mathcal{V}_{o1} \times \mathcal{V}_{a1},\ \mathcal{W}_1 = \mathcal{W}_{o1} \times \mathcal{W}_{a1},\ \mathcal{V}_2 = \mathcal{V}_{o2} \times \mathcal{V}_{a2},$$

$$\mathcal{H} = \mathcal{H}_1 \times \mathcal{H}_2,\ \mathcal{V} = \mathcal{V}_1 \times \mathcal{V}_2,\ \mathcal{W} = \mathcal{W}_1 \times \mathcal{V}_2,$$

where $\mathbf{n}$ is the unit outward vector normal to the boundary.

Remark 2. *1) The semi-norm and the norm defined on H^1-space are equivalent in $\mathcal{W}_{p1}$ and in $\mathcal{V}_{pi}$ ($p = o,a$ and $i = 1,2$). Then the norms (semi-norms) and inner products for these spaces are denoted by $\|\ .\ \|$ and $((.,.))$.*
2) The spaces $\mathcal{W}_{p1}$ and $\mathcal{V}_{p1}$ ($p = o,a$) are equipped with the following norm:

$$\| \mathbf{u}^{(o)} \|_W^2 = \| \mathbf{u}^{(o)} \|^2 + \| \int_{-H}^{z} \mathrm{div}(\mathbf{u}^{(o)})dz' \|^2,\ \ \forall \mathbf{u}^{(o)} \in \mathcal{W}_{o1}\ (respectively\ \mathbf{u}^{(o)} \in \mathcal{V}_{o1}),$$

$$\| \mathbf{u}^{(a)} \|_W^2 = \| \mathbf{u}^{(a)} \|^2 + \| \int_{z}^{L} \mathrm{div}(\mathbf{u}^{(a)})dz' \|^2,\ \ \forall \mathbf{u}^{(a)} \in \mathcal{W}_{a1}\ (respectively\ \mathbf{u}^{(a)} \in \mathcal{V}_{a1}).$$

3) The spaces $\mathcal{V}_p$ ($p = o,a$) are equipped with the following norm:

$$\| X \|_W^2 = \| \mathbf{u}^{(p)} \|_W^2 + \| \mathcal{T}^{(p)} \|^2,\ \ \forall X = (\mathbf{u}^{(p)}, \mathcal{T}^{(p)}) \in \mathcal{V}_p\ (p = o,a).$$

4) Hereafter we shall us the following notation: $\mathbf{u} = (\mathbf{u}^{(o)}, \mathbf{u}^{(a)})$ and $\mathcal{T} = (\mathcal{T}^{(o)}, \mathcal{T}^{(a)})$. □

We now define the following forms (for $p = o, a$).

$$a_{11}^{(p)}(\mathbf{u}^{(p)}, \mathbf{v}^{(p)}) = \nu_{1h}^{(p)}(\nabla \mathbf{u}^{(p)}, \nabla \mathbf{v}^{(p)}) + (\nu_{1v}^{(p)} \frac{\partial \mathbf{u}^{(p)}}{\partial z}, \frac{\partial \mathbf{v}^{(p)}}{\partial z}),$$

$$a_{12}^{(p)}(\mathbf{u}^{(p)}, \mathbf{v}^{(p)}) = \nu_{1h}^{(p)}(\nabla W^{(p)}(\mathbf{u}^{(p)}), \nabla W^{(p)}(\mathbf{v}^{(p)})) + (\nu_{1v}^{(p)} \mathrm{div}(\mathbf{u}^{(p)}), \mathrm{div}(\mathbf{v}^{(p)})),$$

$$a_{1}^{(p)}(\mathbf{u}^{(p)}, \mathbf{v}^{(p)}) = a_{11}^{(p)}(\mathbf{u}^{(p)}, \mathbf{v}^{(p)}) + a_{12}^{(p)}(\mathbf{u}^{(p)}, \mathbf{v}^{(p)}),$$

$$a_{1I}(\mathbf{u}^{(o)}, \mathbf{u}^{(a)}, \mathbf{v}^{(o)}) = (G_1 \mid \mathbf{u}^{(a)} - \mathbf{u}^{(o)} \mid_2^r (\mathbf{u}^{(a)} - \mathbf{u}^{(o)}), \mathbf{v}^{(o)})_{\Gamma_I},$$

$$a_{2}^{(p)}(\mathcal{T}^{(p)}, \phi^{(p)}) = \nu_{2h}^{(p)}(\nabla \mathcal{T}^{(p)}, \nabla \phi^{(p)}) + (\nu_{2v}^{(p)} \frac{\partial \mathcal{T}^{(p)}}{\partial z}, \frac{\partial \phi^{(p)}}{\partial z}),$$

$$a_{2I}(\mathcal{T}^{(o)}, \mathcal{T}^{(a)}, \phi^{(o)}) = (G_2(\mathcal{T}^{(a)} - \mathcal{T}^{(o)}), \phi^{(o)})_{\Gamma_I},$$

$$a_{3I}(\mathbf{u}^{(o)}, \mathcal{T}^{(o)}, \mathbf{u}^{(a)}, \mathcal{T}^{(a)}, \phi^{(o)}) = (G_3 \mid \mathbf{u}^{(a)} - \mathbf{u}^{(o)} \mid_2^r (\mathcal{T}^{(a)} - \mathcal{T}^{(o)}), \phi^{(o)})_{\Gamma_I},$$

$$a(X, Y) = a_{1}^{(p)}(\mathbf{u}^{(p)}, \mathbf{v}^{(p)}) + a_{2}^{(p)}(\mathcal{T}^{(p)}, \phi^{(p)}), \quad \text{with} \quad X = (\mathbf{u}^{(p)}, \mathcal{T}^{(p)}), \ \ Y = (\mathbf{v}^{(p)}, \phi^{(p)}),$$

$$d^{(p)}(\mathbf{u}^{(p)}, \mathbf{v}^{(p)}) = (\mathbf{F} \wedge_2 \mathbf{u}^{(p)}, \mathbf{v}^{(p)}),$$

$$b_{1}^{(p)}(\mathbf{u}^{(p)}, \mathbf{v}^{(p)}, \mathbf{w}^{(p)}) = ((\mathbf{u}^{(p)}.\nabla)\mathbf{v}^{(p)}, \mathbf{w}^{(p)}) + (W^{(p)}(\mathbf{u}^{(p)}) \frac{\partial \mathbf{v}^{(p)}}{\partial z}, \mathbf{w}^{(p)}),$$

$$b_{2}^{(p)}(\mathbf{u}^{(p)}, \mathcal{T}^{(p)}, \phi^{(p)}) = ((\mathbf{u}^{(p)}.\nabla)\mathcal{T}^{(p)}, \phi^{(p)}) + (W^{(p)}(\mathbf{u}^{(p)}) \frac{\partial \mathcal{T}^{(p)}}{\partial z}, \phi^{(p)}),$$

$$b_{01}^{(p)}(\mathbf{u}_0^{(p)}, \mathbf{v}^{(p)}, \mathbf{w}^{(p)}) = ((U_0^{(p)}.\nabla_3)\mathbf{v}^{(p)}, \mathbf{w}^{(p)}),$$

$$b_{02}^{(p)}(\mathbf{u}_0^{(p)}, \mathcal{T}^{(p)}, \phi^{(p)}) = ((U_0^{(p)}.\nabla_3)\mathcal{T}^{(p)}, \phi^{(p)}),$$

$$c^{(o)}(\phi^{(o)}, \mathbf{v}^{(o)}) = \alpha^{(o)}(\phi^{(o)}, \int_{-H}^{z} \mathrm{div}(\mathbf{v}^{(o)}) dz'),$$

$$c^{(a)}(\phi^{(a)}, \mathbf{v}^{(a)}) = -\alpha^{(a)}(\phi^{(a)}, \int_{z}^{L} \mathrm{div}(\mathbf{v}^{(a)}) dz'),$$

$$l_{1}^{(p)}(\mathbf{u}^{(p)}, \mathbf{v}^{(p)}) = b_{1}^{(p)}(\mathbf{u}^{(p)}, \mathbf{u}_0^{(p)}, \mathbf{v}^{(p)}) + b_{01}^{(p)}(\mathbf{u}_0^{(p)}, \mathbf{u}^{(p)}, \mathbf{v}^{(p)}),$$

$$l_{2}^{(p)}(\mathcal{T}^{(p)}, \phi^{(p)}) = b_{02}^{(p)}(\mathbf{u}_0^{(p)}, \mathcal{T}^{(p)}, \phi^{(p)}),$$

where $U_0^{(p)} = (\mathbf{u}_0^{(p)}, w_0^{(p)})$. Then the following forms can be introduced as:

$$A_1(\mathbf{u},\mathbf{v}) = a_1^{(o)}(\mathbf{u}^{(o)},\mathbf{v}^{(o)}) + a_1^{(a)}(\mathbf{u}^{(a)},\mathbf{v}^{(a)}),$$

$$A_2(\mathcal{T},\phi) = a_2^{(o)}(\mathcal{T}^{(o)},\phi^{(o)}) + a_2^{(a)}(\mathcal{T}^{(a)},\phi^{(a)}),$$

$$A_{1I}(\mathbf{u},\mathbf{v}) = (G_1 \mid \mathbf{u}^{(a)} - \mathbf{u}^{(o)} |_2^r (\mathbf{u}^{(a)} - \mathbf{u}^{(o)}), \mathbf{v}^{(a)} - \mathbf{v}^{(o)})_{\Gamma_I},$$

$$A_{2I}(\mathbf{u},\mathbf{v}) = (G_2(\mathcal{T}^{(a)} - \mathcal{T}^{(o)}), \phi^{(a)} - \phi^{(o)})_{\Gamma_I},$$

$$A_{3I}(\mathbf{u},\mathcal{T},\phi) = (G_3 \mid \mathbf{u}^{(a)} - \mathbf{u}^{(o)} |_2^r (\mathcal{T}^{(a)} - \mathcal{T}^{(o)}), \phi^{(a)} - \phi^{(o)})_{\Gamma_I},$$

$$L_1(\mathbf{u},\mathbf{v}) = l_1^{(o)}(\mathbf{u}^{(o)},\mathbf{v}^{(o)}) + l_1^{(a)}(\mathbf{u}^{(a)},\mathbf{v}^{(a)}),$$

$$L_2(\mathcal{T},\phi) = l_2^{(o)}(\mathcal{T}^{(o)},\phi^{(o)}) + l_2^{(a)}(\mathcal{T}^{(a)},\phi^{(a)}),$$

$$B_1(\mathbf{u},\mathbf{v},\mathbf{w}) = b_1^{(o)}(\mathbf{u}^{(o)},\mathbf{v}^{(o)},\mathbf{w}^{(o)}) + b_1^{(a)}(\mathbf{u}^{(a)},\mathbf{v}^{(a)},\mathbf{w}^{(a)}),$$

$$B_2(\mathbf{u},\mathcal{T},\phi) = b_2^{(o)}(\mathbf{u}^{(o)},\mathcal{T}^{(o)},\phi^{(o)}) + b_2^{(a)}(\mathbf{u}^{(a)},\mathcal{T}^{(a)},\phi^{(a)}),$$

$$B_{01}(\mathbf{u}_0,\mathbf{v},\mathbf{w}) = b_{01}^{(o)}(\mathbf{u}_0^{(o)},\mathbf{v}^{(o)},\mathbf{w}^{(o)}) + b_{01}^{(a)}(\mathbf{u}_0^{(a)},\mathbf{v}^{(a)},\mathbf{w}^{(a)}),$$

$$B_{02}(\mathbf{u}_0,\mathcal{T},\phi) = b_{02}^{(o)}(\mathbf{u}_0^{(o)},\mathcal{T}^{(o)},\phi^{(o)}) + b_{02}^{(a)}(\mathbf{u}_0^{(a)},\mathcal{T}^{(a)},\phi^{(a)}),$$

$$C(\phi,\mathbf{v}) = c^{(o)}(\phi^{(o)},\mathbf{v}^{(o)}) + c^{(a)}(\phi^{(a)},\mathbf{v}^{(a)}),$$

$$D(\mathbf{u},\mathbf{v}) = d^{(o)}(\mathbf{u}^{(o)},\mathbf{v}^{(o)}) + d^{(a)}(\mathbf{u}^{(a)},\mathbf{v}^{(a)}).$$

2. Variational Formulation and Some Properties

Before giving the variational formulation of the problem (17)-(18), we begin by introduce and prove the two following lemmas.

Lemma 1. *$\forall(\mathbf{u}^{(o)},\mathbf{v}^{(o)}) \in (\mathcal{W}_{o1} \cap H^2(\Omega_o))^2$ and $\forall(\mathbf{u}^{(a)},\mathbf{v}^{(a)}) \in (\mathcal{W}_{a1} \cap H^2(\Omega_a))^2$ we have*

(i) $$(\nabla\mathcal{L}(\mathbf{u}^{(o)}),\mathbf{v}^{(o)}) = \nu_{1h}^{(o)}(\nabla(\int_{-H}^{z} \mathrm{div}(\mathbf{u}^{(o)})dz'), \nabla(\int_{-H}^{z} \mathrm{div}(\mathbf{v}^{(o)})dz'))$$
$$+(\nu_{1v}^{(o)}\mathrm{div}(\mathbf{u}^{(o)}), \mathrm{div}(\mathbf{v}^{(o)})) - (\nu_{1v}^{(o)}\gamma_0(\mathrm{div}(\mathbf{u}^{(o)})), \int_{-H}^{0} \mathrm{div}(\mathbf{v}^{(o)})dz),$$

(ii) $$(\nabla\tilde{\mathcal{L}}(\mathbf{u}^{(a)}),\mathbf{v}^{(a)}) = \nu_{1h}^{(a)}(\nabla(\int_{z}^{L} \mathrm{div}(\mathbf{u}^{(a)})dz'), \nabla(\int_{z}^{L} \mathrm{div}(\mathbf{v}^{(a)})dz'))$$
$$+(\nu_{1v}^{(a)}\mathrm{div}(\mathbf{u}^{(a)}), \mathrm{div}(\mathbf{v}^{(a)})) - (\nu_{1v}^{(a)}\gamma_0(\mathrm{div}(\mathbf{u}^{(a)})), \int_{0}^{L} \mathrm{div}(\mathbf{v}^{(a)})dz).$$

Proof. (i) Since

$$(\nabla\mathcal{L}(\mathbf{u}^{(o)}),\mathbf{v}^{(o)}) = \int_{\Omega_o} \nabla(\int_z^0 L^{(o)}(\mathbf{u}^{(o)})dz').\mathbf{v}^{(o)}d\Omega_o,$$

we obtain by using Green's formula

$$(\nabla\mathcal{L}(\mathbf{u}^{(o)}),\mathbf{v}^{(o)}) = -\int_{\Omega}(\int_z^0 L^{(o)}(\mathbf{u}^{(o)})dz')\ \mathrm{div}(\mathbf{v}^{(o)})\ d\Omega_o$$

$$+\int_{\Gamma_o}(\int_z^0 L(\mathbf{u}^{(o)})dz')(\mathbf{v}^{(o)},0).\mathbf{n}d\Gamma_o.$$

As $\mathbf{v}^{(o)}$ is any element in $\mathcal{W}_{o1}$ we obtain

$$(\nabla\mathcal{L}(\mathbf{u}^{(o)}),\mathbf{v}^{(o)}) = -\int_{\Gamma_l}(\int_{-H}^0 (\mathrm{div}(\mathbf{v}^{(o)})\int_z^0 L^{(o)}(\mathbf{u}^{(o)})dz')dz)d\Gamma_l.$$

By integrating by part with respect to z, we have

$$(\nabla\mathcal{L}(\mathbf{u}^{(o)}),\mathbf{v}^{(o)}) = \int_{\Omega_o}(L^{(o)}(\mathbf{u}^{(o)})\int_{-H}^z \mathrm{div}(\mathbf{v}^{(o)})dz')d\Omega_o.$$

Since

$$L^{(o)}(\mathbf{u}^{(o)}) = -\nu_{1h}^{(o)}\Delta W^{(o)}(\mathbf{u}^{(o)}) - \frac{\partial}{\partial z}(\nu_{1v}^{(o)}\frac{\partial W^{(o)}(\mathbf{u}^{(o)})}{\partial z}),$$

and by applying Green's formula, we can deduce that

$$(\nabla\mathcal{L}(\mathbf{u}^{(o)}),\mathbf{v}^{(o)}) = \nu_{1h}^{(o)}(\nabla(\int_{-H}^z \mathrm{div}(\mathbf{u}^{(o)})dz'),\nabla(\int_{-H}^z \mathrm{div}(\mathbf{v}^{(o)})dz'))$$

$$+(\nu_{1v}^{(o)}\mathrm{div}(\mathbf{u}^{(o)}),\mathrm{div}(\mathbf{v}^{(o)})) - (\nu_{1v}^{(o)}\gamma_0(\mathrm{div}(\mathbf{u}^{(o)})),\int_{-H}^0 \mathrm{div}(\mathbf{v}^{(o)})dz')_{\Gamma_l}.$$

Then the result of (i). The same proof is valid to derive for the result of (ii). □

Remark 3. *If* $(\mathbf{v}^{(o)},\mathbf{v}^{(a)}) \in \mathcal{V}_1$ *we can deduce that*

$$(\nabla\mathcal{L}(\mathbf{u}^{(o)}),\mathbf{v}^{(o)}) = \nu_{1h}^{(o)}(\nabla(\int_{-H}^z \mathrm{div}(\mathbf{u}^{(o)})dz'),\nabla(\int_{-H}^z \mathrm{div}(\mathbf{v}^{(o)})dz'))$$

$$+(\nu_{1v}^{(o)}\mathrm{div}(\mathbf{u}^{(o)}),\mathrm{div}(\mathbf{v}^{(o)}))$$

and

$$(\nabla\tilde{\mathcal{L}}(\mathbf{u}^{(a)}),\mathbf{v}^{(a)}) = \nu_{1h}^{(a)}(\nabla(\int_z^L \mathrm{div}(\mathbf{u}^{(a)})dz'),\nabla(\int_z^L \mathrm{div}(\mathbf{v}^{(a)})dz'))$$

$$+(\nu_{1v}^{(a)}\mathrm{div}(\mathbf{u}^{(a)}),\mathrm{div}(\mathbf{v}^{(a)})).\quad \square$$

Lemma 2. *Suppose that* $(\mathbf{u}^{(o)}, \mathcal{T}^{(o)}, P^{(o)})$ *and* $(\mathbf{u}^{(a)}, \mathcal{T}^{(a)}, P^{(a)})$ *are sufficiently regular, then the following results hold.*

(i) *For all* $(\mathbf{v}^{(o)}, \mathbf{v}^{(a)}) \in \mathcal{W}_1$, *we have*

$$-\nu_{1h}^{(o)}(\Delta\mathbf{u}^{(o)}, \mathbf{v}^{(o)}) - (\frac{\partial}{\partial z}(\nu_{1v}^{(o)}\frac{\partial \mathbf{u}^{(o)}}{\partial z}), \mathbf{v}^{(o)}) = a_{11}^{(o)}(\mathbf{u}^{(o)}, \mathbf{v}^{(o)}) - a_{1I}(\mathbf{u}^{(o)}, \mathbf{u}^{(a)}, \mathbf{v}^{(o)}),$$

$$-\nu_{1h}^{(a)}(\Delta\mathbf{u}^{(a)}, \mathbf{v}^{(a)}) - (\frac{\partial}{\partial z}(\nu_{1v}^{(a)}\frac{\partial \mathbf{u}^{(a)}}{\partial z}), \mathbf{v}^{(a)}) = a_{11}^{(a)}(\mathbf{u}^{(a)}, \mathbf{v}^{(a)}) + a_{1I}(\mathbf{u}^{(o)}, \mathbf{u}^{(a)}, \mathbf{v}^{(a)}).$$

(ii) *For all* $(\phi^{(o)}, \phi^{(a)}) \in \mathcal{V}_2$ *we have*

$$\begin{aligned}-\nu_{2h}^{(o)}(\Delta\mathcal{T}^{(o)}, \phi^{(o)}) - (\frac{\partial}{\partial z}(\nu_{2v}^{(o)}\frac{\partial \mathcal{T}^{(o)}}{\partial z}), \phi^{(o)}) &= a_2^{(o)}(\mathcal{T}^{(o)}, \phi^{(o)}) \\ &\quad - a_{2I}(\mathbf{u}^{(o)}, \mathcal{T}^{(o)}, \mathbf{u}^{(a)}, \mathcal{T}^{(a)}, \phi^{(o)}) \\ &\quad - a_{3I}(\mathcal{T}^{(o)}, \mathcal{T}^{(a)}, \phi^{(o)}) - (F, \phi^{(o)})_{\Gamma_I},\end{aligned}$$

$$\begin{aligned}-\nu_{2h}^{(a)}(\Delta\mathcal{T}^{(a)}, \phi^{(a)}) - (\frac{\partial}{\partial z}(\nu_{2v}^{(a)}\frac{\partial \mathcal{T}^{(a)}}{\partial z}), \phi^{(a)}) &= a_2^{(a)}(\mathcal{T}^{(a)}, \phi^{(a)}) \\ &\quad + a_{2I}(\mathbf{u}^{(o)}, \mathcal{T}^{(o)}, \mathbf{u}^{(a)}, \mathcal{T}^{(a)}, \phi^{(a)}) \\ &\quad + a_{3I}(\mathcal{T}^{(o)}, \mathcal{T}^{(a)}, \phi^{(a)}) + (F, \phi^{(a)})_{\Gamma_I}.\end{aligned}$$

(iii) *For all* $(\mathbf{v}^{(o)}, \mathbf{v}^{(a)}) \in \mathcal{W}_1$, *we have*

$$(\nabla P^{(o)}, \mathbf{v}^{(o)}) = -(P^{(o)}, \int_{-H}^{0} \operatorname{div}(\mathbf{v}^{(o)})dz)_{\Gamma_I}, \; (\nabla P^{(a)}, \mathbf{v}^{(a)}) = -(P^{(a)}, \int_{0}^{L} \operatorname{div}(\mathbf{v}^{(o)})dz)_{\Gamma_I},$$

(iv) *For all* $(\mathbf{v}^{(o)}, \mathbf{v}^{(a)}) \in \mathcal{V}_1$, *we have*

$$(\nabla P^{(o)}, \mathbf{v}^{(o)}) = 0, \quad (\nabla P^{(a)}, \mathbf{v}^{(a)}) = 0.$$

(v) *For all* $(\mathbf{v}^{(o)}, \mathbf{v}^{(a)}) \in \mathcal{W}_1$, *we have*

$$(\nabla(\int_{z}^{0} \mathcal{T}^{(o)}dz'), \mathbf{v}^{(o)}) = -(\int_{z}^{0} \mathcal{T}^{(o)}dz', \operatorname{div}(\mathbf{v}^{(o)})) = (\mathcal{T}^{(o)}, \int_{-H}^{z} \operatorname{div}(\mathbf{v}^{(o)})dz'),$$

$$(\nabla(\int_{0}^{z} \mathcal{T}^{(a)}dz'), \mathbf{v}^{(a)}) = -(\int_{0}^{z} \mathcal{T}^{(a)}dz', \operatorname{div}(\mathbf{v}^{(a)})) = -(\mathcal{T}^{(a)}, \int_{z}^{L} \operatorname{div}(\mathbf{v}^{(a)})dz').$$

Proof. Results (i) and (ii) are deduced from the definition of the spaces $\mathcal{W}_{o1}, \mathcal{W}_{a1}, \mathcal{V}_{o2}$ and $\mathcal{V}_{a2}$, and from the boundary conditions satisfied by $(\mathbf{u}^{(o)}, \mathcal{T}^{(o)}, \mathbf{u}^{(a)}, \mathcal{T}^{(a)})$.
(iii) According to Green's formula and boundary conditions verified by $\mathbf{v}^{(o)} \in \mathcal{W}_{o1}$ and $\mathbf{v}^{(a)} \in \mathcal{W}_{a1}$, we then obtain

$$(\nabla P^{(o)}, \mathbf{v}^{(o)}) = -(P^{(o)}, \mathrm{div}(\mathbf{v}^{(o)}))$$

and

$$(\nabla P^{(a)}, \mathbf{v}^{(a)}) = -(P^{(a)}, \mathrm{div}(\mathbf{v}^{(a)})).$$

Since $P^{(o)}$ and $P^{(a)}$ are independent on the variable z, we can deduce the result (iii).
(iv) If $\mathbf{v}^{(o)} \in \mathcal{V}_{o1}$ and $\mathbf{v}^{(a)} \in \mathcal{V}_{a1}$, we have $\int_{-H}^{0} \mathrm{div}(\mathbf{v}^{(o)})dz' = 0$ and $\int_{0}^{L} \mathrm{div}(\mathbf{v}^{(a)})dz' = 0$, and therefore (according to the result (iii)), $(\nabla P^{(o)}, \mathbf{v}^{(o)}) = (\nabla P^{(a)}, \mathbf{v}^{(a)}) = 0$.
(v) According to Green's formula and boundary conditions verified by $\mathbf{v}^{(o)} \in \mathcal{W}_{o1}$, we then have $(\nabla(\int_{z}^{0} \mathcal{T}^{(o)}dz'), \mathbf{v}^{(o)}) = -(\int_{z}^{0} \mathcal{T}^{(o)}dz', \mathrm{div}(\mathbf{v}^{(o)}))$. By integrating by part with respect to z gives

$$(\nabla(\int_{z}^{0} \mathcal{T}^{(o)}dz'), \mathbf{v}^{(o)}) = (\mathcal{T}^{(o)}, \int_{-H}^{z} \mathrm{div}(\mathbf{v}^{(o)})dz').$$

The same proof is valid for $\mathcal{T}^{(a)}$. This completes the proof. □

According to Lemmas 1 and 2, the initial problem (17)(18) with the initial-boundary-interface conditions (12)(13)(14) satisfied by the perturbation $(\mathbf{u}^{(o)}, \mathcal{T}^{(o)}, \mathbf{u}^{(a)}, \mathcal{T}^{(a)})$ of the mean flow admits the two following equivalent weak formulations.

The first formulation (that will be used in the rest of the chapter):

find $(\mathbf{u} = (\mathbf{u}^{(o)}, \mathbf{u}^{(a)}), \mathcal{T} = (\mathcal{T}^{(o)}, \mathcal{T}^{(a)})) \in L^2(0,T,\mathcal{V}_1) \times L^2(0,T,\mathcal{V}_2)$, satisfying (for all $(\mathbf{v}, \phi) \in \mathcal{V}$):

$$\begin{aligned}
&(\frac{\partial \mathbf{u}}{\partial t}, \mathbf{v}) + A_2(\mathbf{u},\mathbf{v}) + A_{1I}(\mathbf{u},\mathbf{v}) + L_1(\mathbf{u},\mathbf{v}) + D(\mathbf{u},\mathbf{v}) \\
&\qquad\qquad + C(\mathcal{T},\mathbf{v}) + B_1(\mathbf{u},\mathbf{u},\mathbf{v}) = (\mathbf{f},\mathbf{v}), \\
&(\frac{\partial \mathcal{T}}{\partial t}, \phi) + A_2(\mathcal{T},\phi) + A_{2I}(\mathcal{T},\phi) + A_{3I}(\mathbf{u},\mathcal{T},\phi) + L_2(\mathcal{T},\phi) \\
&\quad + B_2(\mathbf{u},\mathcal{T},\phi) + c^{(a)}(\phi^{(a)}, \mathbf{u}^{(a)}) + B_2(\mathbf{u},\mathcal{T}_0,\phi) = (F, \phi^{(o)} - \phi^{(a)})_{\Gamma_I} + (\mathbf{g},\phi), \\
&\mathbf{u}(0) = \mathbf{0}, \mathcal{T}(0) = 0,
\end{aligned} \tag{20}$$

where $\mathbf{f} = (0, \mathbf{f}^{(a)})$ and $\mathbf{g} = (0, g^{(a)})$

and the second formulation:

find $((\mathbf{u}^{(o)},\mathbf{u}^{(a)}),(\mathcal{T}^{(o)},\mathcal{T}^{(a)}))\in L^2(0,T,\mathcal{V}_1)\times L^2(0,T,\mathcal{V}_2)$ satisfying:

$$
\begin{aligned}
&(\frac{\partial \mathbf{u}^{(o)}}{\partial t},\mathbf{v}^{(o)})+a_1^{(o)}(\mathbf{u}^{(o)},\mathbf{v}^{(o)})-a_{1I}(\mathbf{u}^{(o)},\mathbf{u}^{(a)},\mathbf{v}^{(o)})\\
&\qquad +l_1^{(o)}(\mathbf{u}^{(o)},\mathbf{v}^{(o)})+d^{(o)}(\mathbf{u}^{(o)},\mathbf{v}^{(o)})+c^{(o)}(\mathcal{T}^{(o)},\mathbf{v}^{(o)})\\
&\qquad\quad +b_1^{(o)}(\mathbf{u}^{(o)},\mathbf{u}^{(o)},\mathbf{v}^{(o)})=0,\quad \forall \mathbf{v}^{(o)}\in\mathcal{V}_{o1},\\
&(\frac{\partial \mathcal{T}^{(o)}}{\partial t},\phi^{(o)})+a^{(o)}2(\mathcal{T}^{(o)},\phi^{(o)})-a_{2I}(\mathbf{u}^{(o)},\mathcal{T}^{(o)},\mathbf{u}^{(a)},\mathcal{T}^{(a)},\phi^{(o)})\\
&\quad -a_{3I}(\mathcal{T}^{(o)},\mathcal{T}^{(a)},\phi^{(o)})+l_2^{(o)}(\mathcal{T}^{(o)},\phi^{(o)})\\
&\quad +b_2^{(o)}(\mathbf{u}^{(o)},\mathcal{T}^{(o)},\phi^{(o)})+b_2^{(o)}(\mathbf{u}^{(o)},\mathcal{T}_0^{(o)},\phi^{(o)})=(F,\phi^{(o)})_{\Gamma_I},\quad \forall\phi^{(o)}\in\mathcal{V}_{o2},\\
&(\frac{\partial \mathbf{u}^{(a)}}{\partial t},\mathbf{v}^{(a)})+a_1^{(a)}(\mathbf{u}^{(a)},\mathbf{v}^{(a)})+a_{1I}(\mathbf{u}^{(o)},\mathbf{u}^{(a)},\mathbf{v}^{(o)})\\
&\qquad +l_1^{(a)}(\mathbf{u}^{(a)},\mathbf{v}^{(a)})+d^{(a)}(\mathbf{u}^{(a)},\mathbf{v}^{(a)})+c^{(a)}(\mathcal{T}^{(a)},\mathbf{v}^{(a)})\\
&\qquad\quad +b_1^{(a)}(\mathbf{u}^{(a)},\mathbf{u}^{(a)},\mathbf{v}^{(a)})=(\mathbf{f}^{(a)},\mathbf{v}^{(a)}),\quad \forall \mathbf{v}^{(a)}\in\mathcal{V}_{a1},\\
&(\frac{\partial \mathcal{T}^{(a)}}{\partial t},\phi^{(a)})+a_2^{(a)}(\mathcal{T}^{(a)},\phi^{(a)})-a_{2I}(\mathbf{u}^{(o)},\mathcal{T}^{(o)},\mathbf{u}^{(a)},\mathcal{T}^{(a)},\phi^{(a)})\\
&\qquad +a_{3I}(\mathcal{T}^{(o)},\mathcal{T}^{(a)},\phi^{(a)})+l_2^{(a)}(\mathcal{T}^{(a)},\phi^{(a)})\\
&\qquad +b_2^{(a)}(\mathbf{u}^{(a)},\mathcal{T}^{(a)},\phi^{(a)})+b_2^{(a)}(\mathbf{u}^{(a)},\mathcal{T}_0^{(a)},\phi^{(a)})\\
&\qquad +c^{(a)}(\phi^{(a)},\mathbf{u}^{(a)})=-(F,\phi^{(a)})_{\Gamma_I}+(g^{(a)},\phi^{(a)}),\quad \forall\phi^{(a)}\in\mathcal{V}_{a2},\\
&\mathbf{u}^{(o)}(0)=\mathbf{u}^{(a)}(0)=\mathbf{0},\\
&\mathcal{T}^{(o)}(0)=\mathcal{T}^{(a)}(0)=0.
\end{aligned}
\tag{21}
$$

In the following propositions, we are dealing with some properties of the forms $a_i^{(o)}$ (respectively $a_i^{(a)}$), (for $i=1,2$), $a^{(o)}$ (respectively $a^{(a)}$), $d^{(o)}$ (respectively $d^{(a)}$), $b_i^{(o)}$ (respectively $b_i^{(a)}$), (for $i=1,2$) and $l_i^{(o)}$ (respectively $l^{(a)}$), (for $i=1,2$).

The propositions 1-3 can be proved like the similar ones in Belmiloudi (2000b):

Proposition 1. *For all* $((\mathbf{u}^{(o)},\mathbf{u}^{(a)}),(\mathbf{v}^{(o)},\mathbf{v}^{(a)}),(\mathbf{w}^{(o)},\mathbf{w}^{(a)}),(\mathcal{T}^{(o)},\mathcal{T}^{(a)}),(\phi^{(o)},\phi^{(a)}))$ *in* $\mathcal{W}_1\times\mathcal{V}_1^2\times\mathcal{V}_2^2$, *we have (for* $p=o,a$*)*

(1) $b_1^{(p)}(\mathbf{u}^{(p)},\mathbf{v}^{(p)},\mathbf{v}^{(p)})=0$ *and* $b_2^{(p)}(\mathbf{u}^{(p)},\mathcal{T}^{(p)},\mathcal{T}^{(p)})=0,\quad$ *if* $\mathbf{u}^{(p)}\in\mathcal{V}_{p1}$,

(2) $b_{01}^{(p)}(\mathbf{u}_0^{(p)},\mathbf{v}^{(p)},\mathbf{v}^{(p)})=0$ *and* $b_{02}^{(p)}(\mathbf{u}_0^{(p)},\phi^{(p)},\phi^{(p)})=0$,

(3) $l_1^{(p)}(\mathbf{u}^{(p)},\mathbf{u}^{(p)}) = b_1^{(p)}(\mathbf{u}^{(p)},\mathbf{u}_0^{(p)},\mathbf{u}^{(p)})$ *and* $l_2^{(p)}(\mathcal{T}^{(p)},\mathcal{T}^{(p)}) = 0$,

(4) $| b_1^{(p)}(\mathbf{u}^{(p)},\mathbf{v}^{(p)},\mathbf{w}^{(p)}) | \leq C_1 \| \mathbf{u}^{(p)} \|_W \| \mathbf{v}^{(p)} \| \| \| \mathbf{w}^{(p)} \|_{L^4}$,

$$| b_1^{(p)}(\mathbf{u}^{(p)},\mathbf{v}^{(p)},\mathbf{w}^{(p)}) | \leq C_1 \| \mathbf{u}^{(p)} \|_W \| \mathbf{v}^{(p)} \| \| \| \mathbf{w}^{(p)} \|_{L^2}^{1/4} \| \mathbf{w}^{(p)} \|^{3/4},$$

$$| b_2^{(p)}(\mathbf{u}^{(p)},\mathcal{T}^{(p)},\mathcal{T}^{(p)}) | \leq C_2 \| \mathbf{u}^{(p)} \|_W \| \mathcal{T}^{(p)} \| \| \| \phi^{(p)} \|_{L^4},$$

$$| b_2^{(p)}(\mathbf{u}^{(p)},\mathcal{T}^{(p)},\phi^{(p)}) | \leq C_2 \| \mathbf{u}^{(p)} \|_W \| \mathcal{T}^{(p)} \| \| \| \phi^{(p)} \|_{L^2}^{1/4} \| \phi^{(p)} \|^{3/4},$$

(5) $| b_{01}^{(p)}(\mathbf{u}_0^{(p)},\mathbf{v}^{(p)},\mathbf{w}^{(p)}) | \leq C_1 \| U_0^{(p)} \| \| \| \mathbf{v}^{(p)} \| \| \| \mathbf{w}^{(p)} \|_{L^4}$,

$$| b_{01}^{(p)}(\mathbf{u}_0^{(p)},\mathbf{v}^{(p)},\mathbf{w}^{(p)}) | \leq C_1 \| U_0^{(p)} \| \| \| \mathbf{v}^{(p)} \| \| \| \mathbf{w}^{(p)} \|_{L^2}^{1/4} \| \mathbf{w}^{(p)} \|^{3/4},$$

$$| b_{02}^{(p)}(\mathbf{u}_0^{(p)},\mathcal{T}^{(p)},\phi^{(p)}) | \leq C_2 \| U_0^{(p)} \| \| \| \mathcal{T}^{(p)} \| \| \| \phi^{(p)} \|_{L^4},$$

$$| b_{02}^{(p)}(\mathbf{u}_0^{(p)},\mathcal{T}^{(p)},\phi^{(p)}) | \leq C_2 \| U_0^{(p)} \| \| \| \mathcal{T}^{(p)} \| \| \| \phi^{(p)} \|_{L^2}^{1/4} \| \phi^{(p)} \|^{3/4}.$$

Proposition 2. *For all* $((\mathbf{u}^{(o)},\mathbf{u}^{(a)}),(\mathbf{w}^{(o)},\mathbf{w}^{(a)}),(\mathcal{T}^{(o)},\mathcal{T}^{(a)}))$ *in* $\mathcal{W}_1^2 \times \mathcal{V}_2$, *and* $((\mathbf{v}^{(o)},\mathbf{v}^{(a)}),(\phi^{(o)},\phi^{(a)}))$ *in* $(\mathcal{W}_1 \cap H^2) \times (\mathcal{V}_2 \cap H^2)$, *we have (for* $p = o,a$*)*

$$| b_1^{(p)}(\mathbf{u}^{(p)},\mathbf{v}^{(p)},\mathbf{w}^{(p)}) | \leq C_1 \| \mathbf{u}^{(p)} \|_W \| \mathbf{v}^{(p)} \|_{H^{3/2}} \| \mathbf{w}^{(p)} \|_{L^2},$$

$$| b_1^{(p)}(\mathbf{u}^{(p)},\mathbf{v}^{(p)},\mathbf{w}^{(p)}) | \leq C_1 \| \mathbf{u}^{(p)} \|_W \| \mathbf{v}^{(p)} \|_{H^2}^{1/2} \| \mathbf{v}^{(p)} \|^{1/2} \| \mathbf{w}^{(p)} \|_{L^2},$$

$$| b_2^{(p)}(\mathbf{u}^{(p)},\mathcal{T}^{(p)},\phi^{(p)}) | \leq C_2 \| \mathbf{u}^{(p)} \|_W \| \mathcal{T}^{(p)} \|_{H^{3/2}} \| \phi^{(p)} \|_{L^2},$$

$$| b_2^{(p)}(\mathbf{u}^{(p)},\mathcal{T}^{(p)},\phi^{(p)}) | \leq C_2 \| \mathbf{u}^{(p)} \|_W \| \mathcal{T}^{(p)} \|_{H^2}^{1/2} \| \mathcal{T}^{(p)} \|^{1/2} \| \phi^{(p)} \|_{L^2}.$$

Proposition 3. *If the mean circulation* $(U_0^{(o)},U_0^{(a)},\mathcal{T}_0^{(o)},\mathcal{T}_0^{(a)})$ *is given such that*

$$(U_0^{(o)},U_0^{(a)}),\ (\mathcal{T}_0^{(o)},\mathcal{T}_0^{(a)}),\ \nabla_3(U_0^{(o)},U_0^{(a)}) \text{ and } \nabla_3(\mathcal{T}_0^{(o)},\mathcal{T}_0^{(a)}) \text{ be in } L^\infty,$$

then we have (for $p = o,a$*)*

$$| b_{01}^{(p)}(\mathbf{u}_0^{(p)},\mathbf{u}^{(p)},\mathbf{v}^{(p)}) | \leq c_0 \| U_0^{(p)} \|_{L^\infty} \| \mathbf{u}^{(p)} \| \| \mathbf{v}^{(p)} \|_{L^2}, \text{ for all } (\mathbf{u}^{(p)},\mathbf{v}^{(p)}) \in \mathcal{V}_{p1} \times \mathcal{H}_{p1},$$

$$| b_{02}^{(p)}(\mathbf{u}_0^{(p)},\mathcal{T}^{(p)},\phi^{(p)}) | \leq c_0 \| U_0^{(p)} \|_{L^\infty} \| \mathcal{T}^{(p)} \| \| \phi^{(p)} \|_{L^2}, \text{ for all } (\mathcal{T}^{(p)},\phi^{(p)}) \in \mathcal{V}_{p2} \times \mathcal{H}_{p2},$$

$$| b_1^{(p)}(\mathbf{u}^{(p)},\mathbf{u}_0^{(p)},\mathbf{v}^{(p)}) | \leq c_0 \| \nabla_3 U_0^{(p)} \|_{L^\infty} \| \mathbf{u}^{(p)} \| \| \mathbf{v}^{(p)} \|_{L^2}, \text{ for all } (\mathbf{u}^{(p)},\mathbf{v}^{(p)}) \in \mathcal{W}_{p1} \times \mathcal{H}_{p1},$$

$$| b_2^{(p)}(\mathbf{u}^{(p)},\mathcal{T}_0^{(p)},\phi^{(p)}) | \leq c_0 \| \nabla_3 \mathcal{T}_0^{(p)} \|_{L^\infty} \| \mathbf{u}^{(p)} \| \| \phi^{(p)} \|_{L^2}, \text{ for all } (\mathbf{u}^{(a)},\phi^{(a)}) \in \mathcal{W}_{p1} \times \mathcal{H}_{p2},$$

$$| l_1^{(p)}(\mathbf{u}^{(p)},\mathbf{v}^{(p)}) | \leq c_0 (\| U_0^{(p)} \|_{L^\infty} + \| \nabla_3 U_0^{(p)} \|_{L^\infty}) \| \mathbf{u}^{(p)} \| \| \mathbf{v}^{(p)} \|_{L^2},$$
$$\text{for all } (\mathbf{u}^{(p)},\mathbf{v}^{(p)}) \in \mathcal{W}_{p1} \times \mathcal{H}_{p1},$$

$$| l_2^{(p)}(\mathcal{T}^{(p)},\phi^{(p)}) | \leq c_0 \| U_0^{(p)} \|_{L^\infty} \| \mathcal{T}^{(p)} \| \| \phi^{(p)} \|_{L^2}, \text{ for all } (\mathcal{T}^{(p)},\phi^{(p)}) \in \mathcal{V}_{p2} \times \mathcal{H}_{p2}.$$

Proposition 4. **(i)** *$a_i^{(o)}, i = 1,2$ (respectively $a_i^{(a)}, i = 1,2$) are bilinear continuous and coercive forms on $\mathcal{W}_{o1}^2$ and $\mathcal{V}_{o2}^2$ (respectively on $\mathcal{W}_{a1}^2$ and $\mathcal{V}_{a2}^2$).*
Indeed

$$a_1^{(o)}(\mathbf{u}^{(a)},\mathbf{u}^{(a)}) \geq \alpha_1^{(o)} \parallel \mathbf{u}^{(o)} \parallel_W^2, \quad \textit{for all } \mathbf{u}^{(o)} \in \mathcal{W}_{o1},$$

$$a_1^{(a)}(\mathbf{u}^{(a)},\mathbf{u}^{(a)}) \geq \alpha_1^{(a)} \parallel \mathbf{u}^{(o)} \parallel_W^2, \quad \textit{for all } \mathbf{u}^{(a)} \in \mathcal{W}_{a1},$$

$$a_2^{(o)}(\mathcal{T}^{(o)},\mathcal{T}^{(o)}) \geq \alpha_2^{(o)} \parallel \mathcal{T}^{(o)} \parallel^2, \quad \textit{for all } \mathcal{T}^{(o)} \in \mathcal{V}_{o2},$$

$$a_2^{(a)}(\mathcal{T}^{(a)},\mathcal{T}^{(a)}) \geq \alpha_2^{(a)} \parallel \mathcal{T}^{(a)} \parallel^2, \quad \textit{for all } \mathcal{T}^{(a)} \in \mathcal{V}_{a2}.$$

(ii) *$d^{(o)}$ (respectively $d^{(a)}$) is a bilinear continuous form on $\mathcal{W}_{o1}^2$ (respectively on $\mathcal{W}_{a1}^2$) and moreover we have*

$$d^{(o)}(\mathbf{v}^{(o)},\mathbf{v}^{(o)}) = d^{(a)}(\mathbf{v}^{(a)},\mathbf{v}^{(a)}) = 0, \quad \textit{for all } (\mathbf{v}^{(o)},\mathbf{v}^{(a)}) \in \mathcal{W}_{o1} \times \mathcal{W}_{a1}.$$

(iii) *The vertical velocities are such that:*

$$\parallel W^{(o)}(\mathbf{u}^{(o)}) \parallel_{L^2} \leq c \parallel \mathbf{u}^{(o)} \parallel \quad \textit{and} \quad \parallel W^{(a)}(\mathbf{u}^{(a)}) \parallel_{L^2} \leq c \parallel \mathbf{u}^{(a)} \parallel.$$

(iv) *For all $(\mathbf{v}^{(o)},\phi^{(o)}) \in \mathcal{V}_o$ and for all $(\mathbf{v}^{(a)},\phi^{(a)}) \in \mathcal{V}_a$ the following estimates hold:*

$$| c^{(o)}(\phi^{(o)},\mathbf{v}^{(o)}) | \leq C_H \parallel \phi^{(o)} \parallel_{L^2} \parallel \mathbf{v}^{(o)} \parallel \quad \textit{and} \quad | c^{(o)}(\phi^{(o)},\mathbf{v}^{(o)}) | \leq C_H \parallel \mathbf{v}^{(o)} \parallel_{L^2} \parallel \phi^{(o)} \parallel,$$

$$| c^{(a)}(\phi^{(a)},\mathbf{v}^{(a)}) | \leq C_L \parallel \phi^{(a)} \parallel_{L^2} \parallel \mathbf{v}^{(a)} \parallel \quad \textit{and} \quad | c^{(a)}(\phi^{(a)},\mathbf{v}^{(a)}) | \leq C_L \parallel \mathbf{v}^{(a)} \parallel_{L^2} \parallel \phi^{(a)} \parallel.$$

Proof. The proof of Proposition 4 is the same on a similar result proved in Belmiloudi (2000b). □

By using the results of Propositions 1-4 we have the following lemmas:

Lemma 3. *For all $(\mathbf{u},\mathbf{v},\mathbf{w},\mathcal{T},\phi) \in \mathcal{W}_1 \times \mathcal{V}_1^2 \times \mathcal{V}_2^2$, the following identities and estimates hold.*

(1) $B_1(\mathbf{u},\mathbf{v},\mathbf{v}) = B_2(\mathbf{u},\mathcal{T},\mathcal{T}) = 0, \quad$ *if* $\mathbf{u} \in \mathcal{V}_1$,

(2) $B_{01}(\mathbf{u}_0,\mathbf{v},\mathbf{v}) = B_{o2}(\mathbf{u}_0,\mathcal{T},\mathcal{T}) = 0$,

(3) $L_1(\mathbf{u},\mathbf{u}) = B_1(\mathbf{u},\mathbf{u}_0,\mathbf{u})$ *and* $L_2(\mathcal{T},\mathcal{T}) = 0$,

(4)

$$| B_1(\mathbf{u},\mathbf{v},\mathbf{w}) | \leq c_1 \parallel \mathbf{u} \parallel_W \parallel \mathbf{v} \parallel \parallel \mathbf{w} \parallel_{L^4},$$
$$| B_1(\mathbf{u},\mathbf{v},\mathbf{w}) | \leq c_1 \parallel \mathbf{u} \parallel_W \parallel \mathbf{v} \parallel \parallel \mathbf{w} \parallel_{L^2}^{1/4} \parallel \mathbf{w} \parallel^{3/4},$$

(5)

$$| B_2(\mathbf{u},\mathcal{T},\phi) | \leq c_2 \parallel \mathbf{u} \parallel_W \parallel \mathcal{T} \parallel \parallel \phi \parallel_{L^4},$$
$$| B_2(\mathbf{u},\mathcal{T},\phi) | \leq c_2 \parallel \mathbf{u} \parallel_W \parallel \mathcal{T} \parallel \parallel \phi \parallel_{L^2}^{1/4} \parallel \phi \parallel^{3/4},$$

(6)

$$| B_{01}(\mathbf{u}_0,\mathbf{v},\mathbf{w}) | \leq c_1 \| U_0 \| \| \mathbf{v} \| \| \mathbf{w} \|_{L^4},$$

$$| B_{01}(\mathbf{u}_0,\mathbf{v},\mathbf{w}) | \leq c_1 \| U_0 \| \| \mathbf{v} \| \| \mathbf{w} \|_{L^2}^{1/4} \| \mathbf{w} \|^{3/4},$$

(7)

$$| B_{02}(\mathbf{u}_0,\mathcal{T},\phi) | \leq c_2 \| U_0 \| \| \mathcal{T} \| \| \phi \|_{L^4},$$

$$| B_2(\mathbf{u}_0,\mathcal{T},\phi) | \leq c_2 \| U_0 \| \| \mathcal{T} \| \| \phi \|_{L^2}^{1/4} \| \phi \|^{3/4}.$$

Lemma 4. *For all* $(\mathbf{u},\mathbf{w},\mathcal{T}) \in \mathcal{W}_1^2 \times \mathcal{V}_2$ *and* $(\mathbf{v},\phi) \in (\mathcal{W}_1 \cap H^2) \times (\mathcal{V}_2 \cap H^2)$*, the following estimates hold.*

(1)

$$| B_1(\mathbf{u},\mathbf{v},\mathbf{w}) | \leq c_1 \| \mathbf{u} \|_W \| \mathbf{v} \|_{H^{3/2}} \| \mathbf{w} \|_{L^2},$$

$$| B_1(\mathbf{u},\mathbf{v},\mathbf{w}) | \leq c_1 \| \mathbf{u} \|_W \| \mathbf{v} \|_{H^2}^{1/2} \| \mathbf{v} \|^{1/2} \| \mathbf{w} \|_{L^2},$$

(2)

$$| B_2(\mathbf{u},\mathcal{T},\phi) | \leq c_2 \| \mathbf{u} \|_W \| \mathcal{T} \|_{H^{3/2}} \| \phi \|_{L^2},$$

$$| B_2(\mathbf{u},\mathcal{T},\phi) | \leq c_2 \| \mathbf{u} \|_W \| \mathcal{T} \|_{H^2}^{1/2} \| \mathcal{T} \|^{1/2} \| \phi \|_{L^2}.$$

Lemma 5. *If the mean circulation* $(U_0,\mathcal{T}_0)$ *is given such that* $U_0, \mathcal{T}_0, \nabla_3 U_0$ *and* $\nabla_3 \mathcal{T}_0 \in L^\infty$*, then there exists a positive constant* c_0

$$| B_{01}(\mathbf{u}_0,\mathbf{u},\mathbf{v}) | \leq c_0 \| U_0 \|_{L^\infty} \| \mathbf{u} \| \| \mathbf{v} \|_{L^2}, \quad \text{for all } (\mathbf{u},\mathbf{v}) \in \mathcal{W}_1 \times \mathcal{H}_1,$$

$$| B_{02}(\mathbf{u}_0,\mathcal{T},\phi) | \leq c_0 \| U_0 \|_{L^\infty} \| \mathcal{T} \| \| \phi \|_{L^2}, \quad \text{for all } (\mathcal{T},\phi) \subset \mathcal{V}_2 \times \mathcal{H}_2,$$

$$| B_1(\mathbf{u},\mathbf{u}_0,\mathbf{v}) | \leq c_0 \| \nabla_3 U_0 \|_{L^\infty} \| \mathbf{u} \| \| \mathbf{v} \|_{L^2}, \quad \text{for all } (\mathbf{u},\mathbf{v}) \in \mathcal{W}_1 \times \mathcal{H}_1,$$

$$| B_2(\mathbf{u},\mathcal{T}_0,\phi) | \leq c_0 \| \nabla_3 \mathcal{T}_0 \|_{L^\infty} \| \mathbf{u} \| \| \phi \|_{L^2}, \quad \text{for all } (\mathbf{u},\phi) \in \mathcal{W}_1 \times \mathcal{H}_2,$$

$$| L_1(\mathbf{u},\mathbf{v}) | \leq c_0(\| U_0 \|_{L^\infty} + \| \nabla_3 U_0 \|_{L^\infty}) \| \mathbf{u} \| \| \mathbf{v} \|_{L^2}, \quad \text{for all } (\mathbf{u},\mathbf{v}) \in \mathcal{W}_1 \times \mathcal{H}_1,$$

$$| L_2(\mathcal{T},\phi) | \leq c_0 \| U_0 \|_{L^\infty} \| \mathcal{T} \| \| \phi \|_{L^2}, \quad \text{for all } (\mathcal{T},\phi) \in \mathcal{V}_2 \times \mathcal{H}_2.$$

Lemma 6. **(i)** A_1 *(respectively* A_2*) is a bilinear continuous and coercive form on* $\mathcal{W}_1^2$ *(respectively on* $\mathcal{V}_2^2$*). Indeed*

$$A_1(\mathbf{u},\mathbf{u}) \geq \alpha_1 \| \mathbf{u} \|_W^2 \text{ and } A_2(\mathcal{T},\mathcal{T}) \geq \alpha_2 \| \mathcal{T} \|^2, \quad \text{for all } \mathbf{u} \in \mathcal{W}_1 \text{ and } \mathcal{T} \in \mathcal{V}_2,$$

$$A(X,X) \geq \alpha \| X \|_W^2, \quad \text{for all } X = (\mathbf{u},\mathcal{T}) \in \mathcal{W}_1 \times \mathcal{V}_2,$$
where $\| X \|_W^2 = \| \mathbf{u} \|_W^2 + \| \mathcal{T} \|^2$,

(ii) *D is a bilinear continuous form on* $\mathcal{W}_1^2$ *and we have the following identity*

$$D(\mathbf{v},\mathbf{v}) = 0, \quad \text{for all } \mathbf{v} \in \mathcal{W}_1.$$

(iii) *For all* $(\mathbf{v},\phi) \in \mathcal{V}_1 \times \mathcal{V}_2$ *we have the following estimates*

$$| C(\phi,\mathbf{v}) | \leq C_{HL} \| \phi \|_{L^2} \| \mathbf{v} \| \quad and \quad | C(\phi,\mathbf{v}) | \leq C_{HL} \| \mathbf{v} \|_{L^2} \| \phi \| .$$

Lemma 7. *For* $\mathbf{u},\mathbf{v},\mathcal{T},\phi$ *sufficiently regular we have, for all* $r \geq 0$

(1) $(G_1(| \mathbf{u} |_2^r \mathbf{u} - | \mathbf{v} |_2^r \mathbf{v}), \mathbf{u}-\mathbf{v})_{\Gamma_I} \geq 0,$

(2) $| (G_1 | \mathbf{u} |_2^r \mathbf{u}, \mathbf{v})_{\Gamma_I} | \leq c \, \| | \mathbf{u} |_2^r \mathbf{u} \|_{L^{\frac{r+2}{r+1}}(\Gamma_I)} \| \mathbf{v} \|_{L^{r+2}(\Gamma_I)} = c \| \mathbf{u} \|_{L^{r+2}(\Gamma_I)}^{r+1} \| \mathbf{v} \|_{L^{r+2}(\Gamma_I)},$

(3) $| (G_3 | \mathbf{u} |_2^r \mathcal{T}, \phi)_{\Gamma_I} | \leq c \, \| | \mathbf{u} |_2^r \mathcal{T} \|_{L^{\frac{r+2}{r+1}}(\Gamma_I)} \| \phi \|_{L^{r+2}(\Gamma_I)}$ *and*

$$\| | \mathbf{u} |_2^r \mathcal{T} \|_{L^{\frac{r+2}{r+1}}(\Gamma_I)} \leq c \| \mathbf{u} \|_{L^{r+2}(\Gamma_I)}^{\frac{r}{2}} \| | \mathbf{u} |_2^r | \mathcal{T} |_2^2 \|_{L^1(\Gamma_I)}^{\frac{1}{2}}.$$

Proof. (1) since

$$(G_1(| \mathbf{u} |_2^r \mathbf{u} - | \mathbf{v} |_2^r \mathbf{v}), \mathbf{u}-\mathbf{v})_{\Gamma_I} = (G_1 | \mathbf{u} |_2^r \mathbf{u}, \mathbf{u})_{\Gamma_I} - (G_1 | \mathbf{u} |_2^r \mathbf{u}, \mathbf{v})_{\Gamma_I}$$
$$-(G_1 | \mathbf{v} |_2^r \mathbf{v}, \mathbf{u})_{\Gamma_I} + (G_1 | \mathbf{v} |_2^r \mathbf{v}, \mathbf{v})_{\Gamma_I}$$

and $| (\mathbf{u},\mathbf{v})_2 | \leq | \mathbf{u} |_2 \, | \mathbf{v} |_2$ we have (according to the positivity of the function G_1)

$$(G_1(| \mathbf{u} |_2^r \mathbf{u} - | \mathbf{v} |_2^r \mathbf{v}), \mathbf{u}-\mathbf{v})_{\Gamma_I} \geq (G_1(| \mathbf{u} |_2^{r+1} - | \mathbf{v} |_2^{r+1}), | \mathbf{u} |_2 - | \mathbf{v} |_2)_{\Gamma_I}.$$

This implies that $(G_1(| \mathbf{u} |_2^r \mathbf{u} - | \mathbf{v} |_2^r \mathbf{v}), \mathbf{u}-\mathbf{v})_{\Gamma_I} \geq 0.$
(2) Since

$$| (G_1 | \mathbf{u} |_2^r \mathbf{u}, \mathbf{v})_{\Gamma_I} | \leq \int_{\Gamma_I} G_1 | \mathbf{u} |_2^{r+1} | \mathbf{v} |_2 \, d\Gamma,$$

by using Young's formula, with $p = \frac{r+2}{r+1}$ and $q = r+2$, and according to the boundedness of the function G_1 we can deduce the result (2) of the lemma.
(3) By using Young's formula (with $p = \frac{r+2}{r+1}, q = r+2$) and according to the boundedness of the function G_3 we have

$$| (G_3 | \mathbf{u} |_2^r \mathcal{T}, \phi)_{\Gamma_I} | \leq c \, \| | \mathbf{u} |_2^r \mathcal{T} \|_{L^{\frac{r+2}{r+1}}(\Gamma_I)} \| \phi \|_{L^{r+2}(\Gamma_I)} .$$

According again to Young's formula (with $p = \frac{2(r+1)}{r}, q = \frac{2(r+1)}{r+2}$), we obtain

$$\| | \mathbf{u} |_2^r \mathcal{T} \|_{L^{\frac{r+2}{r+1}}}^{\frac{r+2}{r+1}} = \int_{\Gamma_I} | \mathbf{u} |_2^{\frac{r(r+2)}{2(r+1)}} (| \mathbf{u} |_2^r | \mathcal{T} |_2^2)^{\frac{r+2}{2(r+1)}} d\Gamma_I$$
$$\leq c \| \mathbf{u} \|_{L^{r+2}(\Gamma_I)}^{\frac{r(r+2)}{2(r+1)}} \| | \mathbf{u} |_2^r | \mathcal{T} |_2^2 \|_{L^1(\Gamma_I)}^{\frac{r+2}{2(r+1)}} .$$

We have then the result (3) of the lemma. This completes the proof. □

We are now going to derive the existence and uniqueness of the solution.

3. Existence and Uniqueness of Solution

3.1. Existence of the Solution

Theorem 1. *For* $U_0 = (\mathbf{u}_0^{(o)}, w_0^{(o)}, \mathbf{u}_0^{(a)}, w_0^{(a)})$, $\mathcal{T}_0 = (\mathcal{T}_0^{(o)}, \mathcal{T}_0^{(a)})$, F, $\mathbf{g}$, $\mathbf{f}$ *given such that* U_0, $\mathcal{T}_0$, $\nabla_3 U_0$, $\nabla_3 \mathcal{T}_0 \in L^\infty$, $\mathbf{f}$, $\mathbf{g} \in L^2(0,T,L^2)$, *and* $F \in L^2(0,T,L^2(\Gamma_I))$, *the variational problem (21) (or (20)) has at least one solution* $X = (\mathbf{u}, \mathcal{T})$ *satisfying*

$$X \in L^2(0,T,\mathcal{V}) \cap L^\infty(0,T,\mathcal{H}),$$

$$|\,\mathbf{u}^{(a)} - \mathbf{u}^{(o)}\,|_2^r\,(\mathbf{u}^{(a)} - \mathbf{u}^{(o)}) \in L^{\frac{r+2}{r+1}}(0,T,L^{\frac{r+2}{r+1}}(\Gamma_I)),$$

$$|\,\mathbf{u}^{(a)} - \mathbf{u}^{(o)}\,|_2^r\,(\mathcal{T}^{(a)} - \mathcal{T}^{(o)}) \in L^{\frac{r+2}{r+1}}(0,T,L^{\frac{r+2}{r+1}}(\Gamma_I)).$$

Proof. To prove the existence of a solution we use the Faedo-Galerkin method.
Set $(\mathbf{v}_1, \ldots, \mathbf{v}_l, \ldots)$ a total and free sequence in $\mathcal{V}_1$ and $(\phi_1, \ldots, \phi_l, \ldots)$ a total and free sequence in $\mathcal{V}_2$ (see e.g., Temam (1997), pp. 255).

Set, for each l, $X_l = (\mathbf{u}_l, \mathcal{T}_l) = \sum_{i=1}^{l} g_{il}(t)(\mathbf{v}_i, \phi_i)$ (the functions g_{il} are scalar functions defined on $[0,T]$) an approximation solution of Problem (21), verifying the following problem (for $k = 1, \ldots, l$):

$$\begin{aligned}
&(\frac{\partial \mathbf{u}_l}{\partial t}, \mathbf{v}_k) + A_1(\mathbf{u}_l, \mathbf{v}_k) + A_{1l}(\mathbf{u}_l, \mathbf{v}_k) + L_1(\mathbf{u}_l, \mathbf{v}_k) + D(\mathbf{u}_l, \mathbf{v}_k)\\
&\qquad + C(\mathcal{T}_l, \mathbf{v}_k) + B_1(\mathbf{u}_l, \mathbf{u}_l, \mathbf{v}_k) = (\mathbf{f}, \mathbf{v}_k),\\
&(\frac{\partial \mathcal{T}_l}{\partial t}, \phi_k) + A_2(\mathcal{T}_l, \phi_k) + A_{2l}(\mathcal{T}_l, \phi_k) + A_{3l}(\mathbf{u}_l, \mathcal{T}_l, \phi_k)\\
&\qquad + L_2(\mathcal{T}_l, \phi_k) + c^{(a)}(\phi_k^{(a)}, \mathbf{u}_l^{(a)}) + B_2(\mathbf{u}_l, \mathcal{T}_l, \phi_k)\\
&\qquad + B_2(\mathbf{u}_l, \mathcal{T}_0, \phi_k) = (F, \phi_k^{(o)} - \phi_k^{(a)})_{\Gamma_I} + (\mathbf{g}, \phi_k),\\
&\mathbf{u}_l(0) = 0, \mathcal{T}_l(0) = 0.
\end{aligned} \tag{22}$$

We can deduce from (22) that

$$\begin{aligned}
&\frac{1}{2}\frac{\partial}{\partial t} |\,\mathbf{u}_l\,|^2 + A_1(\mathbf{u}_l, \mathbf{u}_l) + A_{1l}(\mathbf{u}_l, \mathbf{u}_l) + L_1(\mathbf{u}_l, \mathbf{u}_l) + D(\mathbf{u}_l, \mathbf{u}_l)\\
&\qquad + C(\mathcal{T}_l, \mathbf{u}_l) + B_1(\mathbf{u}_l, \mathbf{u}_l, \mathbf{u}_l) = (\mathbf{f}, \mathbf{u}_l),\\
&\frac{1}{2}\frac{\partial}{\partial t} |\,\mathcal{T}_l\,|^2 + A_2(\mathcal{T}_l, \mathcal{T}_l) + A_{2l}(\mathcal{T}_l, \mathcal{T}_l) + A_{3l}(\mathbf{u}_l, \mathcal{T}_l, \mathcal{T}_l)\\
&\qquad + L_2(\mathcal{T}_l, \mathcal{T}_l) + c^{(a)}(\mathcal{T}_l^{(a)}, \mathbf{u}_l^{(a)}) + B_2(\mathbf{u}_l, \mathcal{T}_l, \mathcal{T}_l)\\
&\qquad + B_2(\mathbf{u}_l, \mathcal{T}_0, \mathcal{T}_l) = (F, \mathcal{T}_l^{(o)} - \mathcal{T}_l^{(a)})_{\Gamma_I} + (\mathbf{g}, \mathcal{T}_l).
\end{aligned}$$

According to Lemma 3 and to the definition of D we obtain

$$\frac{1}{2}\frac{\partial}{\partial t}\mid \mathbf{u}_l\mid^2 + A_1(\mathbf{u}_l,\mathbf{u}_l) + A_{1l}(\mathbf{u}_l,\mathbf{u}_l) = -C(\mathcal{T}_l,\mathbf{u}_l) + B_1(\mathbf{u}_l,\mathbf{u}_l,\mathbf{u}_0) + (\mathbf{f},\mathbf{u}_l),$$

$$\frac{1}{2}\frac{\partial}{\partial t}\mid \mathcal{T}_l\mid^2 + A_2(\mathcal{T}_l,\mathcal{T}_l) + A_{2l}(\mathcal{T}_l,\mathcal{T}_l) + A_{3l}(\mathbf{u}_l,\mathcal{T}_l,\mathcal{T}_l)$$

$$= -c^{(a)}(\mathcal{T}_l^{(a)},\mathbf{u}_l^{(a)}) + B_2(\mathbf{u}_l,\mathcal{T}_l,\mathcal{T}_0)$$

$$+(F,\mathcal{T}_l^{(o)} - \mathcal{T}_l^{(a)})_{\Gamma_l} + (\mathbf{g},\mathcal{T}_l).$$

According to the positivity of $A_{1l}(\mathbf{u}_l,\mathbf{u}_l)$ and $A_{2l}(\mathcal{T}_l,\mathcal{T}_l) + A_{3l}(\mathbf{u}_l,\mathcal{T}_l,\mathcal{T}_l)$, and applying Lemmas 3-6 we have (by using Young's inequality) the estimate

$$\frac{\partial}{\partial t}\mid X_l\mid^2 + \alpha\parallel X_l\parallel_W^2 \leq c_1\mid X_l\mid^2 + c_2\mid \mathbf{f}\mid^2 + c_3\mid \mathbf{g}\mid^2 + c_4\parallel F\parallel^2_{L^2(\Gamma_l)},$$

which implies that X_l is bounded in $L^2(0,T,\mathcal{V})\cap L^\infty(0,T,\mathcal{H})$. Moreover, we have that

$$\begin{aligned}&\int_0^T\Big(\int_{\Gamma_l}\mid \mathbf{u}_l^{(a)} - \mathbf{u}_l^{(o)}\mid_2^{r+2} d\Gamma\Big)dt \leq c,\\ &\int_0^T\Big(\int_{\Gamma_l}\mid \mathbf{u}_l^{(a)} - \mathbf{u}_l^{(o)}\mid_2^{r}\mid \mathcal{T}_l^{(a)} - \mathcal{T}_l^{(o)}\mid_2^2 d\Gamma\Big)dt \leq c.\end{aligned} \tag{23}$$

According to Lemma 7 and by using Young's formula we can deduce that

$$\begin{aligned}&G_1\mid \mathbf{u}_l^{(a)} - \mathbf{u}_l^{(o)}\mid_2^r(\mathbf{u}_l^{(a)} - \mathbf{u}_l^{(o)}) \text{ is bounded in } L^{\frac{r+2}{r+1}}(0,T,L^{\frac{r+2}{r+1}}(\Gamma_l)),\\ &G_3\mid \mathbf{u}_l^{(a)} - \mathbf{u}_l^{(o)}\mid_2^r(\mathcal{T}_l^{(a)} - \mathcal{T}_l^{(o)}) \text{ is bounded in } L^{\frac{r+2}{r+1}}(0,T,L^{\frac{r+2}{r+1}}(\Gamma_l)).\end{aligned} \tag{24}$$

Prove now that $\dfrac{\partial X_l}{\partial t} \in L^{\frac{r+2}{r+1}}(0,T,(\mathcal{V}\cap H^{3/2})')$.
According to (22) we have that, for all $Y = (\mathbf{v},\phi)$ in $\mathcal{V}\cap H^{3/2}$

$$\Big(\frac{\partial \mathbf{u}_l}{\partial t},\mathbf{v}\Big) = -(A_1(\mathbf{u}_l,\mathbf{v}) + A_{1l}(\mathbf{u}_l,\mathbf{v}) + L_1(\mathbf{u}_l,\mathbf{v}) + D(\mathbf{u}_l,\mathbf{v}))$$

$$-(C(\mathcal{T}_l,\mathbf{v}) + B_1(\mathbf{u}_l,\mathbf{u}_l,\mathbf{v})) + (\mathbf{f},\mathbf{v}),$$

$$\Big(\frac{\partial \mathcal{T}_l}{\partial t},\phi\Big) = -(A_2(\mathcal{T}_l,\phi) + A_{2l}(\mathcal{T}_l,\phi) + A_{3l}(\mathbf{u}_l,\mathcal{T}_l,\phi) + L_2(\mathcal{T}_l,\phi))$$

$$-(c^{(a)}(\phi^{(a)},\mathbf{u}_l^{(a)}) + B_2(\mathbf{u}_l,\mathcal{T}_l,\phi))$$

$$-B_2(\mathbf{u}_l,\mathcal{T}_0,\phi) + (F,\phi^{(o)} - \phi^{(a)})_{\Gamma_l} + (\mathbf{g},\phi).$$

According to Lemmas 3-6, we have that

$$|(\frac{\partial X_l}{\partial t},Y)| \leq c_1 \| X_l \|_W \| Y \| + c_2 \| X_l \|_W | X_l | \| Y \|_{H^{3/2}}$$

$$+c_3(|\mathbf{f}|+|\mathbf{g}|+\| F \|_{L^2(\Gamma_l)}) \| Y \|$$

$$+|(G_1 | \mathbf{u}_l^{(a)} - \mathbf{u}_l^{(o)} |_2^r (\mathbf{u}_l^{(a)} - \mathbf{u}_l^{(o)}), \mathbf{v}^{(a)} - \mathbf{v}^{(o)})_{\Gamma_l}|$$

$$+|(G_3 | \mathbf{u}_l^{(a)} - \mathbf{u}_l^{(o)} |_2^r (\mathcal{T}_l^{(a)} - \mathcal{T}_l^{(o)}), \phi^{(a)} - \phi^{(o)})_{\Gamma_l}|.$$

By using Lemma 7 we have then

$$|(\frac{\partial X_l}{\partial t},Y)| \leq c_1 \| X_l \|_W \| Y \| + c_2 \| X_l \|_W | X_l | \| Y \|_{H^{3/2}}$$

$$+c_3(|\mathbf{f}|+|\mathbf{g}|+\| F \|_{L^2(\Gamma_l)}) \| Y \|$$

$$+c_4 \| \mathbf{u}_l^{(a)} - \mathbf{u}_l^{(o)} \|_{L^{r+2}(\Gamma_l)}^{r+1} \| Y \|_{L^{r+2}(\Gamma_l)}$$

$$+c_5 \| \mathbf{u}_l^{(a)} - \mathbf{u}_l^{(o)} \|_{L^{r+2}(\Gamma_l)}^{r/2} \| | \mathbf{u}_l^{(a)} - \mathbf{u}_l^{(o)} |_2^r | \mathcal{T}_l^{(a)} - \mathcal{T}_l^{(o)} |_2^2 \|_{L^1(\Gamma_l)}^{1/2} \| Y \|_{L^{r+2}(\Gamma_l)}.$$

Since $Y \in \mathcal{V} \cap H^{3/2}$, so $\gamma_o Y \in H^1(\Gamma_l)$ and according to Sobolev's inclusion (see Appendix) we have $H^1(\Gamma_l) \subset L^q(\Gamma_l)\ \forall q \geq 2$, we can deduce then

$$\| Y \|_{L^{r+2}(\Gamma_l)} \leq c \| Y \|_{H^1(\Gamma_l)} \leq c \| Y \|_{H^{3/2}}.$$

Applying this result (according to young's formula) we can deduce that

$$\| \frac{\partial X_l}{\partial t} \|_{(\mathcal{V} \cap H^{3/2})'} \leq c_1 \| X_l \|_W + c_2 \| X_l \|_W | X_l |$$

$$+c_3(|\mathbf{f}|+|\mathbf{g}|+\| F \|_{L^2(\Gamma_l)})$$

$$+c_6 \| \mathbf{u}_l^{(a)} - \mathbf{u}_l^{(o)} \|_{L^{r+2}(\Gamma_l)}^{r+1} + c_7 \| | \mathbf{u}_l^{(a)} - \mathbf{u}_l^{(o)} |_2^r | \mathcal{T}_l^{(a)} - \mathcal{T}_l^{(o)} |_2^2 \|_{L^1(\Gamma_l)}^{\frac{r+1}{r+2}}.$$

Since X_l is bounded in $L^\infty(0,T,\mathcal{H}) \cap L^2(0,T,\mathcal{V})$, and according to (23) we obtain that $\dfrac{\partial X_l}{\partial t}$ is bounded in $L^{\frac{r+2}{r+1}}(0,T,(\mathcal{V} \cap H^{3/2})')$.

Since the injection of $\{Z \in L^2(0,T,\mathcal{V})\ \mathit{such\ that}\ \dfrac{\partial Z}{\partial t} \in L^{\frac{r+2}{r+1}}(0,T,(\mathcal{V} \cap H^{3/2})')\}$ into $L^2(0,T,\mathcal{H})$ is compact, then it is possible to extract from X_l a subsequence also denoted by X_l and such that

$$
\begin{aligned}
& X_l \longrightarrow X = (\mathbf{u}, \mathcal{T}) \ strongly \ in \ L^2(0,T,\mathcal{H}), \\
& X_l \rightharpoonup X = (\mathbf{u}, \mathcal{T}) \ weakly \ in \ L^2(0,T,\mathcal{V}), \\
& G_1 \mid \mathbf{u}_l^{(a)} - \mathbf{u}_l^{(o)} \mid_2^r (\mathbf{u}_l^{(a)} - \mathbf{u}_l^{(o)}) \rightharpoonup I_1 \ weakly \ in \ L^{\frac{r+2}{r+1}}(0,T,L^{\frac{r+2}{r+1}}(\Gamma_I)), \\
& G_3 \mid \mathbf{u}_l^{(a)} - \mathbf{u}_l^{(o)} \mid_2^r (\mathcal{T}_l^{(a)} - \mathcal{T}_l^{(o)}) \rightharpoonup I_2 \ weakly \ in \ L^{\frac{r+2}{r+1}}(0,T,L^{\frac{r+2}{r+1}}(\Gamma_I)).
\end{aligned} \tag{25}
$$

We now prove that

$$
I_1 = G_1 \mid \mathbf{u}^{(a)} - \mathbf{u}^{(o)} \mid_2^r (\mathbf{u}^{(a)} - \mathbf{u}^{(o)}) \ and \ I_2 = G_3 \mid \mathbf{u}^{(a)} - \mathbf{u}^{(o)} \mid_2^r (\mathcal{T}^{(a)} - \mathcal{T}^{(o)})
$$

by using the same technique as in Lions et al. (1995).
According to the first and second parts of (25) and to Lions et al. (1968), we have that

$$
\begin{aligned}
X_l &\longrightarrow X \ strongly \ in \ L^2(0,T,H^{1/2+s}) \ \forall\, 0 < s < 1/2 \ and \ then \\
\gamma_0 X_l &\longrightarrow \gamma_0 X \ strongly \ in \ L^2(0,T,H^s(\Gamma_I)) \ \forall\, 0 < s < 1/2.
\end{aligned}
$$

Therefore we can deduce that

$$
\begin{aligned}
& G_1 \mid \mathbf{u}_l^{(a)} - \mathbf{u}_l^{(o)} \mid_2^r (\mathbf{u}_l^{(a)} - \mathbf{u}_l^{(o)}) \longrightarrow G_1 \mid \mathbf{u}^{(a)} - \mathbf{u}^{(o)} \mid_2^r (\mathbf{u}^{(a)} - \mathbf{u}^{(o)}), \\
& \qquad\qquad almost \ everywhere \ (a.e.) \ on \ (0,T) \times \Gamma_I, \\
& G_3 \mid \mathbf{u}_l^{(a)} - \mathbf{u}_l^{(o)} \mid_2^r (\mathcal{T}_l^{(a)} - \mathcal{T}_l^{(o)}) \longrightarrow G_3 \mid \mathbf{u}^{(a)} - \mathbf{u}^{(o)} \mid_2^r (\mathcal{T}^{(a)} - \mathcal{T}^{(o)}), \\
& \qquad\qquad almost \ everywhere \ (a.e.) \ on \ (0,T) \times \Gamma_I.
\end{aligned} \tag{26}
$$

By applying Lemma 1.3 of Lions (1969) (pp. 12), we obtain that (according to (25)-(26))

$$
\begin{aligned}
& G_1 \mid \mathbf{u}_l^{(a)} - \mathbf{u}_l^{(o)} \mid_2^r (\mathbf{u}_l^{(a)} - \mathbf{u}_l^{(o)}) \rightharpoonup G_1 \mid \mathbf{u}^{(a)} - \mathbf{u}^{(o)} \mid_2^r (\mathbf{u}^{(a)} - \mathbf{u}^{(o)}) \\
& \qquad\qquad weakly \ in \ L^{\frac{r+2}{r+1}}(0,T,L^{\frac{r+2}{r+1}}(\Gamma_I)), \\
& G_3 \mid \mathbf{u}_l^{(a)} - \mathbf{u}_l^{(o)} \mid_2^r (\mathcal{T}_l^{(a)} - \mathcal{T}_l^{(o)}) \rightharpoonup G_3 \mid \mathbf{u}^{(a)} - \mathbf{u}^{(o)} \mid_2^r (\mathcal{T}^{(a)} - \mathcal{T}^{(o)}) \\
& \qquad\qquad weakly \ in \ L^{\frac{r+2}{r+1}}(0,T,L^{\frac{r+2}{r+1}}(\Gamma_I)).
\end{aligned} \tag{27}
$$

and therefore

$$
\begin{aligned}
& \int_0^T A_{1I}(\mathbf{u}_l, \mathbf{v})dt \longrightarrow \int_0^T A_{1I}(\mathbf{u}, \mathbf{v})dt, \\
& \int_0^T A_{3I}(\mathbf{u}_l, \mathcal{T}_l, \phi)dt \longrightarrow \int_0^T A_{3I}(\mathbf{u}, \mathcal{T}, \phi)dt, \forall (\mathbf{v}, \phi) \in \mathcal{V}.
\end{aligned}
$$

We can now easily prove that X is solution of Problem (21) and verifies the regularity

$$
\begin{aligned}
& X \in L^2(0,T,\mathcal{V}) \cap L^\infty(0,T,\mathcal{H}), \\
& G_1 \mid \mathbf{u}^{(a)} - \mathbf{u}^{(o)} \mid_2^r (\mathbf{u}^{(a)} - \mathbf{u}^{(o)}) \in L^{\frac{r+2}{r+1}}(0,T,L^{\frac{r+2}{r+1}}(\Gamma_I)), \\
& G_3 \mid \mathbf{u}^{(a)} - \mathbf{u}^{(o)} \mid_2^r (\mathcal{T}^{(a)} - \mathcal{T}^{(o)}) \in L^{\frac{r+2}{r+1}}(0,T,L^{\frac{r+2}{r+1}}(\Gamma_I)).
\end{aligned}
$$

This completes the proof of Theorem. □

3.2. Regularity and Uniqueness of the Solution

In this subsection we study only the case where the function G_3 is null. So the problem (20) becomes:

find $(\mathbf{u} = (\mathbf{u}^{(o)}, \mathbf{u}^{(a)}), \mathcal{T} = (\mathcal{T}^{(o)}, \mathcal{T}^{(a)})) \in L^2(0,T,\mathcal{V}_1) \times L^2(0,T,\mathcal{V}_2)$ satisfying (for all $(\mathbf{v}, \phi) \in \mathcal{V})$

$$(\frac{\partial \mathbf{u}}{\partial t}, \mathbf{v}) + A_2(\mathbf{u},\mathbf{v}) + A_{1I}(\mathbf{u},\mathbf{v}) + L_1(\mathbf{u},\mathbf{v}) + D(\mathbf{u},\mathbf{v})$$

$$+C(\mathcal{T},\mathbf{v}) + B_1(\mathbf{u},\mathbf{u},\mathbf{v}) = (\mathbf{f},\mathbf{v}),$$

$$(\frac{\partial \mathcal{T}}{\partial t}, \phi) + A_2(\mathcal{T},\phi) + A_{2I}(\mathcal{T}\phi) + L_2(\mathcal{T},\phi) + B_2(\mathbf{u},\mathcal{T},\phi) + c^{(a)}(\phi^{(a)}, \mathbf{u}^{(a)}) \tag{28}$$

$$+B_2(\mathbf{u},\mathcal{T}_0,\phi) = (F, \phi^{(o)} - \phi^{(a)})_{\Gamma_I} + (\mathbf{g},\phi),$$

$$\mathbf{u}(0) = \mathbf{0}, \mathcal{T}(0) = 0.$$

From now on we assume the following regularity of external forces $\mathbf{f} = (0, \mathbf{f}^{(a)})$ and $\mathbf{g} = (0, g^{(a)})$

$$(\mathbf{f}^{(a)}, g^{(a)}) \in L^2(0,T,H^1), \frac{\partial}{\partial t}(\mathbf{f}^{(a)}, g^{(a)}) \in L^2(0,T,L^2),$$

$$(\mathbf{f}^{(a)}, g^{(a)})|_{\Gamma_{aE}} = (\mathbf{f}^{(a)}, g^{(a)})|_{\Gamma_{aW}}, (\mathbf{f}^{(a)}, g^{(a)})(t=0) = 0,$$

$$F \in L^2(0,T,H^1(\Gamma_I)), \frac{\partial F}{\partial t} \in L^2(0,T,L^2(\Gamma_I)), \tag{29}$$

$$F|_{\gamma_{IE}} = F|_{\gamma_{IW}} \text{ and } F(t=0) = 0,$$

$$G_i|_{\gamma_{IE}} = G_i|_{\gamma_{IW}}, \text{for } \; i = 1,2.$$

where $\gamma_I = \cup_{i=1,4}$ denotes the boundary of Γ_I: γ_{IE} and γ_{IE} are the western and eastern boundaries, γ_{IS} and γ_{IN} are the southern and northern boundaries.

Remark 4. . *The first and third conditions of (29) imply that* $\mathbf{f} \in C([0,T], L^2)$, $\mathbf{g} \in C([0,T], L^2)$ *and* $F \in C([0,T], L^2(\Gamma_I))$ *(see e.g., Lions (1961)).* □

Theorem 2. *Assume that the external force satisfies conditions (29). Subject to the following conditions:*

(i) $K_\alpha = \alpha - C_B \parallel X_0 \parallel > 0,$

(ii) $K_\alpha^3 > C_B exp(C_0 T/2) G$, *where*

$$G = \parallel S_p \parallel_{L^2(0,T)} + \parallel S \parallel_{L^\infty(0,T)}, \quad S^2(t) = |\, \mathbf{f}\,|^2 + |\, \mathbf{g}\,|^2 + C_{Tr} \parallel F \parallel^2_{L^2(\Gamma_I)},$$

$$S_p^2(t) = |\, \frac{\partial \mathbf{f}}{\partial t}\,|^2 + |\, \frac{\partial \mathbf{g}}{\partial t}\,|^2 + C_{Tr} \parallel \frac{\partial F}{\partial t} \parallel^2_{L^2(\Gamma_I)} \quad \textit{and } X_0 = (\mathbf{u}_0, \mathcal{T}_0),$$

the solution $X = (\mathbf{u}, \mathcal{T})$ *of Problem (28) is in the following space* $\mathcal{M} = \mathcal{M}_1 \times \mathcal{M}_2$ *with* $\mathcal{M} = \{Y$ *such that* $\frac{\partial Y}{\partial t} \in L^2(0,T,\mathcal{V}) \cap L^\infty(0,T,\mathcal{H})$ *and* $\| Y \|_W$ *is bounded in* $[0,T]\}$.

Proof. Let us prove that $\frac{\partial X}{\partial t}$ is also in $L^2(0,T,\mathcal{V}) \cap L^\infty(0,T,\mathcal{H})$.

Let $(\mathbf{v}_i, \phi_i), i = 1, \ldots$ denote a total and free sequence in $\mathcal{V}$ (since $\mathcal{V}$ is separable). For each l, we define the function $X_l = (\mathbf{u}_l, \mathcal{T}_l) = \sum_{i=1}^{l} g_{il}(t)(\mathbf{v}_i, \phi_i)$ (the functions g_{il} are scalar functions defined on $[0,T]$) as the solution of the following approach problem (for $k = 1, \ldots, l$):

$$
\begin{aligned}
&(\frac{\partial \mathbf{u}_l}{\partial t}, \mathbf{v}_k) + A_1(\mathbf{u}_l, \mathbf{v}_k) + A_{1l}(\mathbf{u}_l, \mathbf{v}_k) + L_1(\mathbf{u}_l, \mathbf{v}_k) + D(\mathbf{u}_l, \mathbf{v}_k) \\
&\qquad\qquad + C(\mathcal{T}_l, \mathbf{v}_k) + B_1(\mathbf{u}_l, \mathbf{u}_l, \mathbf{v}_k) = (\mathbf{f}, \mathbf{v}_k), \\
&(\frac{\partial \mathcal{T}_l}{\partial t}, \phi_k) + A_2(\mathcal{T}_l, \phi_k) + A_{2l}(\mathcal{T}_l, \phi_k) + L_2(\mathcal{T}_l, \phi_k) \\
&\qquad\qquad + c^{(a)}(\phi_k^{(a)}, \mathbf{u}_l^{(a)}) + B_2(\mathbf{u}_l, \mathcal{T}_l, \phi_k) \\
&\qquad\qquad + B_2(\mathbf{u}_l, \mathcal{T}_0, \phi_k) = (F, \phi_k^{(o)} - \phi_k^{(a)})_{\Gamma_l} + (\mathbf{g}, \phi_k), \\
&\mathbf{u}_l(0) = 0, \mathcal{T}_l(0) = 0.
\end{aligned} \tag{30}
$$

Set $\mathbf{u}'_l = \frac{\partial \mathbf{u}_l}{\partial t}$, $\mathcal{T}'_l = \frac{\partial \mathcal{T}_l}{\partial t}$, $X'_l = \frac{\partial X_l}{\partial t}$, $\mathbf{f}' = \frac{\partial \mathbf{f}}{\partial t}$, $\mathbf{g}' = \frac{\partial \mathbf{g}}{\partial t}$ and $F' = \frac{\partial F}{\partial t}$. We can deduce from (30) that

$$
\begin{aligned}
&| \mathbf{u}'_l |^2 + A_1(\mathbf{u}_l, \mathbf{u}'_l) + A_{1l}(\mathbf{u}_l, \mathbf{u}'_l) + L_1(\mathbf{u}_l, \mathbf{u}'_l) + D(\mathbf{u}_l, \mathbf{u}'_l) \\
&\qquad\qquad + C(\mathcal{T}_l, \mathbf{u}'_l) + B_1(\mathbf{u}_l, \mathbf{u}_l, \mathbf{u}'_l) = (\mathbf{f}, \mathbf{u}'_l), \\
&| \mathcal{T}'_l |^2 + A_2(\mathcal{T}_l, \mathcal{T}'_l) + A_{2l}(\mathcal{T}_l, \mathcal{T}'_l) + L_2(\mathcal{T}_l, \mathcal{T}'_l) + c^{(a)}(\mathcal{T}_l^{(a)'}, \mathbf{u}_l^{(a)}) \\
&\qquad\qquad + B_2(\mathbf{u}_l, \mathcal{T}_l, \mathcal{T}'_l) + B_2(\mathbf{u}_l, \mathcal{T}_0, \mathcal{T}'_l) = (F, \mathcal{T}_l^{(o)'} - \mathcal{T}_l^{(a)'})_{\Gamma_l} + (\mathbf{g}, \mathcal{T}'_l).
\end{aligned}
$$

Since $\mathbf{u}_l(0) = 0, \mathcal{T}_l(0) = 0$ and $\mathbf{f}(0) = 0$, we obtain

$$
X'_l(0) = 0. \tag{31}
$$

By using (30) we have

$$\frac{\partial \mid \mathbf{u}_l' \mid^2}{2\partial t} + A_1(\mathbf{u}_l', \mathbf{u}_l') + A_{1l}(\mathbf{u}_l', \mathbf{u}_l')$$

$$+ \int_{\Gamma_I} (rG_1 \mid \mathbf{u}_l^{(o)} - \mathbf{u}_l^{(a)} \mid_2^{r-2} (\mathbf{u}_l^{(o)'} - \mathbf{u}_l^{(a)'}, \mathbf{u}_l^{(o)} - \mathbf{u}_l^{(a)})_2^2 d\Gamma_I + L_1(\mathbf{u}_l', \mathbf{u}_l')$$

$$+ D(\mathbf{u}_l', \mathbf{u}_l') + C(\mathcal{T}_l', \mathbf{u}_l') + B_1(\mathbf{u}_l', \mathbf{u}_l, \mathbf{u}_l') + B_1(\mathbf{u}_l, \mathbf{u}_l', \mathbf{u}_l') = (\mathbf{f}', \mathbf{u}_l'),$$

$$\frac{\partial \mid \mathcal{T}_l' \mid^2}{2\partial t} + A_2(\mathcal{T}_l', \mathcal{T}_l') + A_{2l}(\mathcal{T}_l', \mathcal{T}_l') + L_2(\mathcal{T}_l', \mathcal{T}_l')$$

$$+ c^{(a)}(\mathcal{T}_l^{(a)'}, \mathbf{u}_l^{(a)'}) + B_2(\mathbf{u}_l', \mathcal{T}_0, \mathcal{T}_l') + B_2(\mathbf{u}_l', \mathcal{T}_l, \mathcal{T}_l')$$

$$+ B_2(\mathbf{u}_l, \mathcal{T}_l', \mathcal{T}_l') = (F', \mathcal{T}_l^{(o)'} - \mathcal{T}_l^{(a)'})_{\Gamma_I} + (\mathbf{g}', \mathcal{T}_l').$$

According to Lemma 3 we obtain

$$\frac{\partial \mid \mathbf{u}_l' \mid^2}{2\partial t} + A_1(\mathbf{u}_l', \mathbf{u}_l') + A_{1l}(\mathbf{u}_l', \mathbf{u}_l')$$

$$+ \int_{\Gamma_I} (rG_1 \mid \mathbf{u}_l^{(o)} - \mathbf{u}_l^{(a)} \mid_2^{r-2} (\mathbf{u}_l^{(o)'} - \mathbf{u}_l^{(a)'}, \mathbf{u}_l^{(o)} - \mathbf{u}_l^{(a)})_2^2 d\Gamma_I$$

$$= B_1(\mathbf{u}_l', \mathbf{u}_l', \mathbf{u}_0) - C(\mathcal{T}_l', \mathbf{u}_l') + B_1(\mathbf{u}_l', \mathbf{u}_l', \mathbf{u}_l) + (\mathbf{f}', \mathbf{u}_l'),$$

$$\frac{\partial \mid \mathcal{T}_l' \mid^2}{2\partial t} + A_2(\mathcal{T}_l', \mathcal{T}_l') + A_{2l}(\mathcal{T}_l', \mathcal{T}_l') + c^{(a)}(\mathcal{T}_l^{(a)'}, \mathbf{u}_l^{(a)'})$$

$$= B_2(\mathbf{u}_l', \mathcal{T}_l', \mathcal{T}_0) + B_2(\mathbf{u}_l', \mathcal{T}_l', \mathcal{T}_l) + (F', \mathcal{T}_l^{(o)'} - \mathcal{T}_l^{(a)'})_{\Gamma_I} + (\mathbf{g}', \mathcal{T}_l').$$

Applying the properties of the forms $B_i, i = 1,2$, and the coercivity of the form $A_i, i = 1,2$ we obtain the following estimation (according to Lemmas 3 -7)

$$\frac{\partial}{\partial t} \mid X_l' \mid^2 + \alpha \parallel X_l' \parallel_W^2 \leq C_B \parallel X_l' \parallel_W^2 \parallel X_l + X_0 \parallel + (1 + \frac{C_{HL}^2}{\alpha}) \mid X_l' \mid^2$$
$$+ \mid \mathbf{f}' \mid^2 + \mid \mathbf{g}' \mid^2 + C_{Tr} \parallel F' \parallel_{\Gamma_I}^2,$$

and then,

$$\frac{\partial}{\partial t} \mid X_l' \mid^2 + \Psi_l(t) \parallel X_l' \parallel_W^2 \leq C_0 \mid X_l' \mid^2 + S_p^2(t), \tag{32}$$

where

$$\Psi_l(t) = \alpha - C_B \parallel X_l + X_0 \parallel, \ S_p^2(t) = \mid \mathbf{f}' \mid^2 + \mid \mathbf{g}' \mid^2 + C_{Tr} \parallel F' \parallel_{L^2(\Gamma_I)}^2 \text{ and } C_0 = 1 + \frac{C_{HL}^2}{\alpha} > 1.$$

Since $X_l(t=0) = 0$, we have then $\Psi_l(0) = \alpha - C_B \parallel X_0 \parallel$. According to the condition (i) of the proposition, we have $\Psi_l(0) > 0$.

Since Ψ_l is a continuous function, let then T_l be the first time in $[0,T]$ such that $\Psi_l(T_l) = 0$ and $\Psi_l(t) > 0 \ \forall t \in [0, T_l[$.

According to the properties of the forms $(L_i)_{i=1,2}$ and $(B_i)_{i=1,2}$, the equations (30) give,

$$\frac{1}{2}\frac{\partial}{\partial t} \mid \mathbf{u}_l \mid^2 + A_1(\mathbf{u}_l,\mathbf{u}_l) + A_{1l}(\mathbf{u}_l,\mathbf{u}_l) = -C(\mathcal{T}_l,\mathbf{u}_l) + B_1(\mathbf{u}_l,\mathbf{u}_l,\mathbf{u}_0) + (\mathbf{f},\mathbf{u}_l),$$

$$\frac{1}{2}\frac{\partial}{\partial t} \mid \mathcal{T}_l \mid^2 + A_2(\mathcal{T}_l,\mathcal{T}_l) + A_{2l}(\mathcal{T}_l,\mathcal{T}_l) = -c^{(a)}(\mathcal{T}_l^{(a)},\mathbf{u}_l^{(a)}) + B_2(\mathbf{u}_l,\mathcal{T}_l,\mathcal{T}_0)$$
$$+(F,\mathcal{T}_l^{(o)} - \mathcal{T}_l^{(a)})_{\Gamma_l} + (\mathbf{g},\mathcal{T}_l).$$

According to the positivity of $A_{1l}(\mathbf{u}_l,\mathbf{u}_l), A_{2l}(\mathcal{T}_l,\mathcal{T}_l)$, and applying Lemmas 3-6 we have (by using Young's inequality) the estimation

$$\frac{\partial}{\partial t} \mid X_l \mid^2 + \Psi_l(0) \parallel X_l \parallel_W^2 \leq C_0 \mid X_l \mid^2 + S^2(t), \tag{33}$$

where $S^2(t) = \mid \mathbf{f} \mid^2 + \mid \mathbf{g} \mid^2 + C_{Tr} \parallel F \parallel^2_{L^2(\Gamma_l)}$.

Multiplying by $exp(-C_0 t)$ and integrating between 0 and t (since $X_l(0) = 0$), we can prove that

$$\mid X_l \mid^2 (t) \leq \int_0^t exp(C_0 s) S^2(t-s)ds,$$

which implies,

$$\mid X_l \mid^2 (t) \leq \frac{exp(C_0 T) - 1}{C_0} F_\infty^2 \ \forall t \in [0,T], \tag{34}$$

where $F_\infty = \parallel S \parallel_{L^\infty(0,T)}$.

According to (33)-(34) and $C_0 > 1$, we get

$$\Psi_l(0) \parallel X_l \parallel_W^2 \leq exp(C_0 T) F_\infty^2 + 2exp(C_0 T/2) F_\infty \mid X_l' \mid \ \forall t \in [0,T]. \tag{35}$$

Multiplying now (32) by $exp(-C_0 t)$ and integrating between 0 and t (since $X_l'(0) = 0$), we have

$$\mid X_l' \mid^2 (t) \leq exp(C_0 T) F_2^2 \ \forall t \in [0,T_l], \tag{36}$$

where $F_2 = \parallel S_p \parallel_{L^2(0,T)}$.

Set $G = F_\infty + F_2$ and $K_\alpha^2 = \Psi_l(0)$. From the inequalities (35) and (36), we get

$$\parallel X_l \parallel_W \leq \frac{exp(C_0 T/2)}{K_\alpha} G \ \forall t \in [0,T_l]. \tag{37}$$

We deduce from (37) and $\parallel X_l \parallel \leq \parallel X_l \parallel_W$ that

$$K_\alpha \Psi_l(t) \geq K_\alpha(\Psi_l(0) - C_B \parallel X_l \parallel_W) \geq K_\alpha^3 - C_B exp(C_0 T/2) G, \forall t \in [0,T_l].$$

According to the condition (ii) we have that $\Psi_l(t) > 0 \ \forall t \in [0,T_l]$, thus the assumption $\Psi_l(T_l) = 0$ cannot be satisfied and we conclude that $\Psi_l(t) > 0 \ \forall t \in [0,T]$.

We can now deduce from inequality (32) that the sequence X_l' is bounded in $L^2(0,T,\mathcal{V}) \cap L^\infty(0,T,\mathcal{H})$ and we conclude that the limit $X = (\mathbf{u},\mathcal{T})$ solution of problem (28) verifies the regularity $\frac{\partial X}{\partial t} \in L^2(0,T,\mathcal{V}) \cap L^\infty(0,T,\mathcal{H})$.

By using the same technique to obtain (35), we can deduce from (28) that

$$K_\alpha^2 \parallel X \parallel_W^2 (t) \leq exp(C_0 T) F_\infty^2 + 2exp(C_0 T/2) F_\infty \parallel \frac{\partial X}{\partial t} \parallel_{L^\infty(0,T,\mathcal{H})} \ \forall t \in [0,T]$$

and then $\| X \|_W$ is bounded in $[0,T]$.

This completes the proof. □

Proposition 5. *If the external force satisfies the regularity (29) under conditions (i) and (ii) of Theorem 2, and if $U_0, \mathcal{T}_0$ are sufficiently regular, the solution $X = (\mathbf{u}, \mathcal{T})$ of Problem (28) is unique.*

Proof. Let us suppose that there exists two solutions $X_1 = (\mathbf{u}_1, \mathcal{T}_1)$ and $X_2 = (\mathbf{u}_2, \mathcal{T}_2)$.

Set $X = X_1 - X_2 = (\mathbf{u}, \mathcal{T})$, by using (27) we have

$$\frac{1}{2}\frac{\partial}{\partial t} \mid \mathbf{u} \mid^2 + A_1(\mathbf{u},\mathbf{u}) + L_1(\mathbf{u},\mathbf{u}) + D(\mathbf{u},\mathbf{u}) + C(\mathcal{T},\mathbf{u})$$

$$+ B_1(\mathbf{u}_1,\mathbf{u}_1,\mathbf{u}) - B_1(\mathbf{u}_2,\mathbf{u}_2,\mathbf{u}) + A_{1I}(\mathbf{u}_1,\mathbf{u}) - A_{1I}(\mathbf{u}_2,\mathbf{u}) = 0,$$

$$\frac{1}{2}\frac{\partial}{\partial t} \mid \mathcal{T} \mid^2 + A_2(\mathcal{T},\mathcal{T}) + L_2(\mathcal{T},\mathcal{T}) + A_{2I}(\mathcal{T},\mathcal{T}) + c^{(a)}(\mathcal{T}^{(a)},\mathbf{u}^{(a)})$$

$$+ B_2(\mathbf{u},\mathcal{T}_0,\mathcal{T}) + B_2(\mathbf{u}_1,\mathcal{T}_1,\mathcal{T}) - B_2(\mathbf{u}_2,\mathcal{T}_2,\mathcal{T}) = 0.$$

In the same way that to obtain (33) we have

$$\frac{\partial}{\partial t} \mid X \mid^2 + (\alpha - C_B \parallel X_2 + X_0 \parallel) \parallel X \parallel_W^2 + 2(A_{1I}(\mathbf{u}_1,\mathbf{u}) - A_{1I}(\mathbf{u}_2,\mathbf{u})) \leq c_1 \mid X \mid^2.$$

According to Lemma 7 and to Theorem 2 we obain $A_{1I}(\mathbf{u}_1,\mathbf{u}) - A_{1I}(\mathbf{u}_2,\mathbf{u}) \geq 0$ and $\alpha - C_B \parallel X_2 + X_0 \parallel \geq 0$.

We can now deduce that $\frac{\partial}{\partial t} \mid X \mid^2 \leq c_1 \mid X \mid^2$, and then $\frac{\partial}{\partial t}(\mid X \mid^2 exp(-c_1 t)) \leq 0$. By integrating between 0 and t, we obtain (since $X(0) = 0$) that $X = 0$ and we conclude that the solution is unique. □

4. Study of Control Framework

In this section we study the optimal control problem for the linear problem associated to (17)-(18) with nonlinear boundary conditions in the physical case i.e. $r = 1$ (see Pedlosky (1979), pp. 245).

4.1. Formulation of the Optimal Control Problem

If we neglect the nonlinear terms in system (17)-(18), the problem (28) becomes:

Find $(\mathbf{u} = (\mathbf{u}^{(o)}, \mathbf{u}^{(a)}), \mathcal{T} = (\mathcal{T}^{(o)}, \mathcal{T}^{(a)})) \in L^2(0,T,\mathcal{V}_1) \times L^2(0,T,\mathcal{V}_2)$ satisfying

$$\begin{aligned}
&(\frac{\partial \mathbf{u}}{\partial t},\mathbf{v}) + A_1(\mathbf{u},\mathbf{v}) + A_{1I}(\mathbf{u},\mathbf{v}) + L_1(\mathbf{u},\mathbf{v}) + D(\mathbf{u},\mathbf{v}) + C(\mathcal{T},\mathbf{v}) = (\mathbf{f},\mathbf{v}),\\
&(\frac{\partial \mathcal{T}}{\partial t},\phi) + A_2(\mathcal{T},\phi) + A_{2I}(\mathcal{T},\phi) + L_2(\mathcal{T},\phi) + c^{(a)}(\phi^{(a)},\mathbf{u}^{(a)})\\
&\qquad + B_2(\mathbf{u},\mathcal{T}_0,\phi) = (F,\phi^{(o)} - \phi^{(a)})_{\Gamma_I} + (\mathbf{g},\phi)\ \forall(\mathbf{v},\phi) \in \mathcal{V},\\
&\mathbf{u}(0) = \mathbf{0}, \mathcal{T}(0) = 0.
\end{aligned} \tag{38}$$

Theorem 3. *Under the assumptions of Theorem 1, the problem (38) admits one unique solution $X = (\mathbf{u}, \mathcal{T})$ satisfying*

$$X \in L^2(0,T,\mathcal{V}) \cap L^\infty(0,T,\mathcal{H})$$

and

$$| \mathbf{u}^{(a)} - \mathbf{u}^{(o)} |_2 (\mathbf{u}^{(a)} - \mathbf{u}^{(o)}) \in L^{\frac{3}{2}}(0,T,L^{\frac{3}{2}}(\Gamma_I)).$$

Proof. The prove of the existence result is similar to that for the problem (28). We are now going to prove the uniqueness.

Let us suppose that there exist two solutions $X_1 = (\mathbf{u}_1, \mathcal{T}_1)$ and $X_2 = (\mathbf{u}_2, \mathcal{T}_2)$.
Set $X = X_1 - X_2 = (\mathbf{u}, \mathcal{T})$, by using (38) we have

$$\frac{1}{2}\frac{\partial}{\partial t} | \mathbf{u} |^2 + A_1(\mathbf{u},\mathbf{u}) + L_1(\mathbf{u},\mathbf{u}) + D(\mathbf{u},\mathbf{u}) + C(\mathcal{T},\mathbf{u}) + A_{1I}(\mathbf{u}_1,\mathbf{u}) - A_{1I}(\mathbf{u}_2,\mathbf{u}) = 0,$$

$$\frac{1}{2}\frac{\partial}{\partial t} | \mathcal{T} |^2 + A_2(\mathcal{T},\mathcal{T}) + L_2(\mathcal{T},\mathcal{T}) + c^{(a)}(\mathcal{T}^{(a)},\mathbf{u}^{(a)}) + B_2(\mathbf{u},\mathcal{T}_0,\mathcal{T}) + A_{2I}(\mathcal{T},\mathcal{T}) = 0.$$

In the same way that to obtain (33) we have

$$\frac{\partial}{\partial t} | X |^2 + \alpha \| X \|_W^2 + 2(A_{1I}(\mathbf{u}_1,\mathbf{u}) - A_{1I}(\mathbf{u}_2,\mathbf{u})) \leq c_1 | X |^2 .$$

According to Lemma 7 we obain $A_{1I}(\mathbf{u}_1,\mathbf{u}) - A_{1I}(\mathbf{u}_2,\mathbf{u}) \geq 0$.
We can now deduce that $\frac{\partial}{\partial t} | X |^2 \leq c_1 | X |^2$, and then $\frac{\partial}{\partial t}(| X |^2 \, exp(-c_1 t)) \leq 0$. By integrating between 0 and t, we obtain (since $X(0) = 0$) that $X = 0$ and we conclude that the solution is unique. This completes the proof. □

The problem is controlled by the variability of the interface-stress F. The observation is the vorticity of the flow.
The cost function J is defined by

$$J(G,\mathbf{v}) = \frac{C^{(a)}}{2} \| curl(\mathbf{v}^{(a)} - \mathbf{u}_{obs}^{(a)}) \|_{L^2(0,T,L^2)}^2 + \frac{C^{(o)}}{2} \| curl(\mathbf{v}^{(o)} - \mathbf{u}_{obs}^{(o)}) \|_{L^2(0,T,L^2)}^2 + \frac{\gamma}{2} \| G \|_{U_c}^2, \tag{39}$$

where $\gamma > 0$, $C^{(o)} \geq 0$, $C^{(a)} \geq 0$, $C^{(o)} + C^{(a)} > 0$ and $U_c = L^2(0,T,L^2(\Gamma_I))$.
Let K_c be a convex, closed, nonempty subset of U_c. The optimal control problem can be written as follows:

$$\begin{aligned} &Find\ (F,\mathbf{u}) \in U_{ad}\ such\ that \\ &J(F,\mathbf{u}) = \inf_{(G,\mathbf{v}) \in U_{ad}} J(G,\mathbf{v}) \end{aligned} \tag{40}$$

$U_{ad} = \{(G,\mathbf{v}) \in K_c \times \mathcal{N}_1 / J(G,\mathbf{v}) < \infty$, and there exists $\mathcal{T} \in \mathcal{N}_2$ such that $(\mathbf{v},\mathcal{T},G)$ is solution of problem (38), with a forcing equal to G $\}$

is the admissibility set, with

$$\mathcal{N}_1 = \{\mathbf{v} \in L^2(0,T,\mathcal{V}_1) \cap L^\infty(0,T,\mathcal{H}_1) \text{ such that } \mid \mathbf{v}^{(a)} - \mathbf{v}^{(o)} \mid_2 (\mathbf{v}^{(a)} - \mathbf{v}^{(o)}) \in L^{\frac{3}{2}}(0,T,L^{\frac{3}{2}}(\Gamma_I))\}$$

and

$$\mathcal{N}_2 = L^2(0,T,\mathcal{V}_2) \cap L^\infty(0,T,\mathcal{H}_2).$$

We are now going to study the existence of a solution to the optimal control problem (40).

4.2. Existence of An Optimal Solution

Theorem 4. *Under the assumptions of Theorem 1, the optimal control problem (40) has at least one solution $(F,\mathbf{u}) \in U_{ad}$.*

Proof. Let $(F_k,\mathbf{u}_k)$ be a minimizing sequence in U_{ad} for the function J such that

$$\liminf_{k\to\infty} J(F_k,\mathbf{u}_k) = \inf_{(G,\mathbf{v})\in U_{ad}} J(G,\mathbf{v})$$

Setting $(\mathbf{v},\phi) = (\mathbf{u}_k,\mathcal{T}_k)$ in (38) gives

$$(\frac{\partial \mathbf{u}_k}{\partial t},\mathbf{v}) + A_2(\mathbf{u}_k,\mathbf{v}) + A_{1I}(\mathbf{u}_k,\mathbf{v}) + L_1(\mathbf{u}_k,\mathbf{v}) + D(\mathbf{u}_k,\mathbf{v}) + C(\mathcal{T}_k,\mathbf{v}) = (\mathbf{f},\mathbf{v}),$$

$$(\frac{\partial \mathcal{T}_k}{\partial t},\phi) + A_2(\mathcal{T}_k,\phi) + A_{2I}(\mathcal{T}_k,\phi) + L_2(\mathcal{T}_k,\phi) + c^{(a)}(\phi^{(a)},\mathbf{u}_k^{(a)})$$

$$+ B_2(\mathbf{u}_k,\mathcal{T}_0,\phi) = (F_k,\phi^{(o)} \quad \phi^{(a)})_{\Gamma_I} + (\mathbf{g},\phi), \quad \forall(\mathbf{v},\phi) \in \mathcal{V},$$

$$\mathbf{u}_k(0) = \mathbf{0}, \mathcal{T}_k(0) = 0.$$

Since $J(F_k,\mathbf{u}_k)$ is bounded for a minimizing sequence $(F_k,\mathbf{u}_k) \in U_{ad}$, then F_k is uniformly bounded in U_c, according to the definition of J.
Then by using the same techniques as to obtain the existence result in Theorem 1 we have that
$X_k = (\mathbf{u}_k,\mathcal{T}_k)$ is uniformly bounded in $L^2(0,T,\mathcal{V}) \cap L^\infty(0,T,\mathcal{H})$, $\frac{\partial X_k}{\partial t}$ is uniformly bounded in $L^{3/2}(0,T,(\mathcal{V} \cap H^{3/2})')$ and $\int_0^T (\int_{\Gamma_I} G_1 \mid \mathbf{u}_k^{(a)} - \mathbf{u}_k^{(o)} \mid_2^3 d\Gamma_I)dt$ is uniformly bounded. Moreover these result makes it possible to extract from X_k a subsequence also denoted by X_k and such that

$F_k \rightharpoonup F$ *weakly in* U_c,

$X_k \longrightarrow X = (\mathbf{u},\mathcal{T})$ *strongly in* $L^2(0,T,\mathcal{H})$,

$X_k \rightharpoonup X = (\mathbf{u},\mathcal{T})$ *weakly in* $L^2(0,T,\mathcal{V})$,

$\mid \mathbf{u}_k^{(a)} - \mathbf{u}_k^{(o)} \mid_2 (\mathbf{u}_k^{(a)} - \mathbf{u}_k^{(o)}) \rightharpoonup \mid \mathbf{u}^{(a)} - \mathbf{u}^{(o)} \mid_2 (\mathbf{u}^{(a)} - \mathbf{u}^{(o)})$ *weakly in* $L^{\frac{3}{2}}(0,T,L^{\frac{3}{2}}(\Gamma_I))$, where (X,F) is solution of problem (38), verifies the following regularity results

$$X \in L^2(0,T,\mathcal{V}) \cap L^\infty(0,T,\mathcal{H}) \text{ and } \mid \mathbf{u}^{(a)} - \mathbf{u}^{(o)} \mid_2 (\mathbf{u}^{(a)} - \mathbf{u}^{(o)}) \in L^{\frac{3}{2}}(0,T,L^{\frac{3}{2}}(\Gamma_I)).$$

Using the sequential weak lower semi-continuity of the function J we conclude that

$$J(F,\mathbf{u}) \leq \liminf_{k\to\infty} J(F_k,\mathbf{u}_k),$$

and then $J(F,\mathbf{u}) = \inf_{(G,\mathbf{v})\in U_{ad}} J(G,\mathbf{v})$, which achieves the proof. □

Remark 5. *The question of the uniqueness is not treated; in general in nonlinear problems the answer is expected to be negative.* □

In order to characterize the optimal control, we introduce a cost function J_ε which is a penalization of J corresponding to a linearization of problem (38). We then proceed as follows. First, we define penalized control problems corresponding to the linearization of initial control problem. Then the optimal penalized control is characterized by a set a linear systems: direct and adjoint problems, linked by an inequality. Finally by passing to the limit (when ε tends to 0) in the characterization of the penalized problems we obtain the optimality conditions.

Let $(F,\mathbf{u})$ be the solution of control problem (40). The function J_ε is defined by

$$\begin{aligned} J_\varepsilon(G,\mathbf{w}) = & \frac{C^{(a)}}{2} \| curl(\mathbf{U}^{(a)}(G,\mathbf{w}) - \mathbf{u}^{(a)}_{obs}) \|^2_{L^2(0,T,L^2)} \\ & + \frac{C^{(o)}}{2} \| curl(\mathbf{U}^{(o)}(G,\mathbf{w}) - \mathbf{u}^{(o)}_{obs}) \|^2_{L^2(0,T,L^2)} \\ & + \frac{1}{2\varepsilon}\int_0^T A_1(\mathbf{U}(G,\mathbf{w}) - \mathbf{w}, \mathbf{U}(G,\mathbf{w}) - \mathbf{w})dt \\ & + \frac{1}{2} \| \mathbf{U}(G,\mathbf{w}) - \mathbf{u} \|^2_{L^2(0,T,\mathcal{H}_1)} + \frac{1}{2}\gamma \| G \|^2_{\mathcal{U}_c} + \frac{1}{2} \| G - F \|^2_{\mathcal{U}_c}, \end{aligned} \tag{41}$$

where $(\mathbf{U}(G,\mathbf{w}),\tau(G,\mathbf{w}))$ is solution of the following problem:

$$\begin{aligned} & (\frac{\partial \mathbf{U}}{\partial t},\mathbf{v}) + A_1(\mathbf{U},\mathbf{v}) + L_1(\mathbf{U},\mathbf{v}) + D(\mathbf{U},\mathbf{v}) + C(\tau,\mathbf{v}) \\ & \quad + \int_{\Gamma_I} G_1(\mid \mathbf{w}^{(a)} - \mathbf{w}^{(o)} \mid_2 (\mathbf{U}^{(a)} - \mathbf{U}^{(o)}), \mathbf{v}^{(a)} - \mathbf{v}^{(o)}) = (\mathbf{f},\mathbf{v}), \\ & (\frac{\partial \tau}{\partial t},\phi) + A_2(\tau,\phi) + A_{2I}(\tau,\phi) + L_2(\tau,\phi) + c^{(a)}(\phi^{(a)},\mathbf{U}^{(a)}) \\ & \quad + B_2(\mathbf{U},\mathcal{T}_0,\phi) = (G,\phi^{(o)} - \phi^{(a)})_{\Gamma_I} + (\mathbf{g},\phi), \quad \forall (\mathbf{v},\phi) \in \mathcal{V}, \\ & (\mathbf{U},\tau)(t=0) = (\mathbf{0},0). \end{aligned} \tag{42}$$

Lemma 8. $\mathbf{U}(F,\mathbf{u}) = \mathbf{u}$ *and* $\tau(F,\mathbf{u}) = \mathcal{T}$, *where* $(\mathbf{u},\mathcal{T})$ *is the solution of Problem (38), the interface forcing being equal to* F.

Proof. Let $\mathbf{u}_1 = \mathbf{U}(F,\mathbf{u}) - \mathbf{u}$ and $\mathcal{T}_1 = \tau(F,\mathbf{u}) - \mathcal{T}$.

Subtracting equations (38) from (42) gives

$$\begin{aligned}
&(\frac{\partial \mathbf{u}_1}{\partial t},\mathbf{v}) + A_1(\mathbf{u}_1,\mathbf{v}) + L_1(\mathbf{u}_1,\mathbf{v}) + D(\mathbf{u}_1,\mathbf{v}) + C(\mathcal{T}_1,\mathbf{v})\\
&\qquad + \int_{\Gamma_I} G_1(|\,\mathbf{u}^{(a)} - \mathbf{u}^{(o)}\,|_2\,(\mathbf{u}_1^{(a)} - \mathbf{u}_1^{(o)}),\mathbf{v}^{(a)} - \mathbf{v}^{(o)})_2 d\Gamma_I = 0,\\
&(\frac{\partial \mathcal{T}_1}{\partial t},\phi) + A_2(\mathcal{T}_1,\phi) + A_{2I}(\mathcal{T}_1,\phi) + L_2(\mathcal{T}_1,\phi) + c^{(a)}(\phi^{(a)},\mathbf{u}_1^{(a)})\\
&\qquad + B_2(\mathbf{u}_1,\mathcal{T}_0,\phi) = 0, \quad \forall(\mathbf{v},\phi) \in \mathcal{V},\\
&(\mathbf{u}_1,\mathcal{T}_1)(t=0) = (\mathbf{0},0).
\end{aligned} \tag{43}$$

Setting $(\mathbf{v},\phi) = (\mathbf{u}_1,\mathcal{T}_1)$ in (43) and applying Lemmas 3-6 we have the following estimation

$$\frac{\partial\,|\,Y\,|^2}{\partial t} + 2\alpha \parallel Y \parallel_W^2 + 2\int_{\Gamma_I} G_1\,|\,\mathbf{u}^{(a)} - \mathbf{u}^{(o)}\,|_2|\,\mathbf{u}_1^{(a)} - \mathbf{u}_1^{(o)}\,|_2^2\,d\Gamma_I \leq c_1\,|\,Y\,|^2,$$

where $Y = (\mathbf{u}_1,\mathcal{T}_1)$.

We can now deduce that $\frac{\partial}{\partial t}(|\,Y\,|^2\,exp(-c_1 t)) \leq 0$. By integrating between 0 and t, we obtain (since $Y(0) = 0$) that $Y = 0$ and then $\mathbf{u}_1 = \mathbf{0}$ and $\mathcal{T}_1 = 0$. □

The control problem associated with direct problem (42) and the cost functional J_ε defined by (41) is then the following:

$$\begin{aligned}
&\textit{Find } (F_\varepsilon,\mathbf{w}_\varepsilon) \in U_{ad}^\varepsilon \textit{ such that}\\
&J_\varepsilon(F_\varepsilon,\mathbf{w}_\varepsilon) = \inf_{(G,\mathbf{v})\in U_{ad}^\varepsilon} J_\varepsilon(G,\mathbf{v}),
\end{aligned} \tag{44}$$

where the admissibility set is $U_{ad}^\varepsilon = K_c \times \mathcal{N}_1$, and K_c is a convex, closed, nonempty subset of U_c.

4.3. Study of Penalized Control Problems

In this section we prove the existence of an optimal control solution of problem (44). We then characterize this control by means of direct problem (42) and its adjoint.

Proposition 6. *For any* $\varepsilon > 0$ *and under assumptions of Theorem 1, the control problem (44) has at least one solution* $(F_\varepsilon,\mathbf{w}_\varepsilon)$ *in* U_{ad}^ε.

Proof. This result is proved in the same way as in Theorem 4. □

From now on, $(F_\varepsilon, \mathbf{w}_\varepsilon)$ denotes the optimal control solution of problem (44). We set: $\mathbf{U}_\varepsilon = \mathbf{U}(F_\varepsilon, \mathbf{w}_\varepsilon), \mathcal{T}_\varepsilon = \tau(F_\varepsilon, \mathbf{w}_\varepsilon)$. $(\mathbf{U}_\varepsilon, \mathcal{T}_\varepsilon)$ is the solution of the following problem:

$$
\begin{aligned}
&(\frac{\partial \mathbf{U}_\varepsilon}{\partial t}, \mathbf{v}) + A_1(\mathbf{U}_\varepsilon, \mathbf{v}) + L_1(\mathbf{U}_\varepsilon, \mathbf{v}) + D(\mathbf{U}_\varepsilon, \mathbf{v}) + C(\mathcal{T}_\varepsilon, \mathbf{v}) \\
&\qquad + \int_{\Gamma_I} G_1(|\, \mathbf{w}_\varepsilon^{(a)} - \mathbf{w}_\varepsilon^{(o)} \,|_2\, (\mathbf{U}_\varepsilon{}^{(a)} - \mathbf{U}_\varepsilon{}^{(o)}), \mathbf{v}^{(a)} - \mathbf{v}^{(o)})_2 d\Gamma_I = (\mathbf{f}, \mathbf{v}), \\
&(\frac{\partial \mathcal{T}_\varepsilon}{\partial t}, \phi) + A_2(\mathcal{T}_\varepsilon, \phi) + A_{2I}(\mathcal{T}_\varepsilon, \phi) + L_2(\mathcal{T}_\varepsilon, \phi) + c^{(a)}(\phi^{(a)}, \mathbf{U}_\varepsilon{}^{(a)}) \\
&\qquad + B_2(\mathbf{U}_\varepsilon, \mathcal{T}_0, \phi) = (F_\varepsilon, \phi^{(a)} - \phi^{(o)})_{\Gamma_I} + (\mathbf{g}, \phi)\ \forall (\mathbf{v}, \phi) \in \mathcal{V}, \\
&(\mathbf{U}_\varepsilon, \mathcal{T}_\varepsilon)(t=0) = (\mathbf{0}, 0).
\end{aligned} \tag{45}
$$

We are now going to characterize the optimal control $(F_\varepsilon, \mathbf{w}_\varepsilon)$.

Theorem 5. *Let $(F, \mathbf{u})$ be a solution of control problem (40) and $(F_\varepsilon, \mathbf{w}_\varepsilon)$ be a solution of control problem (44). Then there exist $(\mathbf{U}_\varepsilon, \mathcal{T}_\varepsilon)$ solution of (45) and $(\mathbf{R}_\varepsilon, \mathcal{S}_\varepsilon)$ such that (for all $(\mathbf{v}, \phi) \in \mathcal{V}$)*

$$
\begin{aligned}
&-(\frac{\partial \mathbf{R}_\varepsilon}{\partial t}, \mathbf{v}) + A_1(\mathbf{R}_\varepsilon, \mathbf{v}) + L_1(\mathbf{v}, \mathbf{R}_\varepsilon) + D(\mathbf{v}, \mathbf{R}_\varepsilon) + B_2(\mathbf{v}, \mathcal{T}_0, \mathcal{S}_\varepsilon) + c^{(a)}(\mathcal{S}_\varepsilon^{(a)}, \mathbf{v}^{(a)}) \\
&\qquad + \int_{\Gamma_I} G_1\left(|\, \mathbf{w}_\varepsilon^{(a)} - \mathbf{w}_\varepsilon^{(o)} \,|_2\, (\mathbf{R}_\varepsilon{}^{(a)} - \mathbf{R}_\varepsilon{}^{(o)}), \mathbf{v}^{(a)} - \mathbf{v}^{(o)}\right)_2 d\Gamma_I \\
&\qquad + \int_{\Gamma_I} G_1\Big(|\, \mathbf{w}_\varepsilon^{(a)} - \mathbf{w}_\varepsilon^{(o)} \,|_2^{-1}\, (\mathbf{w}_\varepsilon^{(a)} - \mathbf{w}_\varepsilon^{(o)})(\mathbf{U}_\varepsilon{}^{(a)} - \mathbf{U}_\varepsilon{}^{(o)}, \mathbf{R}_\varepsilon{}^{(a)} - \mathbf{R}_\varepsilon{}^{(o)})_2 \\
&\qquad\qquad , \mathbf{v}^{(a)} - \mathbf{v}^{(o)}\Big)_2 d\Gamma_I \\
&\qquad = (\mathbf{U}_\varepsilon - \mathbf{u}, \mathbf{v}) + C^{(a)}(curl(\mathbf{U}_\varepsilon{}^{(a)} - \mathbf{u}_{obs}^{(a)}), curl(\mathbf{v}^{(a)})) \\
&\qquad\quad + C^{(o)}(curl(\mathbf{U}_\varepsilon{}^{(o)} - \mathbf{u}_{obs}^{(o)}), curl(\mathbf{v}^{(o)})), \\
&-(\frac{\partial \mathcal{S}_\varepsilon}{\partial t}, \phi) + A_2(\mathcal{S}_\varepsilon, \phi) + A_{2I}(\mathcal{S}_\varepsilon, \phi) + L_2(\phi, \mathcal{S}_\varepsilon) + C(\phi, \mathbf{R}_\varepsilon) = 0, \\
&(\mathbf{R}_\varepsilon, \mathcal{S}_\varepsilon)(t=T) = (\mathbf{0}, 0)
\end{aligned} \tag{46}
$$

and

$$
\left(\gamma_0(\mathcal{S}_\varepsilon^{(a)} - \mathcal{S}_\varepsilon^{(o)}) + \gamma F_\varepsilon + F_\varepsilon - F,\ G - F_\varepsilon\right)_{U_c} \geq 0, \quad \forall G \in K_c,
$$

where γ_0 is the trace function on Γ_I.

Proof. The partial derivative of (45), at point $(F_\varepsilon, \mathbf{w}_\varepsilon)$, with respect to G at constant $\mathbf{w}$ is

$(\forall(\mathbf{v},\phi)\in\mathcal{V})$

$$\begin{aligned}
&(\frac{\partial \mathbf{U}_G}{\partial t},\mathbf{v})+A_1(\mathbf{U}_G,\mathbf{v})+L_1(\mathbf{U}_G,\mathbf{v})+D(\mathbf{U}_G,\mathbf{v})+C(\mathcal{T}_G,\mathbf{v})\\
&\qquad+\int_{\Gamma_I}G_1(\mid \mathbf{w}_\varepsilon^{(a)}-\mathbf{w}_\varepsilon^{(o)}\mid_2(\mathbf{U}_G^{(a)}-\mathbf{U}_G^{(o)}),\mathbf{v}^{(a)}-\mathbf{v}^{(o)})_2 d\Gamma_I=0,\\
&(\frac{\partial \mathcal{T}_G}{\partial t},\phi)+A_2(\mathcal{T}_G,\phi)+A_{2I}(\mathcal{T}_G,\phi)+L_2(\mathcal{T}_G,\phi)+c^{(a)}(\phi^{(a)},\mathbf{U}_G^{(a)})\\
&\qquad\qquad+B_2(\mathbf{U}_G,\mathcal{T}_0,\phi)=(G,\phi^{(a)}-\phi^{(o)})_{\Gamma_I},\\
&(\mathbf{U}_G,\mathcal{T}_G)(t=0)=(\mathbf{0},0),
\end{aligned}\tag{47}$$

and the partial derivative of (45), at point $(F_\varepsilon,\mathbf{w}_\varepsilon)$, with respect to $\mathbf{w}$ at constant G is $(\forall(\mathbf{v},\phi)\in\mathcal{V})$

$$\begin{aligned}
&(\frac{\partial \mathbf{U}_w}{\partial t},\mathbf{v})+A_1(\mathbf{U}_w,\mathbf{v})+L_1(\mathbf{U}_w,\mathbf{v})+D(\mathbf{U}_w,\mathbf{v})+C(\mathcal{T}_w,\mathbf{v})\\
&\quad+\int_{\Gamma_I}G_1\left(\mid \mathbf{w}_\varepsilon^{(a)}-\mathbf{w}_\varepsilon^{(o)}\mid_2(\mathbf{U}_w^{(a)}-\mathbf{U}_w^{(o)}),\mathbf{v}^{(a)}-\mathbf{v}^{(o)}\right)_2 d\Gamma_I\\
&\quad+\int_{\Gamma_I}G_1\left(\mid \mathbf{w}_\varepsilon^{(a)}-\mathbf{w}_\varepsilon^{(o)}\mid_2^{-1}(\mathbf{w}_\varepsilon^{(a)}-\mathbf{w}_\varepsilon^{(o)},\mathbf{w}^{(a)}-\mathbf{w}^{(o)})_2(\mathbf{U}_\varepsilon{}^{(a)}-\mathbf{U}_\varepsilon{}^{(o)})\right.\\
&\qquad\qquad\left.,\mathbf{v}^{(a)}-\mathbf{v}^{(o)}\right)_2 d\Gamma_I=0,\\
&(\frac{\partial \mathcal{T}_w}{\partial t},\phi)+A_2(\mathcal{T}_w,\phi)+A_{2I}(\mathcal{T}_w,\phi)+L_2(\mathcal{T}_w,\phi)\\
&\qquad+c^{(a)}(\phi^{(a)},\mathbf{U}_w^{(a)})+B_2(\mathbf{U}_w,\mathcal{T}_0,\phi)=0,\\
&(\mathbf{U}_w,\mathcal{T}_w)(t=0)=(\mathbf{0},0).
\end{aligned}\tag{48}$$

The existence of $(\mathbf{U}_\varepsilon,\mathcal{T}_\varepsilon)$, solution of the direct problem, follows from the definition of J_ε. On the other hand we have to prove the existence of $(\mathbf{R}_\varepsilon,\mathcal{S}_\varepsilon)$ verifying the adjoint problem. First $(\mathbf{R}_\varepsilon,\mathcal{S}_\varepsilon)$ will denote the solution of the following problem:

$$\begin{aligned}
&-(\frac{\partial \mathbf{R}_\varepsilon}{\partial t},\mathbf{v})+A_1(\mathbf{R}_\varepsilon,\mathbf{v})+L_1(\mathbf{v},\mathbf{R}_\varepsilon)+D(\mathbf{v},\mathbf{R}_\varepsilon)+B_2(\mathbf{v},\mathcal{T}_0,\mathcal{S}_\varepsilon)+c^{(a)}(\mathcal{S}_\varepsilon^{(a)},\mathbf{v}^{(a)})\\
&\qquad+\int_{\Gamma_I}G_1(\mid \mathbf{w}_\varepsilon^{(a)}-\mathbf{w}_\varepsilon^{(o)}\mid_2(\mathbf{R}_\varepsilon{}^{(a)}-\mathbf{R}_\varepsilon{}^{(o)}),\mathbf{v}^{(a)}-\mathbf{v}^{(o)})_2 d\Gamma_I\\
&\quad=(\mathbf{U}_\varepsilon-\mathbf{u},\mathbf{v})+\frac{1}{\varepsilon}A_1(\mathbf{U}_\varepsilon-\mathbf{w}_\varepsilon,\mathbf{v})+C^{(a)}(curl(\mathbf{U}_\varepsilon{}^{(a)}-\mathbf{u}_{obs}^{(a)}),curl(\mathbf{v}^{(a)}))\\
&\qquad+C^{(o)}(curl(\mathbf{U}_\varepsilon{}^{(o)}-\mathbf{u}_{obs}^{(o)}),curl(\mathbf{v}^{(o)})),\\
&-(\frac{\partial \mathcal{S}_\varepsilon}{\partial t},\phi)+A_2(\mathcal{S}_\varepsilon,\phi)+A_{2I}(\mathcal{S}_\varepsilon,\phi)+L_2(\phi,\mathcal{S}_\varepsilon)+C(\phi,\mathbf{R}_\varepsilon)=0\,,\quad\forall(\mathbf{v},\phi)\in\mathcal{V},\\
&(\mathbf{R}_\varepsilon,\mathcal{S}_\varepsilon)(t=T)=(\mathbf{0},0).
\end{aligned}\tag{49}$$

This linear problem has one unique solution $(\mathbf{R}_\varepsilon,\mathcal{S}_\varepsilon)$ in $L^2(0,T,\mathcal{V})$.

Setting $(\mathbf{v},\phi) = (\mathbf{R}_\varepsilon, \mathcal{S}_\varepsilon)$ in equations (48) and integrating with respect to time gives

$$\int_0^T (\frac{\partial \mathbf{U}_w}{\partial t}, \mathbf{R}_\varepsilon)dt + \int_0^T A_1(\mathbf{U}_w, \mathbf{R}_\varepsilon)dt + \int_0^T L_1(\mathbf{U}_w, \mathbf{R}_\varepsilon)dt + \int_0^T D(\mathbf{U}_w, \mathbf{R}_\varepsilon)dt$$

$$+ \int_0^T C(\mathcal{T}_w, \mathbf{R}_\varepsilon)dt + \int_0^T (\int_{\Gamma_I} G_1 \left(\mid \mathbf{w}_\varepsilon^{(a)} - \mathbf{w}_\varepsilon^{(o)} \mid_2 (\mathbf{U}_w^{(a)} - \mathbf{U}_w^{(o)}), \mathbf{R}_\varepsilon{}^{(a)} - \mathbf{R}_\varepsilon{}^{(o)} \right)_2 d\Gamma_I)dt$$

$$+ \int_0^T (\int_{\Gamma_I} G_1 \left(\mid \mathbf{w}_\varepsilon^{(a)} - \mathbf{w}_\varepsilon^{(o)} \mid_2^{-1} (\mathbf{w}_\varepsilon^{(a)} - \mathbf{w}_\varepsilon^{(o)}, \mathbf{w}^{(a)} - \mathbf{w}^{(o)})_2 (\mathbf{U}_\varepsilon{}^{(a)} - \mathbf{U}_\varepsilon{}^{(o)}) \right.$$

$$\left. , \mathbf{R}_\varepsilon{}^{(a)} - \mathbf{R}_\varepsilon{}^{(o)} \right)_2 d\Gamma_I)dt = 0,$$

$$\int_0^T (\frac{\partial \mathcal{T}_w}{\partial t}, \mathcal{S}_\varepsilon)dt + \int_0^T A_2(\mathcal{T}_w, \mathcal{S}_\varepsilon)dt + \int_0^T A_{2I}(\mathcal{T}_w, \mathcal{S}_\varepsilon)dt + \int_0^T L_2(\mathcal{T}_w, \mathcal{S}_\varepsilon)dt$$

$$+ \int_0^T c^{(a)}(\mathcal{S}_\varepsilon^{(a)}, \mathbf{U}_w^{(a)})dt + \int_0^T B_2(\mathbf{U}_w, \mathcal{T}_0, \mathcal{S}_\varepsilon)dt = 0.$$

After integration by parts we obtain (according to the third part of (48) and of (49))

$$- \int_0^T (\frac{\partial \mathbf{R}_\varepsilon}{\partial t}, \mathbf{U}_w)dt + \int_0^T A_1(\mathbf{R}_\varepsilon, \mathbf{U}_w)dt + \int_0^T L_1(\mathbf{U}_w, \mathbf{R}_\varepsilon)dt + \int_0^T D(\mathbf{U}_w, \mathbf{R}_\varepsilon)dt$$

$$+ \int_0^T C(\mathcal{T}_w, \mathbf{R}_\varepsilon)dt + \int_0^T (\int_{\Gamma_I} G_1 (\mid \mathbf{w}_\varepsilon^{(a)} - \mathbf{w}_\varepsilon^{(o)} \mid_2 (\mathbf{R}_\varepsilon{}^{(a)} - \mathbf{R}_\varepsilon{}^{(o)}), \mathbf{U}_w^{(a)} - \mathbf{U}_w^{(o)})_2 d\Gamma_I)dt$$

$$+ \int_0^T (\int_{\Gamma_I} G_1 \left(\mid \mathbf{w}_\varepsilon^{(a)} - \mathbf{w}_\varepsilon^{(o)} \mid_2^{-1} (\mathbf{w}_\varepsilon^{(a)} - \mathbf{w}_\varepsilon^{(o)}, \mathbf{w}^{(a)} - \mathbf{w}^{(o)})_2 (\mathbf{R}_\varepsilon{}^{(a)} - \mathbf{R}_\varepsilon{}^{(o)}) \right.$$

$$\left. , \mathbf{U}_\varepsilon{}^{(a)} - \mathbf{U}_\varepsilon{}^{(o)} \right)_2 d\Gamma_I)dt = 0,$$

$$- \int_0^T (\frac{\partial \mathcal{S}_\varepsilon}{\partial t}, \mathcal{T}_w)dt + \int_0^T A_2(\mathcal{S}_\varepsilon, \mathcal{T}_w)dt + \int_0^T A_{2I}(\mathcal{T}_w, \mathcal{S}_\varepsilon)dt + \int_0^T L_2(\mathcal{T}_w, \mathcal{S}_\varepsilon)dt$$

$$+ \int_0^T c^{(a)}(\mathcal{S}_\varepsilon^{(a)}, \mathbf{U}_w^{(a)})dt + \int_0^T B_2(\mathbf{U}_w, \mathcal{T}_0, \mathcal{S}_\varepsilon)dt = 0.$$

Since $(\mathbf{R}_\varepsilon, \mathcal{S}_\varepsilon)$ verifies (49), this implies

$$- \int_0^T c^{(a)}(\mathcal{S}_\varepsilon^{(a)}, \mathbf{U}_w^{(a)})dt - \int_0^T B_2(\mathbf{U}_w, \mathcal{T}_0, \mathcal{S}_\varepsilon)dt + \int_0^T C(\mathcal{T}_w, \mathbf{R}_\varepsilon)dt$$

$$+ \int_0^T (\int_{\Gamma_I} G_1 \left(\mid \mathbf{w}_\varepsilon^{(a)} - \mathbf{w}_\varepsilon^{(o)} \mid_2^{-1} (\mathbf{w}_\varepsilon^{(a)} - \mathbf{w}_\varepsilon^{(o)}, \mathbf{w}^{(a)} - \mathbf{w}^{(o)})_2 (\mathbf{R}_\varepsilon{}^{(a)} - \mathbf{R}_\varepsilon{}^{(o)}) \right.$$

$$\left. , \mathbf{U}_\varepsilon{}^{(a)} - \mathbf{U}_\varepsilon{}^{(o)} \right)_2 d\Gamma_I)dt$$

$$+ \int_0^T C^{(a)}(curl(\mathbf{U}_\varepsilon{}^{(a)} - \mathbf{u}_{obs}^{(a)}), curl(\mathbf{U}_w^{(a)}))dt + \int_0^T C^{(o)}(curl(\mathbf{U}_\varepsilon{}^{(o)} - \mathbf{u}_{obs}^{(o)}), curl(\mathbf{U}_w^{(o)}))dt$$

$$+ \int_0^T (\mathbf{U}_\varepsilon - \mathbf{u}, \mathbf{U}_w)dt + \int_0^T \frac{1}{\varepsilon} A_1(\mathbf{U}_\varepsilon - \mathbf{w}_\varepsilon, \mathbf{U}_w)dt = 0,$$

$$\int_0^T c^{(a)}(\mathcal{S}_\varepsilon^{(a)}, \mathbf{U}_w^{(a)})dt + \int_0^T B_2(\mathbf{U}_w, \mathcal{T}_0, \mathcal{S}_\varepsilon)dt - \int_0^T C(\mathcal{T}_w, \mathbf{R}_\varepsilon)dt = 0.$$

Hence,

$$\begin{aligned}
&\int_0^T (\mathbf{U}_\varepsilon - \mathbf{u}, \mathbf{U}_w)dt + \int_0^T \frac{1}{\varepsilon} A_1(\mathbf{U}_\varepsilon - \mathbf{w}_\varepsilon, \mathbf{U}_w)dt \\
&\quad + \int_0^T C^{(a)}(curl(\mathbf{U}_\varepsilon{}^{(a)} - \mathbf{u}^{(a)}_{obs}), curl(\mathbf{U}^{(a)}_w))dt \\
&\quad + \int_0^T C^{(o)}(curl(\mathbf{U}_\varepsilon{}^{(o)} - \mathbf{u}^{(o)}_{obs}), curl(\mathbf{U}^{(o)}_w))dt \\
&+ \int_0^T (\int_{\Gamma_I} G_1 \Big(\mid \mathbf{w}^{(a)}_\varepsilon - \mathbf{w}^{(o)}_\varepsilon \mid_2^{-1} (\mathbf{w}^{(a)}_\varepsilon - \mathbf{w}^{(o)}_\varepsilon, \mathbf{w}^{(a)} - \mathbf{w}^{(o)})_2 (\mathbf{R}_\varepsilon{}^{(a)} - \mathbf{R}_\varepsilon{}^{(o)}) \\
&\qquad , \mathbf{U}_\varepsilon{}^{(a)} - \mathbf{U}_\varepsilon{}^{(o)} \Big)_2 d\Gamma_I)dt = 0.
\end{aligned} \tag{50}$$

The partial derivative of J_ε, at point $(F_\varepsilon, \mathbf{w}_\varepsilon)$, with respect to $\mathbf{w}$ at constant G is

$$\begin{aligned}
\frac{\partial J_\varepsilon}{\partial \mathbf{w}}(F_\varepsilon, \mathbf{w}_\varepsilon).\mathbf{w} = &\int_0^T C^{(a)}(curl(\mathbf{U}_\varepsilon{}^{(a)} - \mathbf{u}^{(a)}_{obs}), curl(\mathbf{U}^{(a)}_w))dt \\
&+ \int_0^T C^{(o)}(curl(\mathbf{U}_\varepsilon{}^{(o)} - \mathbf{u}^{(o)}_{obs}), curl(\mathbf{U}^{(o)}_w))dt \\
&+ \frac{1}{\varepsilon}\int_0^T A_1(\mathbf{U}_\varepsilon - \mathbf{w}_\varepsilon, \mathbf{U}_w - \mathbf{w})dt + \int_0^T (\mathbf{U}_\varepsilon - \mathbf{u}, \mathbf{U}_w)dt.
\end{aligned}$$

The control problem (44) includes no constraint concerning $\mathbf{w}$, so

$$\frac{\partial J_\varepsilon}{\partial \mathbf{w}}(F_\varepsilon, \mathbf{w}_\varepsilon).\mathbf{w} = 0, \ \forall \mathbf{w} \in L^2(0, T, \mathcal{V}_1).$$

From this and (50) we deduce

$$\begin{aligned}
&\int_0^T (\int_{\Gamma_I} G_1 \Big(\mid \mathbf{w}^{(a)}_\varepsilon - \mathbf{w}^{(o)}_\varepsilon \mid_2^{-1} (\mathbf{w}^{(a)}_\varepsilon - \mathbf{w}^{(o)}_\varepsilon, \mathbf{w}^{(a)} - \mathbf{w}^{(o)})_2 (\mathbf{R}_\varepsilon{}^{(a)} - \mathbf{R}_\varepsilon{}^{(o)}) \\
&\qquad , \mathbf{U}_\varepsilon{}^{(a)} - \mathbf{U}_\varepsilon{}^{(o)} \Big)_2 d\Gamma_I)dt \\
&\qquad = -\int_0^T \frac{1}{\varepsilon} A_1(\mathbf{w}, \mathbf{U}_\varepsilon - \mathbf{w}_\varepsilon)dt.
\end{aligned}$$

In particular we have, for all $(\mathbf{v}, \varphi) \in \mathcal{V}_1 \times \mathcal{D}(0, T)$

$$\begin{aligned}
&\int_0^T (\int_{\Gamma_I} G_1 \Big(\mid \mathbf{w}^{(a)}_\varepsilon - \mathbf{w}^{(o)}_\varepsilon \mid_2^{-1} (\mathbf{R}_\varepsilon{}^{(a)} - \mathbf{R}_\varepsilon{}^{(o)}, \mathbf{U}_\varepsilon{}^{(a)} - \mathbf{U}_\varepsilon{}^{(o)})_2 (\mathbf{w}^{(a)}_\varepsilon - \mathbf{w}^{(o)}_\varepsilon) \\
&\qquad , \mathbf{v}^{(a)} - \mathbf{v}^{(o)} \Big)_2 \varphi(t) d\Gamma_I)dt \\
&\qquad = -\int_0^T \frac{1}{\varepsilon} A_1(\mathbf{v}\varphi(t), \mathbf{U}_\varepsilon - \mathbf{w}_\varepsilon)dt.
\end{aligned}$$

So, in the distribution sense

$$\int_{\Gamma_I} G_1 \Big(| \mathbf{w}_\varepsilon^{(a)} - \mathbf{w}_\varepsilon^{(o)} |_2^{-1} (\mathbf{R}_\varepsilon{}^{(a)} - \mathbf{R}_\varepsilon{}^{(o)}, \mathbf{U}_\varepsilon{}^{(a)} - \mathbf{U}_\varepsilon{}^{(o)})_2 (\mathbf{w}_\varepsilon^{(a)} - \mathbf{w}_\varepsilon^{(o)}) , \mathbf{v}^{(a)} - \mathbf{v}^{(o)} \Big)_2 d\Gamma_I = -\frac{1}{\varepsilon} A_1(\mathbf{v}, \mathbf{U}_\varepsilon - \mathbf{w}_\varepsilon), \quad \forall \mathbf{v} \in \mathcal{V}_1. \tag{51}$$

Since $(\mathbf{R}_\varepsilon, \mathcal{S}_\varepsilon)$ is solution of problem (49), we obtain by using (51) that $(\mathbf{R}_\varepsilon, \mathcal{S}_\varepsilon)$ verifies the adjoint problem (46) occurring in Theorem 5. We now are going to prove the inequality which characterizes the optimal control F_ε.
Setting $(\mathbf{v}, \phi) = (\mathbf{R}_\varepsilon, \mathcal{S}_\varepsilon)$ in equations (47) and integrating with respect to time gives

$$-\int_0^T (\frac{\partial \mathbf{R}_\varepsilon}{\partial t}, \mathbf{U}_G) dt + \int_0^T A_1(\mathbf{R}_\varepsilon, \mathbf{U}_G) dt + \int_0^T L_1(\mathbf{U}_G, \mathbf{R}_\varepsilon) dt + \int_0^T D(\mathbf{U}_G, \mathbf{R}_\varepsilon) dt$$

$$+ \int_0^T C(\mathcal{T}_G, \mathbf{R}_\varepsilon) dt + \int_0^T (\int_{\Gamma_I} G_1(| \mathbf{w}_\varepsilon^{(a)} - \mathbf{w}_\varepsilon^{(o)} |_2 (\mathbf{R}_\varepsilon{}^{(a)} - \mathbf{R}_\varepsilon{}^{(o)}), \mathbf{U}_G^{(a)} - \mathbf{U}_G^{(o)})_2 d\Gamma_I) dt = 0,$$

$$-\int_0^T (\frac{\partial \mathcal{S}_\varepsilon}{\partial t}, \mathcal{T}_G) dt + \int_0^T A_2(\mathcal{S}_\varepsilon, \mathcal{T}_G) dt + \int_0^T A_{2I}(\mathcal{T}_G, \mathcal{S}_\varepsilon) dt + \int_0^T L_2(\mathcal{T}_G, \mathcal{S}_\varepsilon) dt$$

$$+ \int_0^T c^{(a)}(\mathcal{S}_\varepsilon^{(a)}, \mathbf{U}_G^{(a)}) dt + \int_0^T B_2(\mathbf{U}_G, \mathcal{T}_0, \mathcal{S}_\varepsilon) dt = \int_0^T (G, \mathcal{S}_\varepsilon^{(a)} - \mathcal{S}_\varepsilon^{(o)})_{\Gamma_I} dt.$$

Since $(\mathbf{R}_\varepsilon, \mathcal{S}_\varepsilon)$ is solution of problem (46), we obtain

$$\begin{aligned} -\int_0^T (\int_{\Gamma_I} G_1 \Big(| \mathbf{w}_\varepsilon^{(a)} - \mathbf{w}_\varepsilon^{(o)} |_2^{-1} (\mathbf{R}_\varepsilon{}^{(a)} - \mathbf{R}_\varepsilon{}^{(o)}, \mathbf{U}_\varepsilon{}^{(a)} - \mathbf{U}_\varepsilon{}^{(o)})_2 (\mathbf{w}_\varepsilon^{(a)} - \mathbf{w}_\varepsilon^{(o)}) , \mathbf{U}_G^{(a)} - \mathbf{U}_G^{(o)} \Big)_2 d\Gamma_I) dt \\ + \int_0^T (\mathbf{U}_\varepsilon - \mathbf{u}, \mathbf{U}_G) dt \\ + \int_0^T C^{(a)}(curl(\mathbf{U}_\varepsilon{}^{(a)} - \mathbf{u}_{obs}^{(a)}), curl(\mathbf{U}_G^{(a)})) dt \\ + \int_0^T C^{(o)}(curl(\mathbf{U}_\varepsilon{}^{(o)} - \mathbf{u}_{obs}^{(o)}), curl(\mathbf{U}_G^{(o)})) dt \\ = \int_0^T (G, \mathcal{S}_\varepsilon^{(a)} - \mathcal{S}_\varepsilon^{(o)})_{\Gamma_I} dt. \end{aligned} \tag{52}$$

The partial derivative of J_ε, at point $(F_\varepsilon, \mathbf{w}_\varepsilon)$, with respect to G at constant $\mathbf{w}$ is

$$\frac{\partial J_\varepsilon}{\partial G}(F_\varepsilon, \mathbf{w}_\varepsilon).G = \int_0^T C^{(a)}(curl(\mathbf{U}_\varepsilon{}^{(a)} - \mathbf{u}_{obs}^{(a)}), curl(\mathbf{U}_G^{(a)}))dt$$

$$+\int_0^T C^{(o)}(curl(\mathbf{U}_\varepsilon{}^{(o)} - \mathbf{u}_{obs}^{(o)}), curl(\mathbf{U}_G^{(o)}))dt$$

$$+\frac{1}{\varepsilon}\int_0^T A_1(\mathbf{U}_\varepsilon - \mathbf{w}_\varepsilon, \mathbf{U}_G)dt + \int_0^T (\mathbf{U}_\varepsilon - \mathbf{u}, \mathbf{U}_G)dt$$

$$+\gamma(F_\varepsilon, G)_{U_c} + (F_\varepsilon - F, G)_{U_c}.$$

Applying relationship (51)-(52), we deduce

$$\frac{\partial J_\varepsilon}{\partial G}(F_\varepsilon, \mathbf{w}_\varepsilon).G = \int_0^T (G, \mathcal{S}_\varepsilon^{(a)} - \mathcal{S}_\varepsilon^{(o)})_{\Gamma_I} dt + \gamma(F_\varepsilon, G)_{U_c} + (F_\varepsilon - F, G)_{U_c}.$$

Control problem (44) includes a constraint on G: $G \in K_c \subset U_c$. Therefore the optimal control F_ε has to verify the following inequality

$$\frac{\partial J_\varepsilon}{\partial G}(F_\varepsilon, \mathbf{w}_\varepsilon).(G - F_\varepsilon) \geq 0, \quad \forall G \in K_c$$

and we have

$$(\gamma_0(\mathcal{S}_\varepsilon^{(a)} - \mathcal{S}_\varepsilon^{(o)}) + \gamma F_\varepsilon + F_\varepsilon - F, G - F_\varepsilon)_{U_c} \geq 0, \quad \forall G \in K_c.$$

This achieves the proof of Theorem 5. □

4.4. Optimality Conditions for the Nonlinear Control Problem

The penalized control problem (44) has a solution which is characterized by means of the direct and adjoint penalized equations.

We are now going to characterize the optimal control $(F, \mathbf{u})$, solution of problem (40). $(\mathbf{u}, \mathcal{T})$ denotes the solution of direct problem (38). The adjoint problem is obtained by passing to the limit in the adjoint penalized equations introduced in Theorem 5.

Theorem 6. *Let* $(F_\varepsilon, \mathbf{w}_\varepsilon)$ *be a solution of control problem (44).*
The following convergence results are verified, as $\varepsilon \longrightarrow 0$

(i) F_ε *converges to* F *in* U_c *strongly,*

(ii) $\lim_{\varepsilon \longrightarrow 0} \frac{1}{\varepsilon} \parallel \mathbf{U}_\varepsilon - \mathbf{w}_\varepsilon \parallel^2_{L^2(0,T,H^1)} = 0,$

(iii) $X_\varepsilon \rightharpoonup X$ *weakly in* $L^2(0,T,H^1)$,

(iv) $X_\varepsilon \longrightarrow X$ *strongly in* $L^2(0,T,L^2)$,

(v) $\mathbf{w}_\varepsilon \rightharpoonup \mathbf{u}$ *weakly in* $L^2(0,T,H^1)$,

(vi) $\mathbf{w}_\varepsilon \longrightarrow \mathbf{u}$ *strongly in* $L^2(0,T,L^2)$,

(vii) $| \mathbf{w}_\varepsilon^{(a)} - \mathbf{w}_\varepsilon^{(o)} |_2 (\mathbf{U}_\varepsilon{}^{(a)} - \mathbf{U}_\varepsilon{}^{(o)}) \rightharpoonup | \mathbf{u}^{(a)} - \mathbf{u}^{(o)} |_2 (\mathbf{u}^{(a)} - \mathbf{u}^{(o)})$
weakly in $L^{\frac{3}{2}}(0,T,L^{\frac{3}{2}}(\Gamma_I))$,

(viii) $J_\varepsilon(\mathbf{w}_\varepsilon, F_\varepsilon)$ *converges to* $J(\mathbf{u},F)$,
where, $X_\varepsilon = (\mathbf{U}_\varepsilon, \mathcal{T}_\varepsilon)$ *and* $X = (\mathbf{u}, \mathcal{T})$.

Proof. $(F,\mathbf{u}) \in U_{ad}$ denotes a solution of control problem (40). $(F_\varepsilon, \mathbf{w}_\varepsilon) \in U^\varepsilon_{ad}$ denotes a solution of the penalized control problem (44). Since $J_\varepsilon(F,\mathbf{u}) = J(F,\mathbf{u})$ and $U_{ad} \subset U^\varepsilon_{ad}$, we have

$$J_\varepsilon(F_\varepsilon, \mathbf{w}_\varepsilon) \leq J(F,\mathbf{u}).$$

According to the definition of J_ε, this implies that

$$\begin{aligned}
&\| F_\varepsilon - F \|^2_{U_c} \leq 2J(F,\mathbf{u}),\\
&\int_0^T A_1(\mathbf{U}_\varepsilon - \mathbf{w}_\varepsilon, \mathbf{U}_\varepsilon - \mathbf{w}_\varepsilon)dt \leq 2\varepsilon J(F,\mathbf{u}),\\
&\| \mathbf{U}_\varepsilon - \mathbf{u} \|^2_{L^2(0,T,H_1)} \leq 2J(F,\mathbf{u}),\\
&C^{(a)} \| curl(\mathbf{U}_\varepsilon{}^{(a)} - \mathbf{u}_{obs}) \|^2_{L^2(0,T,L^2)} \leq 4J(F,\mathbf{u}),\\
&C^{(o)} \| curl(\mathbf{U}_\varepsilon{}^{(o)} - \mathbf{u}_{obs}) \|^2_{L^2(0,T,L^2)} \leq 4J(F,\mathbf{u}).
\end{aligned} \tag{53}$$

We deduce from (53) the following results:

$$\begin{aligned}
&\| F_\varepsilon \|^2_{U_c} \leq 2 \| F \|^2_{U_c} + 4J(\mathbf{u},F),\\
&\lim_{\varepsilon\to 0} \int_0^T A_1(\mathbf{U}_\varepsilon - \mathbf{w}_\varepsilon, \mathbf{U}_\varepsilon - \mathbf{w}_\varepsilon)dt = 0, \quad \| \mathbf{U}_\varepsilon \|_{L^2(0,T,\mathcal{H}_1)} \leq c,\\
&C^{(a)} \| curl(\mathbf{U}_\varepsilon{}^{(a)}) \|_{L^2(0,T,L^2)} \leq c \text{ and } C^{(o)} \| curl(\mathbf{U}_\varepsilon{}^{(o)}) \|_{L^2(0,T,L^2)} \leq c.
\end{aligned} \tag{54}$$

where c is a positive constant.
The relations (54) imply that $\mathbf{U}_\varepsilon$ converges towards $\mathbf{w}_\varepsilon$ strongly in $L^2(0,T,\mathcal{V}_1)$ and that $\mathbf{U}_\varepsilon$ is uniformly bounded in $L^2(0,T,\mathcal{H}_1)$. For ε sufficiently small, $\| \mathbf{U}_\varepsilon - \mathbf{w}_\varepsilon \|_{L^2(0,T,\mathcal{V}_1)}$ is uniformly bounded. Therefore there exists a positive constant $c > 0$ independent of ε such that

$$\begin{aligned}
&\| \mathbf{w}_\varepsilon \|_{L^2(0,T,\mathcal{V}_1)} \leq c(1 + \| \mathbf{U}_\varepsilon \|_{L^2(0,T,\mathcal{V}_1)})\\
&\textit{and}\\
&\| \mathbf{w}_\varepsilon \|_{L^2(0,T,\mathcal{H}_1)} \leq c(1 + \| \mathbf{U}_\varepsilon \|_{L^2(0,T,\mathcal{H}_1)}).
\end{aligned} \tag{55}$$

Since $\mathbf{U}_\varepsilon$ is uniformly bounded in $L^2(0,T,\mathcal{H}_1)$, we have

$$\| \mathbf{w}_\varepsilon \|_{L^2(0,T,\mathcal{H}_1)} \leq c. \tag{56}$$

Setting $(\mathbf{v},\phi) = (\mathbf{U}_\varepsilon, \mathcal{T}_\varepsilon)$ in (45), gives

$$\frac{\partial \mid \mathbf{U}_\varepsilon \mid^2}{2\partial t} + A_1(\mathbf{U}_\varepsilon,\mathbf{U}_\varepsilon) + L_1(\mathbf{U}_\varepsilon,\mathbf{U}_\varepsilon) + D(\mathbf{U}_\varepsilon,\mathbf{U}_\varepsilon) + C(\mathcal{T}_\varepsilon,\mathbf{U}_\varepsilon)$$

$$+\int_{\Gamma_I} G_1 \mid \mathbf{w}_\varepsilon^{(a)} - \mathbf{w}_\varepsilon^{(o)} \mid_2 \mid \mathbf{U}_\varepsilon{}^{(a)} - \mathbf{U}_\varepsilon{}^{(o)} \mid_2^2 d\Gamma_I = (\mathbf{f},\mathbf{v}),$$

$$\frac{\partial \mid \mathcal{T}_\varepsilon \mid^2}{2\partial t} + A_2(\mathcal{T}_\varepsilon,\mathcal{T}_\varepsilon) + A_{2I}(\mathcal{T}_\varepsilon,\mathcal{T}_\varepsilon) + L_2(\mathcal{T}_\varepsilon,\mathcal{T}_\varepsilon) + c^{(a)}(\mathcal{T}_\varepsilon^{(a)},\mathbf{U}_\varepsilon{}^{(a)})$$

$$+B_2(\mathbf{U}_\varepsilon,\mathcal{T}_0,\mathcal{T}_\varepsilon) = (F_\varepsilon,\mathcal{T}_\varepsilon^{(a)} - \mathcal{T}_\varepsilon^{(o)})_{\Gamma_I} + (\mathbf{g},\mathcal{T}_\varepsilon).$$

F_ε is uniformly bounded in U_c, then by applying the same technique as in proof of Theorem 1, we deduce that sequence X_ε is uniformly bounded in $L^2(0,T,\mathcal{V}) \cap L^\infty(0,T,\mathcal{H})$ and $\int_0^T (\int_{\Gamma_I} G_1 \mid \mathbf{w}_\varepsilon^{(a)} - \mathbf{w}_\varepsilon^{(o)} \mid\mid \mathbf{U}_\varepsilon{}^{(a)} - \mathbf{U}_\varepsilon{}^{(o)} \mid_2^2 d\Gamma_I)dt$ is uniformly bounded. By using (55)-(56) we deduce that $\mathbf{w}_\varepsilon$ is uniformly bounded in $L^2(0,T,\mathcal{V}_1) \cap L^\infty(0,T,\mathcal{H}_1)$.

Prove now that $\frac{\partial X_\varepsilon}{\partial t}$ is uniformly bounded in $L^1(0,T,\mathcal{V}')$.

According to Lemmas 3-6, we have that for all $Y = (\mathbf{v},\phi) \in \mathcal{V}$

$$\begin{aligned}\mid (\frac{\partial X_\varepsilon}{\partial t}, Y) \mid &\leq c_1 \parallel X_\varepsilon \parallel_W \parallel Y \parallel \\ &+c_2(\mid \mathbf{f} \mid + \mid \mathbf{g} \mid + \parallel F_\varepsilon \parallel_{L^2(\Gamma_I)}) \parallel Y \parallel \\ &+ \mid (G_1 \mid \mathbf{w}_\varepsilon^{(a)} - \mathbf{w}_\varepsilon^{(o)} \mid_2 (\mathbf{U}_\varepsilon{}^{(a)} - \mathbf{U}_\varepsilon{}^{(o)}), \mathbf{v}^{(a)} - \mathbf{v}^{(o)})_{\Gamma_I} \mid .\end{aligned}$$

By using Lemma 7 we have then

$$\begin{aligned}\mid (\frac{\partial X_\varepsilon}{\partial t}, Y) \mid &\leq c_1 \parallel X_\varepsilon \parallel_W \parallel Y \parallel \\ &+c_2(\mid \mathbf{f} \mid + \mid \mathbf{g} \mid + \parallel F_\varepsilon \parallel_{L^2(\Gamma_I)}) \parallel Y \parallel \\ &+c_3 \parallel \mathbf{w}_\varepsilon^{(a)} - \mathbf{w}_\varepsilon^{(o)} \parallel_{L^3(\Gamma_I)}^{1/2} \parallel\mid \mathbf{w}_\varepsilon^{(a)} - \mathbf{w}_\varepsilon^{(o)} \mid_2^r \mid \mathbf{U}_\varepsilon{}^{(a)} - \mathbf{U}_\varepsilon{}^{(o)} \mid_2^2 \parallel_{L^1(\Gamma_I)}^{1/2} \parallel Y \parallel_{L^3(\Gamma_I)}.\end{aligned}$$

Since $Y \in \mathcal{V}$, so $\gamma_o Y \in H^{1/2}(\Gamma_I)$ and according to Sobolev's inclusion (according to Theorem A.(a) and to Theorem B where $s_2 = 1/2$ and $p_2 = 2$ of the Appendix) we have $H^{1/2}(\Gamma_I) \subset L^q(\Gamma_I)\ \forall 4 \geq q \geq 2$, we can deduce then $\parallel Y \parallel_{L^3(\Gamma_I)} \leq c \parallel Y \parallel_{H^{1/2}(\Gamma_I)} \leq c \parallel Y \parallel_{H^1}$.

Applying this result and Sobolev's inclusion we can deduce that

$$\begin{aligned}\parallel \frac{\partial X_\varepsilon}{\partial t} \parallel_{\mathcal{V}'} &\leq c_1 \parallel X_\varepsilon \parallel_W + c_4 \parallel \mathbf{w}_\varepsilon \parallel \\ &+c_2(\mid \mathbf{f} \mid + \mid \mathbf{g} \mid + \parallel F_\varepsilon \parallel_{L^2(\Gamma_I)}) \\ &+c_5 \parallel\mid \mathbf{w}_\varepsilon^{(a)} - \mathbf{w}_\varepsilon^{(o)} \mid_2 \mid \mathbf{U}_\varepsilon{}^{(a)} - \mathbf{U}_\varepsilon{}^{(o)} \mid_2^2 \parallel_{L^1(\Gamma_I)}.\end{aligned}$$

Since X_ε and $\mathbf{w}_\varepsilon$ are uniformly bounded in $L^\infty(0,T,\mathcal{H}) \cap L^2(0,T,\mathcal{V})$, F_ε is uniformly bounded in U_c, and according to the boundedness of $\int_0^T (\int_{\Gamma_I} G_1 \mid \mathbf{w}_\varepsilon^{(a)} - \mathbf{w}_\varepsilon^{(o)} \mid\mid \mathbf{U}_\varepsilon{}^{(a)} - \mathbf{U}_\varepsilon{}^{(o)} \mid_2^2 d\Gamma_I)dt$ we obtain that $\frac{\partial X_\varepsilon}{\partial t}$ is uniformly bounded in $L^1(0,T,\mathcal{V}')$.

Now we can use the end of proof of Theorem 1. We can extract from $(F_\varepsilon, X_\varepsilon, \mathbf{w}_\varepsilon)$ a subsequence also denoted by $(F_\varepsilon, X_\varepsilon, \mathbf{w}_\varepsilon)$, which verifies

$$\begin{aligned}
& F_\varepsilon \rightharpoonup F_0 \text{ weakly in } U_c, \\
& X_\varepsilon \rightharpoonup X_0 \text{ weakly in } L^2(0,T,H^1), \\
& X_\varepsilon \longrightarrow X_0 \text{ strongly in } L^2(0,T,L^2), \\
& \mathbf{w}_\varepsilon \rightharpoonup \mathbf{U}_0 \text{ weakly in } L^2(0,T,H^1), \\
& \mathbf{w}_\varepsilon \longrightarrow \mathbf{U}_0 \text{ strongly in } L^2(0,T,L^2), \\
& |\, \mathbf{w}_\varepsilon^{(a)} - \mathbf{w}_\varepsilon^{(o)} \,|_2 \, (\mathbf{U}_\varepsilon{}^{(a)} - \mathbf{U}_\varepsilon{}^{(o)}) \rightharpoonup |\, \mathbf{U}_0^{(a)} - \mathbf{U}_0^{(o)} \,|_2 \, (\mathbf{U}_0^{(a)} - \mathbf{U}_0^{(o)}) \\
& \qquad\qquad\qquad\qquad \text{weakly in } L^{\frac{3}{2}}(0,T,L^{\frac{3}{2}}(\Gamma_I)),
\end{aligned} \tag{57}$$

where $X_0 = (\mathbf{U}_0, \tau_0)$ is solution of (38), F_0 being the interface forcing.

Now applying the property of weak lower semicontinuity of J, we obtain

$$\begin{aligned}
& J(F_0, \mathbf{U}_0) \leq \liminf_{\varepsilon\to 0} J(F_\varepsilon, \mathbf{U}_\varepsilon), \\
& \text{and} \\
& \frac{1}{2}(\| \mathbf{U}_0 - \mathbf{u} \|^2_{L^2(0,T,L^2(\Omega))} + \| F_0 - F \|^2_{U_c}) \\
& \qquad \leq \liminf_{\varepsilon\to 0} \frac{1}{2}(\| \mathbf{U}_\varepsilon - \mathbf{u} \|^2_{L^2(0,T,L^2(\Omega))} + \| F_\varepsilon - F \|^2_{U_c}).
\end{aligned} \tag{58}$$

From (58), we can deduce that

$$J(F_0, \mathbf{U}_0) \leq \liminf_{\varepsilon\to 0} J_\varepsilon(F_\varepsilon, \mathbf{w}_\varepsilon).$$

Since $(F_\varepsilon, \mathbf{w}_\varepsilon)$ (respectively $(F, \mathbf{u})$) is the optimal control solution of (40) (respectively of (44)), we thus have

$$J(F_0, \mathbf{U}_0) \leq \liminf_{\varepsilon\to 0} J_\varepsilon(F_\varepsilon, \mathbf{w}_\varepsilon) \leq J(F, \mathbf{u}) \leq J(F_0, \mathbf{U}_0).$$

We conclude that $J(F_0, \mathbf{U}_0) = J(F, \mathbf{u})$, and $(F_0, \mathbf{U}_0)$ is solution of control problem (40).

Applying Lemma 8, we also have

$$J(F, \mathbf{u}) \leq \liminf_{\varepsilon\to 0} J_\varepsilon(F_\varepsilon, \mathbf{w}_\varepsilon) \leq \limsup_{\varepsilon\to 0} J_\varepsilon(F_\varepsilon, \mathbf{w}_\varepsilon) \leq J(F, \mathbf{u}),$$

which implies that $\lim_{\varepsilon\to 0} J_\varepsilon(F_\varepsilon, \mathbf{w}_\varepsilon) = J(F, \mathbf{u}) = J(F_0, \mathbf{U}_0)$.

Since $\mathbf{U}_\varepsilon$ converges towards $\mathbf{U}_0$ in $L^2(0,T,L^2)$ strongly, therefore

$$0 \leq \limsup_{\varepsilon\to 0} (\frac{1}{2}(\frac{1}{\varepsilon}\int_0^T A_1(\mathbf{U}_\varepsilon - \mathbf{w}_\varepsilon - \mathbf{U}_\varepsilon - \mathbf{w}_\varepsilon)dt + \| F_\varepsilon - F \|^2_{U_c})) + \frac{1}{2} \| \mathbf{U}_0 - \mathbf{u} \|^2_{L^2(0,T,L^2)}$$

$$\leq \limsup_{\varepsilon\to 0} J_\varepsilon(F_\varepsilon, \mathbf{w}_\varepsilon) - \liminf_{\varepsilon\to 0} J(F_\varepsilon, \mathbf{U}_\varepsilon)$$

According to (58) this implies

$$\lim_{\varepsilon\to 0}\frac{1}{\varepsilon}\int_0^T A_1(\mathbf{U}_\varepsilon-\mathbf{w}_\varepsilon,\mathbf{U}_\varepsilon-\mathbf{w}_\varepsilon)dt=0,$$

$$\lim_{\varepsilon\to 0}\| F_\varepsilon - F\|^2_{U_c},$$

$$(\mathbf{U}_0,\tau_0,F_0)=(\mathbf{u},\mathcal{T},F).$$

This completes the proof. □

These convergence results will allow us to pass to the limit in the adjoint penalized problem and thus to characterize the optimal control $(F,\mathbf{u})$.

Theorem 7. *Let $(F,\mathbf{u})$ be a solution of control problem (40) and $(\mathbf{u},\mathcal{T})$ be the solution of direct problem (38). Then there exists $(\mathbf{R},\mathcal{S})\in L^2(0,T,\mathcal{V})\cap L^\infty(0,T,\mathcal{H})$ solution of the following adjoint problem*

$$-(\frac{\partial\mathbf{R}}{\partial t},\mathbf{v})+A_1(\mathbf{R},\mathbf{v})+L_1(\mathbf{v},\mathbf{R})+D(\mathbf{v},\mathbf{R})+B_2(\mathbf{v},\mathcal{T},\mathcal{S})+B_2(\mathbf{v},\mathcal{T}_0,\mathcal{S})+c^{(a)}(\mathcal{S}^{(a)},\mathbf{v}^{(a)})$$

$$+\int_{\Gamma_I} G_1(|\,\mathbf{u}^{(a)}-\mathbf{u}^{(o)}\,|_2\,(\mathbf{R}^{(a)}-\mathbf{R}^{(o)}),\mathbf{v}^{(a)}-\mathbf{v}^{(o)})_2 d\Gamma_I$$

$$+\int_{\Gamma_I} G_1\left(|\,\mathbf{u}^{(a)}-\mathbf{u}^{(o)}\,|_2^{-1}\,(\mathbf{u}^{(a)}-\mathbf{u}^{(o)},\mathbf{R}^{(a)}-\mathbf{R}^{(o)})_2(\mathbf{u}^{(a)}-\mathbf{u}^{(o)}),\mathbf{v}^{(a)}-\mathbf{v}^{(o)}\right)_2 d\Gamma_I$$

$$=C^{(a)}(curl(\mathbf{u}^{(a)}-\mathbf{u}^{(a)}_{obs}),curl(\mathbf{v}^{(a)}))+C^{(o)}(curl(\mathbf{u}^{(o)}-\mathbf{u}^{(o)}_{obs}),curl(\mathbf{v}^{(o)})),$$

$$-(\frac{\partial\mathcal{S}}{\partial t},\phi)+A_2(\mathcal{S},\phi)+A_{2I}(\mathcal{S},\phi)+L_2(\phi,\mathcal{S})+C(\phi,\mathbf{R})=0,\quad \forall(\mathbf{v},\phi)\in\mathcal{V},$$

$$(\mathbf{R},\mathcal{S})(t=T)=(\mathbf{0},0),$$

and the optimal solution F satisfies the following inequality

$$(\gamma_0(\mathcal{S}^{(a)}-\mathcal{S}^{(o)})+\gamma F,\ G-F)_{U_c}\geq 0,\quad \forall G\in K_c.$$

Moreover the gradient of the cost J at F in direction G is given by

$$\frac{\partial J}{\partial G}(F,\mathbf{u}).G=\int_0^T\int_{\Gamma_I}(\gamma_0(\mathcal{S}^{(a)}-\mathcal{S}^{(o)})+\gamma F)G\,d\Gamma dt.$$

Proof. $(\mathbf{R}_\varepsilon,\mathcal{S}_\varepsilon)$ being the solution of the adjoint penalized problem, we introduce the sequence d_ε by setting

$$d_\varepsilon=(\int_0^T\int_{\Gamma_I}|\,\mathbf{U}_\varepsilon{}^{(a)}-\mathbf{U}_\varepsilon{}^{(o)}\,|_2\,|\,\mathbf{R}_\varepsilon{}^{(a)}-\mathbf{R}_\varepsilon{}^{(o)}\,|_2^2\,d\Gamma_I dt)^{-1}$$

if $|\,\mathbf{U}_\varepsilon{}^{(a)}-\mathbf{U}_\varepsilon{}^{(o)}\,|_2\,|\,\mathbf{R}_\varepsilon{}^{(a)}-\mathbf{R}_\varepsilon{}^{(o)}\,|_2^2$ is not bounded in $L^1(0,T,L^1(\Gamma_I))$ and

$d_\varepsilon=1$ if $|\,\mathbf{U}_\varepsilon{}^{(a)}-\mathbf{U}_\varepsilon{}^{(o)}\,|_2\,|\,\mathbf{R}_\varepsilon{}^{(a)}-\mathbf{R}_\varepsilon{}^{(o)}\,|_2^2$ is bounded in $L^1(0,T,L^1(\Gamma_I))$.

Remark 6. *The sequence d_ε converges towards d, with $d=0$ or $d=1$.* □

Let $\tilde{\mathbf{R}}_\varepsilon = d_\varepsilon \mathbf{R}_\varepsilon$ and $\tilde{\mathcal{S}}_\varepsilon = d_\varepsilon \mathcal{S}_\varepsilon$. We multiply the systeme (46) by the value d_ε and we thus obtain the following system

$$
\begin{aligned}
&-(\frac{\partial \tilde{\mathbf{R}}_\varepsilon}{\partial t}, \mathbf{v}) + A_1(\tilde{\mathbf{R}}_\varepsilon, \mathbf{v}) + L_1(\mathbf{v}, \tilde{\mathbf{R}}_\varepsilon) + D(\mathbf{v}, \tilde{\mathbf{R}}_\varepsilon) + B_2(\mathbf{v}, \mathcal{T}_0, \tilde{\mathcal{S}}_\varepsilon) \\
&+c^{(a)}(\tilde{\mathcal{S}}_\varepsilon^{(a)}, \mathbf{v}^{(a)}) + \int_{\Gamma_I} G_1(\mid \mathbf{w}_\varepsilon^{(a)} - \mathbf{w}_\varepsilon^{(o)} \mid_2 (\tilde{\mathbf{R}}_\varepsilon^{(a)} - \tilde{\mathbf{R}}_\varepsilon^{(o)}), \mathbf{v}^{(a)} - \mathbf{v}^{(o)})_2 d\Gamma_I \\
&+ \int_{\Gamma_I} G_1 \Big(\mid \mathbf{w}_\varepsilon^{(a)} - \mathbf{w}_\varepsilon^{(o)} \mid_2^{-1} (\mathbf{U}_\varepsilon{}^{(a)} - \mathbf{U}_\varepsilon{}^{(o)}, \tilde{\mathbf{R}}_\varepsilon^{(a)} - \tilde{\mathbf{R}}_\varepsilon^{(o)})_2 (\mathbf{w}_\varepsilon^{(a)} - \mathbf{w}_\varepsilon^{(o)}) \\
&\qquad\qquad\qquad\qquad\qquad\qquad , \mathbf{v}^{(a)} - \mathbf{v}^{(o)} \Big)_2 d\Gamma_I \\
&= d_\varepsilon(\mathbf{U}_\varepsilon - \mathbf{u}, \mathbf{v}) + d_\varepsilon C^{(a)}(curl(\mathbf{U}_\varepsilon{}^{(a)} - \mathbf{u}_{obs}^{(a)}), curl(\mathbf{v}^{(a)})) \\
&\qquad + d_\varepsilon C^{(o)}(curl(\mathbf{U}_\varepsilon{}^{(o)} - \mathbf{u}_{obs}^{(o)}), curl(\mathbf{v}^{(o)})), \\
&-(\frac{\partial \tilde{\mathcal{S}}_\varepsilon}{\partial t}, \phi) + A_2(\tilde{\mathcal{S}}_\varepsilon, \phi) + A_{2I}(\tilde{\mathcal{S}}_\varepsilon, \phi) + L_2(\phi, \tilde{\mathcal{S}}_\varepsilon) + C(\phi, \tilde{\mathbf{R}}_\varepsilon) = 0\,, \quad \forall (\mathbf{v}, \phi) \in \mathcal{V}, \\
&(\tilde{\mathbf{R}}_\varepsilon, \tilde{\mathcal{S}}_\varepsilon)(t = T) = (\mathbf{0}, 0)
\end{aligned}
\tag{59}
$$

and the inequality

$$
(\gamma_0(\tilde{\mathcal{S}}_\varepsilon^{(a)} - \tilde{\mathcal{S}}_\varepsilon^{(o)}) + d_\varepsilon(\gamma F_\varepsilon + F_\varepsilon - F),\ G - F_\varepsilon)_{U_c} \geq 0, \quad \forall G \in K_c.
$$

Setting $(\mathbf{v}, \phi) = (\tilde{\mathbf{R}}_\varepsilon, \tilde{\mathcal{S}}_\varepsilon)$ in (59), we obtain

$$
\begin{aligned}
&-\frac{\partial \mid \tilde{\mathbf{R}}_\varepsilon \mid^2}{2\partial t} + A_1(\tilde{\mathbf{R}}_\varepsilon, \tilde{\mathbf{R}}_\varepsilon) + L_1(\tilde{\mathbf{R}}_\varepsilon, \tilde{\mathbf{R}}_\varepsilon) + D(\tilde{\mathbf{R}}_\varepsilon, \tilde{\mathbf{R}}_\varepsilon) + B_2(\tilde{\mathbf{R}}_\varepsilon, \mathcal{T}_0, \tilde{\mathcal{S}}_\varepsilon) \\
&+c^{(a)}(\tilde{\mathcal{S}}_\varepsilon^{(a)}, \tilde{\mathbf{R}}_\varepsilon^{(a)}) + \int_{\Gamma_I} G_1(\mid \mathbf{w}_\varepsilon^{(a)} - \mathbf{w}_\varepsilon^{(o)} \mid_2 \mid \tilde{\mathbf{R}}_\varepsilon^{(a)} - \tilde{\mathbf{R}}_\varepsilon^{(o)} \mid_2^2 d\Gamma_I \\
&+ \int_{\Gamma_I} G_1(\mid \mathbf{w}_\varepsilon^{(a)} - \mathbf{w}_\varepsilon^{(o)} \mid_2^{-1} (\mathbf{U}_\varepsilon{}^{(a)} - \mathbf{U}_\varepsilon{}^{(o)}, \tilde{\mathbf{R}}_\varepsilon^{(a)} - \tilde{\mathbf{R}}_\varepsilon^{(o)})_2 (\mathbf{w}_\varepsilon^{(a)} - \mathbf{w}_\varepsilon^{(o)}), \tilde{\mathbf{R}}_\varepsilon^{(a)} - \tilde{\mathbf{R}}_\varepsilon^{(o)})_2 d\Gamma_I \\
&= d_\varepsilon(\mathbf{U}_\varepsilon - \mathbf{u}, \tilde{\mathbf{R}}_\varepsilon) + d_\varepsilon C^{(a)}(curl(\mathbf{U}_\varepsilon{}^{(a)} - \mathbf{u}_{obs}^{(a)}), curl(\tilde{\mathbf{R}}_\varepsilon)) \\
&\qquad + d_\varepsilon C^{(o)}(curl(\mathbf{U}_\varepsilon{}^{(o)} - \mathbf{u}_{obs}^{(o)}), curl(\tilde{\mathbf{R}}_\varepsilon)), \\
&-(\frac{\partial \tilde{\mathcal{S}}_\varepsilon}{\partial t}, \tilde{\mathcal{S}}_\varepsilon) + A_2(\tilde{\mathcal{S}}_\varepsilon, \tilde{\mathcal{S}}_\varepsilon) + A_{2I}(\tilde{\mathcal{S}}_\varepsilon, \tilde{\mathcal{S}}_\varepsilon) + L_2(\tilde{\mathcal{S}}_\varepsilon, \tilde{\mathcal{S}}_\varepsilon) + C(\tilde{\mathcal{S}}_\varepsilon, \tilde{\mathbf{R}}_\varepsilon) = 0.
\end{aligned}
$$

According to Lemma 3, to the defintion of D and to the relation (51) we obtain

$$-\frac{\partial \mid \tilde{\mathbf{R}}_\varepsilon \mid^2}{2\partial t} + A_1(\tilde{\mathbf{R}}_\varepsilon, \tilde{\mathbf{R}}_\varepsilon) + \int_{\Gamma_I} G_1 \mid \mathbf{w}_\varepsilon^{(a)} - \mathbf{w}_\varepsilon^{(o)} \mid_2 \mid \tilde{\mathbf{R}}_\varepsilon^{(a)} - \tilde{\mathbf{R}}_\varepsilon^{(o)} \mid_2^2 d\Gamma_I$$

$$= -B_1(\tilde{\mathbf{R}}_\varepsilon, \mathbf{u}_0, \tilde{\mathbf{R}}_\varepsilon) - B_2(\tilde{\mathbf{R}}_\varepsilon, \mathcal{T}_0, \tilde{\mathcal{S}}_\varepsilon) - c^{(a)}(\tilde{\mathcal{S}}_\varepsilon^{(a)}, \tilde{\mathbf{R}}_\varepsilon^{(a)})$$

$$-\int_{\Gamma_I} G_1 \left(\mid \mathbf{w}_\varepsilon^{(a)} - \mathbf{w}_\varepsilon^{(o)} \mid_2^{-1} (\mathbf{U}_\varepsilon{}^{(a)} - \mathbf{U}_\varepsilon{}^{(o)}, \tilde{\mathbf{R}}_\varepsilon^{(a)} - \tilde{\mathbf{R}}_\varepsilon^{(o)})_2 (\mathbf{w}_\varepsilon^{(a)} - \mathbf{w}_\varepsilon^{(o)}), \tilde{\mathbf{R}}_\varepsilon^{(a)} - \tilde{\mathbf{R}}_\varepsilon^{(o)} \right)_2 d\Gamma_I$$

$$+ d_\varepsilon(\mathbf{U}_\varepsilon - \mathbf{u}, \tilde{\mathbf{R}}_\varepsilon) + d_\varepsilon C^{(a)}(curl(\mathbf{U}_\varepsilon{}^{(a)} - \mathbf{u}_{obs}^{(a)}), curl(\tilde{\mathbf{R}}_\varepsilon))$$

$$+ d_\varepsilon C^{(o)}(curl(\mathbf{U}_\varepsilon{}^{(o)} - \mathbf{u}_{obs}^{(o)}), curl(\tilde{\mathbf{R}}_\varepsilon)),$$

$$-\frac{\partial \mid \tilde{\mathcal{S}}_\varepsilon \mid^2}{2\partial t} + A_2(\tilde{\mathcal{S}}_\varepsilon, \tilde{\mathcal{S}}_\varepsilon) + A_{2I}(\tilde{\mathcal{S}}_\varepsilon, \tilde{\mathcal{S}}_\varepsilon) = -C(\tilde{\mathcal{S}}_\varepsilon, \tilde{\mathbf{R}}_\varepsilon).$$

Adding these two equations gives

$$-\frac{\partial \mid \tilde{\mathbf{R}}_\varepsilon \mid^2}{2\partial t} - \frac{\partial \mid \tilde{\mathcal{S}}_\varepsilon \mid^2}{2\partial t} + A_1(\tilde{\mathbf{R}}_\varepsilon, \tilde{\mathbf{R}}_\varepsilon) + A_2(\tilde{\mathcal{S}}_\varepsilon, \tilde{\mathcal{S}}_\varepsilon)$$

$$+ \int_{\Gamma_I} G_1 \mid \mathbf{w}_\varepsilon^{(a)} - \mathbf{w}_\varepsilon^{(o)} \mid_2 \mid \tilde{\mathbf{R}}_\varepsilon^{(a)} - \tilde{\mathbf{R}}_\varepsilon^{(o)} \mid_2^2 d\Gamma_I + A_{2I}(\tilde{\mathcal{S}}_\varepsilon, \tilde{\mathcal{S}}_\varepsilon)$$

$$= -B_1(\tilde{\mathbf{R}}_\varepsilon, \mathbf{u}_0, \tilde{\mathbf{R}}_\varepsilon) - B_2(\tilde{\mathbf{R}}_\varepsilon, \mathcal{T}_0, \tilde{\mathcal{S}}_\varepsilon) - c^{(a)}(\tilde{\mathcal{S}}_\varepsilon^{(a)}, \tilde{\mathbf{R}}_\varepsilon^{(a)}) - C(\tilde{\mathcal{S}}_\varepsilon, \tilde{\mathbf{R}}_\varepsilon) + d_\varepsilon((\mathbf{U}_\varepsilon - \mathbf{u}, \tilde{\mathbf{R}}_\varepsilon)$$

$$+ C^{(a)}(curl(\mathbf{U}_\varepsilon{}^{(a)} - \mathbf{u}_{obs}^{(a)}), curl(\tilde{\mathbf{R}}_\varepsilon)) + C^{(o)}(curl(\mathbf{U}_\varepsilon{}^{(o)} - \mathbf{u}_{obs}^{(o)}), curl(\tilde{\mathbf{R}}_\varepsilon)))$$

$$-\int_{\Gamma_I} G_1 \left(\mid \mathbf{w}_\varepsilon^{(a)} - \mathbf{w}_\varepsilon^{(o)} \mid^{-1} (\mathbf{U}_\varepsilon{}^{(a)} - \mathbf{U}_\varepsilon{}^{(o)}, \tilde{\mathbf{R}}_\varepsilon^{(a)} - \tilde{\mathbf{R}}_\varepsilon^{(o)}) (\mathbf{w}_\varepsilon^{(a)} - \mathbf{w}_\varepsilon^{(o)}), \tilde{\mathbf{R}}_\varepsilon^{(a)} - \tilde{\mathbf{R}}_\varepsilon^{(o)} \right)_2 d\Gamma_I.$$

Applying Lemmas 3-6, and according to the positivity of $A_{2I}(\tilde{\mathcal{S}}_\varepsilon, \tilde{\mathcal{S}}_\varepsilon)$, the boundedness of $X_\varepsilon = (\mathbf{U}_\varepsilon, \mathcal{T}_\varepsilon)$ in $L^2(0,T,\mathcal{V}) \cap L^\infty(0,T,\mathcal{H})$, we have the estimations

$$-\frac{\partial \mid Y_\varepsilon \mid^2}{\partial t} + \alpha \parallel Y_\varepsilon \parallel_W^2 + \int_{\Gamma_I} G_1 \mid \mathbf{w}_\varepsilon^{(a)} - \mathbf{w}_\varepsilon^{(o)} \mid_2 \mid \tilde{\mathbf{R}}_\varepsilon^{(a)} - \tilde{\mathbf{R}}_\varepsilon^{(o)} \mid_2^2 d\Gamma_I$$

$$\leq c_1 d_\varepsilon^2 + c_2 \mid Y_\varepsilon \mid^2 + c_3 \parallel \mathbf{U}_\varepsilon \parallel^2 + c_4 \parallel \mathbf{u}_{obs} \parallel^2 \tag{60}$$

$$+ c_5 \int_{\Gamma_I} \mid \mathbf{U}_\varepsilon{}^{(a)} - \mathbf{U}_\varepsilon{}^{(o)} \mid \mid \tilde{\mathbf{R}}_\varepsilon^{(a)} - \tilde{\mathbf{R}}_\varepsilon^{(o)} \mid_2^2 d\Gamma_I,$$

which implies that (d_ε is bounded) Y_ε is uniformly bounded in $L^2(0,T,\mathcal{V}) \cap L^\infty(0,T,\mathcal{H})$. Moreover we have that

$$\int_0^T \left(\int_{\Gamma_I} G_1 \mid \mathbf{w}_\varepsilon^{(a)} - \mathbf{w}_\varepsilon^{(o)} \mid_2 \mid \tilde{\mathbf{R}}_\varepsilon^{(a)} - \tilde{\mathbf{R}}_\varepsilon^{(o)} \mid_2^2 d\Gamma_I\right) dt \; is \; uniformly \; bounded. \tag{61}$$

By using the boundedness of d_ε we obtain that

$$\int_0^T \left(\int_{\Gamma_I} G_1 \mid \mathbf{U}_\varepsilon{}^{(a)} - \mathbf{U}_\varepsilon{}^{(o)} \mid_2 \mid \tilde{\mathbf{R}}_\varepsilon^{(a)} - \tilde{\mathbf{R}}_\varepsilon^{(o)} \mid_2^2 d\Gamma_I\right)dt \text{ is uniformly bounded.} \quad (62)$$

According to Lemma 7, to the uniformly boundness of $\mathbf{U}_\varepsilon$ and $\mathbf{w}_\varepsilon$ in $L^2(0,T,\mathcal{V}_1)$, and to (61)-(62) (by using Young's formula) we can deduce that

$$\begin{aligned} &G_1 \mid \mathbf{w}_\varepsilon^{(a)} - \mathbf{w}_\varepsilon^{(o)} \mid_2 (\tilde{\mathbf{R}}_\varepsilon^{(a)} - \tilde{\mathbf{R}}_\varepsilon^{(o)}) \text{ is uniformly bounded} \\ &\qquad\qquad \text{in } L^{3/2}(0,T,L^{3/2}(\Gamma_I)), \\ &G_1(\mathbf{U}_\varepsilon{}^{(a)} - \mathbf{U}_\varepsilon{}^{(o)}, \tilde{\mathbf{R}}_\varepsilon^{(a)} - \tilde{\mathbf{R}}_\varepsilon^{(o)})_2 \frac{(\mathbf{w}_\varepsilon^{(a)} - \mathbf{w}_\varepsilon^{(o)})}{\mid \mathbf{w}_\varepsilon^{(a)} - \mathbf{w}_\varepsilon^{(o)} \mid_2} \text{ is uniformly bounded} \\ &\qquad\qquad \text{in } L^{3/2}(0,T,L^{3/2}(\Gamma_I)). \end{aligned} \quad (63)$$

Prove now that $\dfrac{\partial Y_\varepsilon}{\partial t} \in L^1(0,T,\mathcal{V}')$.

Using (59), we have that for all $Z = (\mathbf{v},\phi)$ in $\mathcal{V}$

$$\begin{aligned} -(\frac{\partial \tilde{\mathbf{R}}_\varepsilon}{\partial t},\mathbf{v}) &= -A_1(\tilde{\mathbf{R}}_\varepsilon,\mathbf{v}) - L_1(\mathbf{v},\tilde{\mathbf{R}}_\varepsilon) - D(\mathbf{v},\tilde{\mathbf{R}}_\varepsilon) - B_2(\mathbf{v},\mathcal{T}_0,\tilde{\mathcal{S}}_\varepsilon) \\ &\quad -c^{(a)}(\tilde{\mathcal{S}}_\varepsilon^{(a)},\mathbf{v}^{(a)}) - \int_{\Gamma_I} G_1(\mid \mathbf{w}_\varepsilon^{(a)} - \mathbf{w}_\varepsilon^{(o)} \mid_2 (\tilde{\mathbf{R}}_\varepsilon^{(a)} - \tilde{\mathbf{R}}_\varepsilon^{(o)}),\mathbf{v}^{(a)} - \mathbf{v}^{(o)})_2 d\Gamma_I \\ &\quad - \int_{\Gamma_I} G_1\left(\mid \mathbf{w}_\varepsilon^{(a)} - \mathbf{w}_\varepsilon^{(o)} \mid_2^{-1} (\mathbf{U}_\varepsilon{}^{(a)} - \mathbf{U}_\varepsilon{}^{(o)}, \tilde{\mathbf{R}}_\varepsilon^{(a)} - \tilde{\mathbf{R}}_\varepsilon^{(o)})_2(\mathbf{w}_\varepsilon^{(a)} - \mathbf{w}_\varepsilon^{(o)}),\mathbf{v}^{(a)} - \mathbf{v}^{(o)}\right)_2 d\Gamma_I \\ &\quad + d_\varepsilon(\mathbf{U}_\varepsilon - \mathbf{u},\mathbf{v}) + d_\varepsilon C^{(a)}(curl(\mathbf{U}_\varepsilon{}^{(a)} - \mathbf{u}_{obs}^{(a)}),curl(\mathbf{v}^{(a)})) \\ &\quad + d_\varepsilon C^{(o)}(curl(\mathbf{U}_\varepsilon{}^{(o)} - \mathbf{u}_{obs}^{(o)}),curl(\mathbf{v}^{(o)})), \\ -(\frac{\partial \tilde{\mathcal{S}}_\varepsilon}{\partial t},\phi) &= -A_2(\tilde{\mathcal{S}}_\varepsilon,\phi) - A_{2I}(\tilde{\mathcal{S}}_\varepsilon,\phi) - L_2(\phi,\tilde{\mathcal{S}}_\varepsilon) - C(\phi,\tilde{\mathbf{R}}_\varepsilon), \\ (\tilde{\mathbf{R}}_\varepsilon,\tilde{\mathcal{S}}_\varepsilon)(t=T) &= (\mathbf{0},0). \end{aligned}$$

According to Lemmas 3-6 and to the boundedness of d_ε, we have that

$$\begin{aligned} \mid (\frac{\partial Y_\varepsilon}{\partial t},Z) \mid &\le c_0 \parallel Y_\varepsilon \parallel_W \parallel Z \parallel + c_1 \parallel \mathbf{U}_\varepsilon \parallel \parallel Z \parallel \\ &\quad + \mid (G_1 \mid \mathbf{w}_\varepsilon^{(a)} - \mathbf{w}_\varepsilon^{(o)} \mid_2 (\tilde{\mathbf{R}}_\varepsilon^{(a)} - \tilde{\mathbf{R}}_\varepsilon^{(o)}),\mathbf{v}^{(a)} - \mathbf{v}^{(o)})_{\Gamma_I} \mid \\ &\quad + \mid (G_1(\mid \mathbf{w}_\varepsilon^{(a)} - \mathbf{w}_\varepsilon^{(o)} \mid_2^{-1} (\mathbf{U}_\varepsilon{}^{(a)} - \mathbf{U}_\varepsilon{}^{(o)}, \tilde{\mathbf{R}}_\varepsilon^{(a)} - \tilde{\mathbf{R}}_\varepsilon^{(o)})_2(\mathbf{w}_\varepsilon^{(a)} - \mathbf{w}_\varepsilon^{(o)}),\mathbf{v}^{(a)} - \mathbf{v}^{(o)})_{\Gamma_I} \mid. \end{aligned}$$

By using Lemma 7 we have then

$$\mid (\frac{\partial Y_\varepsilon}{\partial t},Z) \mid \le c_0 \parallel Y_\varepsilon \parallel_W \parallel Z \parallel + c_1 \parallel \mathbf{U}_\varepsilon \parallel \parallel Z \parallel$$

$$+c_2 \parallel \mathbf{w}_\varepsilon^{(a)} - \mathbf{w}_\varepsilon^{(o)} \parallel_{L^3(\Gamma_I)}^{1/2} \parallel | \mathbf{w}_\varepsilon^{(a)} - \mathbf{w}_\varepsilon^{(o)} |_2 | \tilde{\mathbf{R}}_\varepsilon^{(a)} - \tilde{\mathbf{R}}_\varepsilon^{(o)} |_2^2 \parallel_{L^1(\Gamma_I)}^{1/2} \parallel Z \parallel_{L^3(\Gamma_I)}$$

$$+c_3 \parallel \mathbf{U}_\varepsilon{}^{(a)} - \mathbf{U}_\varepsilon{}^{(o)} \parallel_{L^3(\Gamma_I)}^{1/2} \parallel | \mathbf{U}_\varepsilon{}^{(a)} - \mathbf{U}_\varepsilon{}^{(o)} |_2 | \tilde{\mathbf{R}}_\varepsilon^{(a)} - \tilde{\mathbf{R}}_\varepsilon^{(o)} |_2^2 \parallel_{L^1(\Gamma_I)}^{1/2} \parallel Z \parallel_{L^3(\Gamma_I)}.$$

Since $Z \in \mathcal{V}$, so $\gamma_o Z \in H^{1/2}(\Gamma_I)$ and according to Sobolev's inclusion (see Appendix) we have $H^{1/2}(\Gamma_I) \subset L^q(\Gamma_I)$ for all $4 \geq q \geq 2$, we can deduce then $\parallel Z \parallel_{L^3(\Gamma_I)} \leq c \parallel Z \parallel_{H^1}$.

Applying this result and using Sobolev's inclusion, we can deduce that

$$\parallel \frac{\partial Y_\varepsilon}{\partial t} \parallel_{\mathcal{V}'} \leq c_0 \parallel Y_\varepsilon \parallel_W + c_4 \parallel \mathbf{w}_\varepsilon \parallel + c_5 \parallel | \mathbf{w}_\varepsilon^{(a)} - \mathbf{w}_\varepsilon^{(o)} |_2 | \tilde{\mathbf{R}}_\varepsilon^{(a)} - \tilde{\mathbf{R}}_\varepsilon^{(o)} |_2^2 \parallel_{L^1(\Gamma_I)}$$

$$+c_6 \parallel \mathbf{U}_\varepsilon \parallel + c_7 \parallel | \mathbf{U}_\varepsilon{}^{(a)} - \mathbf{U}_\varepsilon{}^{(o)} |_2 | \tilde{\mathbf{R}}_\varepsilon^{(a)} - \tilde{\mathbf{R}}_\varepsilon^{(o)} |_2^2 \parallel_{L^1(\Gamma_I)}.$$

Since $(\mathbf{w}_\varepsilon, \mathbf{U}_\varepsilon)$ is uniformly bounded in $L^\infty(0,T,\mathcal{H}_1) \cap L^2(0,T,\mathcal{V}_1)$, and according to (61)-(62) we obtain that $\frac{\partial Y_\varepsilon}{\partial t}$ is uniformly bounded in $L^1(0,T,\mathcal{V}'')$. Now we can use the end of proof of Theorem 1 (according to (63) and to Theorem 5). We can extract from Y_ε a subsequence also denoted by Y_ε and such that

$$\begin{aligned}
&Y_\varepsilon \longrightarrow Y = (\mathbf{R}, \mathcal{S}) \; strongly\ in\ L^2(0,T,\mathcal{H}),\\
&Y_\varepsilon \rightharpoonup Y = (\mathbf{R}, \mathcal{S}) \; weakly\ in\ L^2(0,T,\mathcal{V}),\\
&G_1 \,| \mathbf{w}_\varepsilon^{(a)} - \mathbf{w}_\varepsilon^{(o)} |_2 (\tilde{\mathbf{R}}_\varepsilon^{(a)} - \tilde{\mathbf{R}}_\varepsilon^{(o)}) \rightharpoonup G_1 \,| \mathbf{u}^{(a)} - \mathbf{u}^{(o)} |_2 (\mathbf{R}^{(a)} - \mathbf{R}^{(o)})\\
&\qquad\qquad \text{weakly in } L^{3/2}(0,T,L^{3/2}(\Gamma_I)),\\
&G_1(\mathbf{U}_\varepsilon{}^{(a)} - \mathbf{U}_\varepsilon{}^{(o)}, \tilde{\mathbf{R}}_\varepsilon^{(a)} - \tilde{\mathbf{R}}_\varepsilon^{(o)}) \frac{(\mathbf{w}_\varepsilon^{(a)} - \mathbf{w}_\varepsilon^{(o)})}{| \mathbf{w}_\varepsilon^{(a)} - \mathbf{w}_\varepsilon^{(o)} |}\\
&\quad \rightharpoonup G_1(\mathbf{u}^{(a)} - \mathbf{u}^{(o)}, \mathbf{R}^{(a)} - \mathbf{R}^{(o)})_2 \frac{(\mathbf{u}^{(a)} - \mathbf{u}^{(o)})}{| \mathbf{u}^{(a)} - \mathbf{u}^{(o)} |_2} \text{ weakly in } L^{3/2}(0,T,L^{3/2}(\Gamma_I)).
\end{aligned} \tag{64}$$

Since $(F_\varepsilon, \mathbf{U}_\varepsilon, \mathcal{T}_\varepsilon, \mathbf{w}_\varepsilon)$ verifies the convergence results proved in Theorem 6, we thus obtain that $Y = (\mathbf{R}, \mathcal{S})$ verifies the following system:

$$\begin{aligned}
&-(\frac{\partial \mathbf{R}}{\partial t}, \mathbf{v}) + A_1(\mathbf{R}, \mathbf{v}) + L_1(\mathbf{v}, \mathbf{R}) + D(\mathbf{v}, \mathbf{R}) + B_2(\mathbf{v}, \mathcal{T}_0, \mathcal{S}) + c^{(a)}(\mathcal{S}^{(a)}, \mathbf{v}^{(a)})\\
&\quad + \int_{\Gamma_I} G_1(| \mathbf{u}^{(a)} - \mathbf{u}^{(o)} |_2 (\mathbf{R}^{(a)} - \mathbf{R}^{(o)}), \mathbf{v}^{(a)} - \mathbf{v}^{(o)})_2 d\Gamma_I\\
&\quad + \int_{\Gamma_I} G_1 \left(| \mathbf{u}^{(a)} - \mathbf{u}^{(o)} |_2^{-1} (\mathbf{u}^{(a)} - \mathbf{u}^{(o)}, \mathbf{R}^{(a)} - \mathbf{R}^{(o)})_2 (\mathbf{u}^{(a)} - \mathbf{u}^{(o)}), \mathbf{v}^{(a)} - \mathbf{v}^{(o)}\right)_2 d\Gamma_I\\
&\quad = dC^{(a)}(curl(\mathbf{u}^{(a)} - \mathbf{u}_{obs}^{(a)}), curl(\mathbf{v}^{(a)})) + dC^{(o)}(curl(\mathbf{u}^{(o)} - \mathbf{u}_{obs}^{(o)}), curl(\mathbf{v}^{(o)})),\\
&-(\frac{\partial \mathcal{S}}{\partial t}, \phi) + A_2(\mathcal{S}, \phi) + A_{2I}(\mathcal{S}, \phi) + L_2(\phi, \mathcal{S}) + C(\phi, \mathbf{R}) = 0,\\
&(\mathbf{R}, \mathcal{S})(t = T) = (\mathbf{0}, 0)
\end{aligned} \tag{65}$$

and the inequality

$$(\gamma_0(\mathcal{S}^{(a)} - \mathcal{S}^{(o)}) + d\gamma F,\ G - F)_{U_c} \geq 0, \quad \forall G \in K_c.$$

We are now going to prove that $d = 1$ in order to achieve the proof of Theorem 7.
Let us suppose that $d = 0$ then $(\mathbf{R}, \mathcal{S})$ verifies the system:

$$\begin{aligned}
&-(\frac{\partial \mathbf{R}}{\partial t}, \mathbf{v}) + A_1(\mathbf{R}, \mathbf{v}) + L_1(\mathbf{v}, \mathbf{R}) + D(\mathbf{v}, \mathbf{R}) + B_2(\mathbf{v}, \mathcal{T}_0, \mathcal{S}) + c^{(a)}(\mathcal{S}^{(a)}, \mathbf{v}^{(a)}) \\
&+ \int_{\Gamma_I} G_1 \left(| \mathbf{u}^{(a)} - \mathbf{u}^{(o)} |_2^{-1} (\mathbf{u}^{(a)} - \mathbf{u}^{(o)}, \mathbf{R}^{(a)} - \mathbf{R}^{(o)})_2 (\mathbf{u}^{(a)} - \mathbf{u}^{(o)}), \mathbf{v}^{(a)} - \mathbf{v}^{(o)} \right)_2 d\Gamma_I \\
&+ \int_{\Gamma_I} G_1 (| \mathbf{u}^{(a)} - \mathbf{u}^{(o)} |_2 (\mathbf{R}^{(a)} - \mathbf{R}^{(o)}), \mathbf{v}^{(a)} - \mathbf{v}^{(o)})_2 d\Gamma_I = 0, \\
&-(\frac{\partial \mathcal{S}}{\partial t}, \phi) + A_2(\mathcal{S}, \phi) + A_{2I}(\mathcal{S}, \phi) + L_2(\phi, \mathcal{S}) + C(\phi, \mathbf{R}) = 0, \\
&(\mathbf{R}, \mathcal{S})(t = T) = (\mathbf{0}, 0).
\end{aligned} \tag{66}$$

Setting $(\mathbf{v}, \phi) = (\mathbf{R}, \mathcal{S})$ in (66) we obtain

$$\begin{aligned}
&-\frac{1}{2} \frac{\partial | \mathbf{R} |^2}{\partial t} + A_1(\mathbf{R}, \mathbf{R}) + L_1(\mathbf{R}, \mathbf{R}) + D(\mathbf{R}, \mathbf{R}) + B_2(\mathbf{R}, \mathcal{T}_0, \mathcal{S}) + c^{(a)}(\mathcal{S}^{(a)}, \mathbf{R}^{(a)}) \\
&+ \int_{\Gamma_I} G_1 | \mathbf{u}^{(a)} - \mathbf{u}^{(o)} |_2 | \mathbf{R}^{(a)} - \mathbf{R}^{(o)} |_2^2 \, d\Gamma_I \\
&+ \int_{\Gamma_I} G_1 \frac{(\mathbf{u}^{(a)} - \mathbf{u}^{(o)}, \mathbf{R}^{(a)} - \mathbf{R}^{(o)})_2^2}{| \mathbf{u}^{(a)} - \mathbf{u}^{(o)} |_2} d\Gamma_I = 0, \\
&-\frac{\partial | \mathcal{S} |^2}{\partial t}, \phi) + A_2(\mathcal{S}, \mathcal{S}) + A_{2I}(\mathcal{S}, \mathcal{S}) + L_2(\mathcal{S}, \mathcal{S}) + C(\mathcal{S}, \mathbf{R}) = 0, \\
&(\mathbf{R}, \mathcal{S})(t = T) = (\mathbf{0}, 0).
\end{aligned}$$

where $Y = (\mathbf{R}, \mathcal{S})$.

According to Lemmas 3-6, the positivity of G_1 and Hölder's inequality we have

$$-\frac{\partial | Y |^2}{\partial t} + \alpha \| Y \|_W^2 \leq c | Y |^2 .$$

By applying Gronwall's lemma and according to $Y(T) = 0$, we can deduce that $Y = 0$.
Since the solution $(\mathbf{R}, \mathcal{S})$ is the limit with $\varepsilon \to 0$ of $(\tilde{\mathbf{R}}_\varepsilon, \tilde{\mathcal{S}}_\varepsilon)$ and the sequence $\int_{\Gamma_I} | \mathbf{U}_\varepsilon{}^{(a)} - \mathbf{U}_\varepsilon{}^{(o)} |_2 | \mathbf{R}_\varepsilon{}^{(a)} - \mathbf{R}_\varepsilon{}^{(o)} |_2^2 \, d\Gamma_I$ is equal to 1, then $\int_{\Gamma_I} | \mathbf{u}^{(a)} - \mathbf{u}^{(o)} |_2 | \mathbf{R}^{(a)} - \mathbf{R}^{(o)} |_2^2$ $d\Gamma_I$ is equal to 1 and we have a contradiction. So we can conclude that $d = 1$.
This achieves the proof of the theorem. □

Remark 7. *In the case where we take into account the final observation, we obtain the same results (i.e. the existence theorem of the optimal control problem and necessary optimality conditions for its solution) using the same method (see e.g., Belmiloudi (2008)). In this case the cost functional can be given as follows:*

$$\begin{aligned}
J(G, \mathbf{v}) = &\frac{C^{(a)}}{2} \| curl(\mathbf{v}^{(a)} - \mathbf{u}_{obs}^{(a)}) \|_{L^2(0,T,L^2)}^2 + \frac{C^{(o)}}{2} \| curl(\mathbf{v}^{(o)} - \mathbf{u}_{obs}^{(o)}) \|_{L^2(0,T,L^2)}^2 \\
&+ \frac{m^{(a)}}{2} \| \mathbf{v}^{(a)}(T) - \eta_{obs}^{(a)} \|_{L^2}^2 + \frac{m^{(o)}}{2} \| \mathbf{v}^{(o)}(T) - \eta_{obs}^{(o)} \|_{L^2}^2 + \frac{\gamma}{2} \| G \|_{U_c}^2,
\end{aligned}$$

where $\gamma > 0$, $C^{(o)} \geq 0$, $C^{(a)} \geq 0$, $m^{(o)} \geq 0$, $m^{(a)} \geq 0$, $C^{(o)} + C^{(a)} + m^{(o)} + m^{(a)} > 0$ *and* $U_c = L^2(0,T,L^2(\Gamma_I))$. $\square$

5. Conclusion and Comments

We have developed this study in order to investigate the effects of long equatorial waves in the tropical Atlantic and Pacific region, taking the interaction between the ocean and atmosphere into account. For this, we have introduced a model to calculate the perturbation in the ocean and in the atmosphere of given mean circulations. At initial time the mean circulations verify nonlinear stationary problems. The mean situation is supposed to be known and resulting from physical observations. We have obtained some theoretical results. First, we have proved the existence and uniqueness of the perturbation. To obtain the uniqueness we have assumed some regularities, in space and time, on the external force and on the mean circulations, and we have imposed the compatibility conditions. Secondly, we have developed a control method in order to calculate the velocity and temperature corresponding to the observed vorticity in the ocean-atmosphere domain Ω and during time T. The control is the interface temperature variability F. The cost function measures the distance between the observed and computed vorticity. The observed vorticity is driven by an interface temperature variability F_r. F is an approximation of the unknown real interface temperature variability F_r. The existence of optimal control is proved by means of minimizing sequences. To characterize the control, we introduce a family of penalized control problems; we obtain a set of equations characterizing these problems, by introducing the adjoint problems. Finally we demonstrate the convergence of the penalized problems and obtain the appropriate necessary optimality conditions from which the solution of the initial nonlinear control problem and states may be determined. These conditions (obtained in a Lagrangian form) correspond to identifying the gradient of the cost functional that is necessary to develop numerical optimization methods (gradient methods, Newton methods, etc.).

This study corresponds to the modelization of tropical instability waves which are illustrated by the "*El Niño*" phenomena. It is easy to deduce from the variability of the currents the characteristics of the excited waves, e.g., by Fourier analysis. So our model makes it possible to study the relationship between mean tropical circulation and interface temperature anomalies, and the oceanic and atmospheric eddies and then to obtain the energy, heat transport and vertical propagation induced by the eddies. The perturbation being made of zonally propagating waves, we construct a cyclic model by assuming the following x-dependence: for each perturbed real function Φ (vector or scalar) , we denote by $\tilde{\Phi}$ the complex function independent of x such that: $\Phi(x,y,z,t) = e^{imx}\tilde{\Phi}(y,z,t)$. The wavelength $\lambda = 2\pi/m$ will be fixed for each numerical run. We can thus obtain a process model focusing on "instability waves" which are characterized by eddies of a time-space scale of 30 days-1000 kms propagating westward along the equator (see e.g., Belmiloudi (2002a).

It is clear that, in accordance with practical applications and available experimental observations, we can consider other observations and controls (which can appear in the boundary conditions, in the initial conditions or in the state system) in order to take into account at best the influence of the temperature fluctuations at the atmosphere-ocean interface on the main phenomena and processes that occur in the tropical Atlantic and Pacific regions, and we obtain similar results by using similar approach as used in this work.

For predicting and acting on phenomena and processes occurring inside and surrounding the atmosphere-ocean domain, we can discussed stabilization, identification and regulation processes with and without randomness in data, parameters and boundary conditions, in order to determine, from some observations, the best optimal prognostic values of different data functions or sources appearing in the state system and in the boundary conditions, in the presence of disturbance and fluctuations. In order to take into account the influence of these noises and fluctuations (e.g. on the turbulent and radiative surface heat fluxes), we can study robust control problems by using the approach developed in Book of Belmiloudi (2008).

Appendix. Some Sobolev Imbedding Theorems and Fundamental Inequalities

We use the standard notation for Sobolev spaces (see Adams (1975)), denoting the norm of $W^{s,p}(\Omega)$ $(s>0,\ p\in[1,\infty])$ by $\|\,.\,\|_{W^{s,p}(\Omega)}$. In the special case $p=2$ we use $H^s(\Omega)$ instead of $W^{s,2}(\Omega)$. The duality pairing of a Banach space X with its dual space X' is given by $<.,.>_{X',X}$. For a Hilbert space Y the inner product is denoted by $(.,.)_Y$. For any pair of real numbers $r,s\geq 0$, we introduce the Sobolev space $H^{r,s}(Q)$ defined by $H^{r,s}(Q)=L^2(0,T,H^r(\Omega))\cap H^s(0,T,L^2(\Omega))$, which is a Hilbert space normed by

$$(\|\, v\,\|^2_{L^2(0,T,H^r(\Omega))}+\|\, v\,\|^2_{H^s(0,T,L^2(\Omega))})^{1/2},$$

where $H^s(0,T,L^2(\Omega))$ denotes the Sobolev space of order s of functions defined on $(0,T)$ and taking values in $L^2(\Omega)$, and defined by, for $\theta\in(0,1), s=(1-\theta)m$, m is an integer, (see e.g. Lions et al. (1968)) $H^s(0,T,L^2(\Omega))=[H^m(0,T,L^2(\Omega)),L^2(Q)]_\theta$, $H^m(0,T,L^2(\Omega))=\{v\in L^2(Q)|\ \dfrac{\partial^j v}{\partial t^j}\in L^2(Q),\forall j=1,m\}$.

Theorem A. *(see Adams (1975)): Let Ω be a domain having the cone property in $I\!R^n$. Let $s>0$ and $1<p<n$. Then there exist the following imbeddings (with continuous injections):*

(a) *If $n>sp$, then $W^{s,p}(\Omega)\longrightarrow L^q(\Omega)$ for $p\leq q\leq \frac{np}{n-sp}$.*

(b) *If $n=sp$, then $W^{s,p}(\Omega)\longrightarrow L^q(\Omega)$ for $p\leq q\leq\infty$.*

(c) *If $n<(s-j)p$ for some nonnegative integer j, then $W^{s,p}(\Omega)\longrightarrow C^j(\Omega)$.*

Theorem B. *(see Grisvard (1985)): Let Ω be a bounded lipschitz domain in $I\!R^n$. Let $0<s_1\leq s_2$ and $1<p_2\leq p_1<n$ such that $s_1-\frac{n}{p_1}=s_2-\frac{n}{p_2}$. Then there exist the following embeddings (with continuous injections):*

$$W^{s_2,p_2}(\Omega)\longrightarrow W^{s_1,p_1}(\Omega).$$

□

Remark 8. *Let $\Omega \subset I\!R^n$, $n \geq 1$, be an open and bounded set with a smooth boundary and q be a nonnegative integer. We have the following results (see e.g. Adams (1975))*

(i) $H^q(\Omega) \subset L^p(\Omega)$, $\forall p \in [1, \frac{2n}{n-2q}]$, *with continuous embedding (with the exception that if* $2q = n$, *then* $p \in [1, +\infty[$ *and if* $2q > n$, *then* $p \in [1, +\infty]$ *).*

(ii) *(Gagliardo-Nirenberg inequalities) There exists* $C > 0$ *such that*

$$\| v \|_{L^p} \leq C \| v \|_{H^q}^{\theta} \| v \|_{L^2}^{1-\theta}, \forall v \in H^q(\Omega),$$

where $0 \leq \theta < 1$ *and* $p = \frac{2n}{n-2\theta q}$ *(with the exception that if* $q - n/2$ *is a nonnegative integer, then* θ *is restricted to 0).* □

Our study involve the following fundamental inequalities, which are repeated here for review:

(i) *Hölder's inequality*

$$\int_D \Pi_{i=1,k} f_i dx \leq \Pi_{i=1,k} \| f_i \|_{L^{q_i}(D)}, \text{ where}$$
$$\| f_i \|_{L^{q_i}(D)} = \left(\int_D | f_i |^{q_i} dx\right)^{1/q_i} \text{ and } \sum_{i=1,k} \frac{1}{q_i} = 1.$$

(ii) *Young's inequality* ($\forall a, b > 0$ and $\varepsilon > 0$)

$$ab \leq \frac{\varepsilon}{p} a^p + \frac{\varepsilon^{-q/p}}{q} b^q, \ for\ p, q \in]1, +\infty[\ and\ \frac{1}{p} + \frac{1}{q} = 1.$$

(iii) *Gronwall's Lemma*

$$If\ \frac{d\Phi}{dt} \leq g(t)\Phi(t) + h(t),\ \forall t \geq 0\ then$$
$$\Phi(t) \leq \Phi(0) exp\left(\int_0^t g(s)ds\right) + \int_0^t h(s) exp\left(\int_s^t g(\tau)d\tau\right)ds,\ \forall t \geq 0.$$

□

References

R.A. Adams, *Sobolev spaces.* Academic Press: New-York, 1975.

A. Belmiloudi, A nonlinear optimal control problem for assimilation of surface data in a Navier-Stokes type equations related to oceanography. *Num. Func. Analy. Opti*, 20 (1999) 1-26.

A. Belmiloudi, *Regularity results and optimal control problems for the perturbation of Boussinesq equations of the ocean.* Num.Func.Analy.Opti, **21** (2000) 623-651.

A. Belmiloudi, Asymptotic Behaviour of the Perturbation of the primitive equations of the ocean with vertical viscosity. *The Canadian Applied Mathematics Quarterly*, 8 (2000) 97-139.

A. Belmiloudi, Existence and characterization of an optimal control for the problem of long waves in a shallow-water model. *SIAM Journal on Control and Optimization*, (published electronically January 31, 2001) 39 (2001) 1558-1584.

A. Belmiloudi, Mathematical analysis and optimal control problems for the perturbation of the primitive equations of the ocean with vertical viscosity. *J. Appl. Anal*, 8 (2002).

A. Belmiloudi, Robin-type boundary control problems for the nonlinear Boussinesq type equations. *J. Math. Anal. Appl*, 273 (2002) 428-456.

A. Belmiloudi, *Stabilization, optimal and robust control. Theory and applications in biological and physical sciences.* Springer-Verlag: London, Berlin (2008).

A. Belmiloudi & F. Brossier, A control method for assimilation of surface data in a linearized Navier-Stokes type problem related to oceanography. *SIAM Journal on Control and Optimization*, 35 (1997) 2183-2197.

A. Belmiloudi & F. Brossier, Regularity results for a Navier-Stokes type problem related to oceanography. *Acta. Appl. Math.*, 48 (1997) 299-316.

A. Belmiloudi & F. Brossier, L. Monier, Mathematical and numerical modellization of large scale oceanic waves. *Math. Meth. Appl. Sci.*, 22 (1999) 967-999.

F. Bryan, Parameter sensitivity of primitive equation ocean general circulation models. *J. Phys. Oceanogr.*, 17 (1987) 970-985.

F. Brossier; L. Monier & P. Delecluse, Simulation of long equatorial waves in the Pacific ocean in relation with sea level oscillations and zonal mean currents. *Oceanologica Acta*, 17 (1994) 461-477.

D. B. Chelton; S. K. Esbensen; N. Schlax; N. Thum & M. H. Freilich, Observations of coupling between surface wind stress and sea surface temperature in the eastern tropical Pacific. *J. Climate*, 14 (2001) 1479-1498.

J. A. Cummings & O.M. Smedstad, Variational Data Assimilation for the Global Ocean. Chapter 13, In: *Data Assimilation for Atmospheric, Oceanic & Hydrologic Applications* (Vol.II), S. Park and L. Xu (eds), Springer-Verlag, Berlin, Heidelberg, (2013) 303-343.

I.Y Gejadze; G.J.M.Copeland & I.M. Navon, Open boundary control problem for Navier-Stokes equations including a free surface: adjoint sensitivity analysis. *Comput. Math. Appl.*, 52 (2006) 1243-1268.

I.Y Gejadze; G.J.M.Copeland & I.M. Navon, Open boundary control problem for Navier-Stokes equations including a free surface: data assimilation. *Comput. Math. Appl.*, 52 (2006) 1269-1288.

M. Grisvard, Elliptic problems in nonsmooth domains. *Monographs and studies in Mathematics*, 24, Pitman, London, (1985).

E.K. Jin et al., Current status of ENSO prediction skill in coupled ocean-atmosphere models. *Clim. Dyn.*, 31 647-664 (2008).

J.L. Lions, *Quelques méthodes de résolution des problèmes aux limites non linéaires.* Dunod, Paris, (1969).

J.L. Lions, *Equations differentielles operationnelles.* Springer Verlag, (1961).

J.L. Lions & E.Magenes, *Problèmes aux limites non homogènes et applications.* tomes 1 et 2, Dunod, (1968).

J.L. Lions; R. Temam& S. Wang, New formulation of the primitive equations of atmosphere and application. *Nonlinearity* 5 (1992) 237-288.

J.L. Lions; R. Temam & S. Wang, Mathematical Theory for the coupled atmosphere-ocean models. *J. Math. Pure. Appl.*, 74 (1995) 105-163.

A.J. Majda,, Introduction to PDEs and Waves for the Atmosphere and Ocean. *Courant Lecture Notes in Mathematics*, 9, American Mathematical Society, (2003).

G.I. Marchuk, *Adjoint Equation and Analysis of Complex systems.* kluwer, dordrecht, the Netherlands, (1995).

E.I. Parmuzin; V.P. Shutyaev & N.A. Diansky, Numerical solution of a variational data assimilation problem for a 3D ocean thermohydrodynamics model with a nonlinear vertical heat exchange. *Russ. J. Numer. Anal. Math. Model.*, 22 (2007) 177-198.

S. Park & L. Xu (eds.), *Data Assimilation for Atmospheric, Oceanic & Hydrologic Applications.* (Vol.II), Springer-Verlag: Berlin, Heidelberg (2013).

J. Pedlosky, *Geophysical fluid dynamics.* Springer-Verlag, (1979).

S.G.H. Philander & R.C. Pacanowski, Parametrization of vertical mixing in numerical models of tropical oceans. *J. Phys. Oceanogr.*, 11 (1981) 1443-1451.

S.G.H. Philander, Instabilities of zonal currents 1. *J. Geophys. Res.*, 81 (1976) 3725-3735.

S.G.H. Philander; W.J. Hurlin & R.C. Pacanowski, Properties of long equatorial waves in models of the seasonal cycle in the tropical Atlantic and Pacific oceans. *J. Geophys. Res.*, 91 (1986) 14207-14211.

A.D. Siegel, A comment on long waves in the Pacific ocean. *J. Phys. Oceanogr.*, 15 (1985) 1881-1883.

S.S. Sritharan (ed.), *Optimal control of viscous.* SIAM, Philadelphia, PA., (1998).

R. Temam, *Navier-Stokes equations.* North Holland, (1977).

C. Wen; Y. Xue & A. Kumar, Ocean-Atmosphere Characteristics of Tropical Instability Waves Simulated in the NCEP Climate Forecast System Reanalysis. *J. Climate* 25 (2012) 6409-6425 .

C. Wunsch, *The ocean circulation inverse problem.* Cambridge University Press, Cambridge, (1996).

In: Partial Differential Equations
Editor: Deborah E. Richards, pp. 71-163
ISBN: 978-1-63482-643-3

Chapter 3

INTEGRATION OF PDE WITH THE HELP OF ANALYSIS OVER OCTONIONS AND CAYLEY-DICKSON ALGEBRAS

Sergey V. Ludkovsky*
Department of Applied Mathematics,
Moscow State Technical University MIREA, Moscow, Russia

Abstract

The chapter is devoted to the integration of partial differential equations (PDE) with the help of non-commutative analysis over octonions and Cayley-Dickson algebras. For this purpose a non-commutative line integration of functions and generalized functions is studied. First, second and higher order PDE with continuous, piecewise continuous and generalized variable coefficients over Cayley-Dickson algebras are investigated. Then solutions of some types of non-linear PDE over Cayley-Dickson algebras are studied.

This technique is based on decompositions of partial differential operators (PDO) into products of lower order PDO over octonions and Cayley-Dickson algebras.

Moreover, PDE of higher order with real and complex coefficients and with variable coefficients with or without boundary conditions are considered. Applications to Sobolev's mixed type PDE are investigated as well. Local, global solutions and problems with boundary conditions are described.

Keywords and Phrases: partial differential operator, partial differential equation, Dirac operator, non-commutative, generalized coefficients, line integration, hypercomplex numbers, Cayley-Dickson algebras

Mathematics Subject Classification 2010: 35G20, 35G30, 35F20, 35F30, 35M10, 35M20, 44A10, 30G35, 32W50, 46F10, 17A05, 17A70

1. Introduction

Partial differential equations (PDE) are very important not only for mathematics, but also for their many-sided applications in different sciences including physics, mechanics, other

*E-mail address: ludkowski@mirea.ru

natural sciences, techniques, economics, etc. [10, 16, 52, 45, 57]. Mainly PDE are considered over fields such as real, complex, or with non-archimedean norms. Recently they are also begun to be studied over Clifford algebras [11, 12, 13].

This chapter is devoted to integration approaches of PDE with the help of analysis over octonions and Cayley-Dickson algebras. It opens new possibilities for finding solutions or integrating PDE and their systems. For this purpose previous results of the author on non-commutative line integration over Cayley-Dickson algebras are used [26, 25, 36]. In these works the super-differentiable functions of the Cayley-Dickson variable z were investigated as well [40, 38, 41]. It is worth to mention that there was studied a rather mild type of differentiation over Cayley-Dickson algebras in terms of word algebras. This permits to work with rather abundant families or modules of functions. Such functions were also called z-holomorphic or z-differentiable, or shortly holomorphic or differentiable, when the Cayley-Dickson algebra is outlined.

Cayley-Dickson algebras have a long history, because quaternions were first introduced by W.R. Hamilton in 1843. He had planned to use them for problems of mechanics and mathematics [15]. Their generalization known as the octonion algebra was introduced by J.T. Graves and A. Cayley in 1843-45. Then Dickson had investigated more general algebras known now as the Cayley-Dickson algebras [2, 5, 18].

Octonions and quaternions are widely used in physics, but mainly algebraically. Maxwell had utilized quaternions to derive his equations of electrodynamics, but then he had rewritten them in real coordinates.

Already Yang and Mills had used them in quantum field theory, but theory of functions over octonions and quaternions in their times was not sufficiently developed to satisfy their needs. They have formulated the problem of developing analysis over octonions and quaternions [14]. This is natural, because quantum fields are frequently non-abelian.

Dirac had used complexified bi-quaternions to solve spin problems of quantum mechanics. He had solved the Klein-Gordon hyperbolic PDE with constant coefficients.

Each Cayley-Dickson algebra $\mathcal{A}_{r+1}$ is obtained from the preceding $\mathcal{A}_r$ with the help of the so called doubling procedure [2, 18, 21]. This gives the family of embedded algebras: $\mathcal{A}_r \hookrightarrow \mathcal{A}_{r+1} \hookrightarrow \dots$. For a unification of notation it is convenient to put: $\mathcal{A}_0 = \mathbf{R}$ for the real field, $\mathcal{A}_1 = \mathbf{C}$ for the complex field, $\mathcal{A}_2 = \mathbf{H}$ denotes the quaternion skew field, $\mathcal{A}_3 = \mathbf{O}$ is the octonion algebra, $\mathcal{A}_4$ denotes the sedenion algebra. The quaternion skew field is associative, but non-commutative. The octonion algebra is the alternative division algebra with the multiplicative norm. The sedenion algebra and Cayley-Dickson algebras of higher order $r \geq 4$ are not division algebras and have not any non-trivial multiplicative norm. Each equation of the form $ax = b$ with non-zero octonion a and any octonion b can be resolved in the octonion algebra: $x = a^{-1}b$, but it may be non-resolvable in Cayley-Dickson algebras of higher order $r \geq 4$ because of divisors of zero. Therefore, in this chapter PDE are considered mainly with octonion or quaternion variables for octonion or quaternion valued functions and with precautions for general Cayley-Dickson algebras.

We recall the doubling procedure for the Cayley-Dickson algebra $\mathcal{A}_{r+1}$ from $\mathcal{A}_r$, because it is frequently used. Each Cayley-Dickson number $z \in \mathcal{A}_{r+1}$ is written in the form $z = \xi + \eta \mathbf{l}$, where $\mathbf{l}^2 = -1$, $\mathbf{l} \notin \mathcal{A}_r$, $\xi, \eta \in \mathcal{A}_r$. The addition of such numbers is componentwise. The conjugate of any Cayley-Dickson number z is given by the formula:

$$z^* := \xi^* - \eta \mathbf{l}. \tag{1}$$

The multiplication in $\mathcal{A}_{r+1}$ is prescribed by the following equation:

$$(\xi + \eta \mathbf{l})(\gamma + \delta \boldsymbol{l}) = (\xi\gamma - \tilde{\delta}\eta) + (\delta\xi + \eta\tilde{\gamma})\boldsymbol{l} \tag{2}$$

for each $\xi, \eta, \gamma, \delta \in \mathcal{A}_r$, $z := \xi + \eta \boldsymbol{l} \in \mathcal{A}_{r+1}$, $\zeta := \gamma + \delta \boldsymbol{l} \in \mathcal{A}_{r+1}$.

The basis of $\mathcal{A}_r$ over $\mathbf{R}$ is denoted by $\mathbf{b}_r := \mathbf{b} := \{1, i_1, ..., i_{2^r-1}\}$, where $i_s^2 = -1$ for each $1 \le s \le 2^r - 1$, $i_{2^r} := \mathbf{l}$ is the additional element of the doubling procedure of $\mathcal{A}_{r+1}$ from $\mathcal{A}_r$, choose $i_{2^r+m} = i_m \mathbf{l}$ for each $m = 1, ..., 2^r - 1$, $i_0 := 1$. This implies that $\xi \mathbf{l} = \mathbf{l}\xi^*$ for each $\xi \in \mathcal{A}_r$, when $1 \le r$.

The chapter is devoted to the integration of PDE with the help of non-commutative analysis over octonions and Cayley-Dickson algebras. For this purpose a non-commutative line integration of functions and generalized functions is studied. First, second and higher order PDE with continuous, piecewise continuous and generalized variable coefficients over Cayley-Dickson algebras are investigated.

Formulas for integrations of such equations are deduced. For this purpose a non-commutative line integration is used. Examples of solutions of PDE are given. This continues and develops further approach of Dirac to integration of PDE. Moreover the approach of over Cayley-Dickson algebras gives the fundamental solution of any first and second order linear PDE with variable differentiable coefficients, in a domain satisfying some mild convexity conditions. These results can be used for solutions of concrete PDE or their systems of different orders with piecewise continuous or generalized coefficients, for example, of Helmholtz' or Klein-Gordon's types, which are important in optics of composite materials or quantum field theory. Then solutions of some types of non-linear PDE over Cayley-Dickson algebras are studied.

This technique is based on decompositions of partial differential operators (PDO) into products of lower order PDO over octonions and Cayley-Dickson algebras. Results on investigations of such decompositions are described. This permits to integrate not only elliptic, but also hyperbolic and parabolic PDE of the second and higher orders developing further Dirac's approach. It is important for many-sided applications of PDE. Solutions of diverse type PDE with the help of non-commutative line integration over octonions are also investigated.

Then multidimensional non-commutative Laplace transforms over octonions are considered as well. Their applications to PDE are discussed.

Moreover, PDE of higher order with real and complex coefficients and with variable coefficients with or without boundary conditions are considered. Applications to Sobolev's mixed type PDE are investigated as well. Applications to non-linear vector PDE such as Korteweg-de-Vries' and Kadomtzev-Petviashvili's and describing stationary non-isothermal flows of incompressible Newtonian liquids are given. Kadomtzev-Petviashvili's and modified Korteweg-de-Vries' equations are used in modeling of physical processes such as non-stationary spread of waves in a material with dispersion, magneto-hydrodynamical waves in a non-collision plasma, heat conductivity of a lattice of anharmonic oscillators. Moreover, PDE describing stationary non-isothermal flows of incompressible Newtonian liquids are also very important. This method is also applicable to a resonance interactions in a non-linear material, which plays very important role in the non-linear optics. Local, global solutions and problems with boundary conditions are described.

2. PDE and Their Systems of Diverse Types over Octonions

2.1. Existence of Solutions

1. Remark. Suppose there is given a system of partial differential equations:

$$\partial^{n_j} u_j/\partial t^{n_j} = F_j(t, x_1, ..., x_n, u_1, ..., u_N, ..., \partial^k u_l/\partial t^{k_0}\partial x_1^{k_1}...\partial x_n^{k_n}, ...), \qquad (1)$$

where $j, l = 1, ..., N$; $k_0 + k_1 + ... + k_n = k \leq n_l$; $k_0 < n_l$, $N, n \in \mathbf{N}$, $0 \leq k \in \mathbf{Z}$, $(x_1, ..., x_n) \in U$, $x_j \in \mathbf{K}$ for each j, U is a domain in $\mathbf{K^n}$, where $\mathbf{K} = \mathbf{H}$ is either the quaternion skew field or $\mathbf{K} = \mathbf{O}$ is the octonion algebra. When a number t is real the F_j are $\mathbf{K}$-valued functions or functionals. In the case $t \in \mathbf{K}$ each F_j is a mapping with values in the space $L_q(\mathbf{K}^{\otimes n_j}; \mathbf{K})$ of all $\mathbf{R}$ n_j-poly-homogeneous $\mathbf{K}$-additive operators with values in $\mathbf{K}$. Each partial derivative $\partial^k u_l/\partial t^{k_0}\partial x_1^{k_1}...\partial x_n^{k_n}$ is the mapping with values in $L_q(\mathbf{K}^{\otimes s}; \mathbf{K})$, where $s = k - k_0$ for $t \in \mathbf{R}$ and $s = k$ for $t \in \mathbf{K}$, $\mathbf{K} = \mathbf{H}$ or $\mathbf{K} = \mathbf{O}$.

For $t \in \mathbf{R}$ there are given initial conditions:

$$(\partial^s u_j/\partial t^s)|_{t=t_0} = \phi_j^{(s)}(x_1, x_2, ..., x_n) \qquad (2)$$

while for $t \in \mathbf{K}$ let

$$(\partial^s u_j/\partial t^s)|_{Re(t)=Re(t_0)} = \phi_j^{(s)}(t; x_1, x_2, ..., x_n) \qquad (2')$$

for each $s = 0, 1, ..., n_j - 1$ and each $(x_1, ..., x_n) \in U$, where $\partial^0 u_j/\partial t^0 := u_j$, in addition each $\phi_j^{(s)}(t; x_1, x_2, ..., x_n)$ is locally analytic by all variables and independent from $Re(t)$ (see also the introduction).

We search a solution of the problem $(1, 2)$ in a neighborhood V of the given point $(t_0, x_1^0, ..., x_n^0)$ either in $\mathbf{R} \times U$ or in $\mathbf{K} \times U$. For our convenience we introduce the notation:

$$(\partial^{k-k_0}\phi_j^{(k_0)}/\partial x_1^{k_1}...\partial x_n^{k_n})|_{(x_1=x_1^0,...,x_n=x_n^0)} = \phi^0_{(j,k_0,k_1,...,k_n)}$$

for each $j = 1, ..., N$,

$$k_0 + k_1 + ... + k_n = k \leq n_j. \qquad (3)$$

Suppose that all functions F_j are analytic by all variables and holomorphic by $x_1, ..., x_n$; $u_1, ..., u_N$, also holomorphic by t, when $t \in \mathbf{K}$, in a neighborhood Y of the point $(t_0, x_1^0, ..., x_n^0, ..., \phi^0_{j,k_0,k_1,...,k_n}, ...)$, where analyticity by operators $A_{l,k} := \partial^k u_l/\partial t^{k_0}\partial x_1^{k_1}...\partial x_n^{k_n}$ is defined relative to their suitable compositions and the convergence of a series of F_j by $\{A_{l,k} : l, k\}$ is supposed to be relative to the operator norm, where $A_{l,k} \in \bigcup_{m=1}^{\infty} L_q(\mathbf{K}^{\otimes m}, \mathbf{K})$. Let also all functions $\phi_j^{(k)}$ be $\mathbf{K}$-holomorphic (i.e. locally x-analytic) in a neighborhood X of $(x_1^0, ..., x_n^0)$.

2. Theorem. *The problem 1*$(1-3)$ *has a holomorphic solution either by all variables* $(t, x_1, ..., x_n)$ *in some neighborhood of* $(t_0, x_1^0, ..., x_n^0)$, *when* $t \in \mathbf{K}$, *or holomorphic by* *the variables* $(x_1, ..., x_n)$ *and (locally) analytic by* t*, when* $t \in \mathbf{R}$*. This solution is unique in such class of functions.*

Proof. To prove this theorem we introduce new functions, when $n_j > 1$, so that $v_{j,p} := \partial^p u_j/\partial t^p$ for each $p = 1, ..., n_j - 1$, $w_{l,k} := \partial^k u_l/\partial t^{k_0}\partial x_1^{k_1}...\partial x_n^{k_n}$ for each $k_0 + k_1 + ... + k_n = k \le n_l$ and $k_0 < n_l$. Then $\partial v_{j,p}/\partial t = v_{j,p+1}$ for each $p < n_j$ and $\partial v_{j,n_j-1}/\partial t = F_j$, also $w_{j,k} = \partial^k v_{l,k_0}/\partial x_1^{k_1}...\partial x_n^{k_n}$ for each $(k_0, ..., k_n)$ and l, where $F_j = F_j(t, x_1, ..., x_n, u_1, ..., u_N, ..., w_{j,k}, ...)$. The initial conditions now become:

$$u_j|_{Re(t)=Re(t_0)} = \phi_j^{(0)}(t; x_1, x_2, ..., x_n),$$

$$v_{j,s}|_{Re(t)=Re(t_0)} = \phi_j^{(s)}(t; x_1, x_2, ..., x_n)$$

for each $s = 1, ..., n_j - 1$ and each $(x_1, ..., x_n) \in U$ and every $t \in \mathbf{K}$ with $Re(t) = Re(t_0)$ in the considered domain. We can take as a neighborhood of t_0 a ball in $\mathbf{K}$ with center at t_0 of a sufficiently small positive radius. In the real case of the variable t it would be an interval in $\mathbf{R}$. For $t \in \mathbf{R}$ we automatically get initial conditions from these general.

If $\partial(v_{j,k-k_0} - \partial^{k-k_0}u/\partial x_1^{k_1}...\partial x_n^{k_n})/\partial t = 0$ for each t and $(x_1, ..., x_n) \in U$, then $(v_{j,k-k_0} - \partial^{k-k_0}u/\partial x_1^{k_1}...\partial x_n^{k_n})$ is independent from t. If in addition, $v_{j,k-k_0}|_{Re(t)=Re(t_0)} = \phi_j^{(k-k_0)}(t; x_1, x_2, ..., x_n)$ for each $(x_1, ..., x_n) \in U$ and each t for the given domain, then $\partial^{k-k_0}u/\partial x_1^{k_1}...\partial x_n^{k_n} = \phi_j^{(k-k_0)}(t; x_1, x_2, ..., x_n)$ in Y due to analyticity of functions.

Therefore, this procedure of the introduction of new functions $v_{j,s}$ reduces the problem $1(1-3)$ to the problem, when $n_j = 1$ for each j. It is possible to simplify further the problem, making substitutions: $w_j(t,x) := u_j(t,x) - \phi_j^{(0)}(t;x)$, $w_{j,s}(t,x) := v_{j,s}(t,x) - \phi_j^{(s)}(t;x)$ for each j, s, where $x = (x_1, ..., x_n)$. For new functions w_j and $w_{j,s}$ the problem has the same form, but with zero initial conditions. Thus without loss of generality consider zero initial conditions and denote w_j and $w_{j,s}$ again by u_j and $v_{j,s}$.

Consider at first the case when each F_j is $\mathbf{R}$-homogeneous and $\mathbf{K}$-additive by each u_j and each $v_{j,s}$, where $n_j = 1$ for each j. Without loss of generality consider $t \in \mathbf{K}$, since when $t \in \mathbf{R}$, then due to the local analyticity by t it is possible to take the extension of each F_j to $F_j \in L_q$ on a neighborhood of t_0 in $\mathbf{K}$. In view of Proposition 2.18 and Remark 2.18.1 in [25, 26] and Theorem 3.2.14 [40] if a solution of the problem exists in the considered class of holomorphic functions u_j, $j = 1, ..., N$, then it is unique.

Consider now series of F_j, u_j, $v_{j,s}$ for each j and s by all their variables $(t - t_0, x_1 - x_1^0, ..., x_n - x_n^0)$ in a neighborhood of $(t_0, x_1^0, ..., x_n^0)$. Expansion coefficients of u_j and $v_{j,s}$ can be evaluated from the system of equations by induction in the order of increasing powers, since $u_j(t,x)|_{Re(t)=Re(t_0)} = 0$ and $v_{j,s}(t,x)|_{Re(t)=Re(t_0)} = 0$ for each j, s and each $x = (x_1, ..., x_n) \in U$ and every t with the prescribed real part $Re(t) = Re(t_0)$ in the considered domain. Now prove the local convergence of the series.

It is possible to make the shift of variables such that one can consider $(t_0, x_1^0, ..., x_n^0) = 0$ without loss of generality. Let the series decomposition $g(t,x) = \sum_{|k|\ge 0}\{(a_k, (t,x)^k)\}_{q(2|k|)}$ be converging at each point $b = (b_0, b_1, ..., b_n)$ with given values $|b_0|, ..., |b_n|$, where $b_l \ne 0$ for each $l = 0, 1, ..., n$. Then there exists a constant $S > 0$

such that $|\sum_{k=(k_0,...,k_n)}\{(a_k,(t,x)^k)\}_{q(2|k|)}| \le S$ for each k, $|k| := k_0 + ... + k_n$. Therefore, this implies the inequality:

$$|\sum_{k=(k_0,...,k_n)}\{(a_k,(\beta_0,\beta_1,...,\beta_n)^k)\}_{q(2|k|)}| \le S|b_0|^{-k_0}...|b_n|^{-k_n}$$

for each $\beta_0 \in \mathbf{K}, ..., \beta_n \in \mathbf{K}$ with $|\beta_0| = 1,...,|\beta_n| = 1$ and each integers $0 \le k_0, ..., k_n < \infty$, where vectors of associators $q(*)$ indicating on orders of brackets or multiplications are important only over $\mathbf{O}$, $a_k = a_k(g)$ (see Formulas 2.13$(ii - iv)$ in [40]). Take the function $Q(t,x) := S(1 - t/|a_0|)^{-1}...(1 - x_n/|a_n|)^{-1}$. Then $Q(t,x)$ is characterized by the condition:

$$|\sum_{k=(k_0,...,k_n)}\{(a_k(g),(\beta_0,\beta_1,...,\beta_n)^k)\}_{q(2|k|)}| \le |\sum_{k=(k_0,...,k_n)}\{(a_k(Q),(\beta_0,\beta_1,...,\beta_n)^k)\}_{q(2|k|)}|$$

for each k and each $\beta_0, ..., \beta_n$ in the algebra $\mathbf{K}$ with the natural restrictions $|\beta_0| = 1,...,|\beta_n| = 1$. The multiplications $z \mapsto \beta_j^p z$ and $z \mapsto z\beta_j^p$ are pseudo-conformal mappings from $\mathbf{K}$ onto $\mathbf{K}$ for each $p \in \mathbf{N}$. Such holomorphic function Q is called the majorant of the holomorphic function g. In the case of variables $A_{j,s} \in L_q$ consider majorants in the class of analytic functions by $A_{j,s}$ and their suitable compositions, where we choose $b_{j,s} \in \mathbf{K}$ such that $|b_{j,s}| := \|A_{j,s}\|$, since we can take $A_{j,s}.(h_1, ..., h_s)$ and $F_j.h_0$ for $F_j \in L_1(\mathbf{K},\mathbf{K})$ for arbitrary $h_0, h_1, ..., h_s \in \mathbf{K}$ of unit norms $|h_0| = 1$, $|h_1| = 1,...,|h_s| = 1$. Then such composite functions while action on $h_0, h_1, ..., h_s$ have also series expansions which are uniformly convergent series by $h_0, h_1, ..., h_s \in B(\mathbf{K}, 0, 1)$.

Let the problem I be the initial problem formulated in §1 and let the problem II be the problem, in which each participating function is substituted by its majorant. Let U_j, $V_{j,s}$ denotes the solution of problem II. Then

$$|\sum_{k=(0,k_1,...,k_n)}\{(a_k(v_{j,s}),(\beta_0,\beta_1,...,\beta_n)^k)\}_{q(2|k|)}| \le |\sum_{k=(0,k_1,...,k_n)}\{(a_k(V_{j,s}),(\beta_0,\beta_1,...,\beta_n)^k)\}_{q(2|k|)}|$$

for each $\beta_0, ..., \beta_n$ in $\mathbf{K}$ with $|\beta_0| = 1,...,|\beta_n| = 1$ and each j and s due to the initial conditions, where $V_{j,0} := U_j$ and $v_{j,0} := u_j$. In the case $k_0 > 0$ coefficients $a_k(v_{j,s})$ or $a_k(V_{j,s})$ are obtained from coefficients $a_m(v_{j,s})$ or $a_m(V_{j,s})$ respectively with the help of addition and multiplication with numbers subjected to the conditions $0 \le m_0 < k_0$. Therefore, by induction these inequalities are true for each k. Thus if problem II has a solution, the problem I also has a solution.

Choose constants $S > 0$ and $a > 0$ such that the function $S[1-(t/\alpha+x_1+...+x_n)/a]^{-1}$ for some $0 < \alpha < 1$ would be the majorant for all terms of the system besides free terms, as well as take $T[1 - (t/\alpha + x_1 + ... + x_n)/a]^{-1}$ as majorant for free terms, since such majorants with $S = S_j$, $a = a_j$, $T = T_j$ can be taken for each F_j. Hence we can take $S = \max_{j=1,...,n} S_j$, $T := \max_{j=1,...,n} T_j$, $a = \min_{j=1,...,n} a_j$, where α we shall choose below. For it problem II can be taken as:

$$\partial U_j/\partial t = [\sum_{j,k}\partial U_j/\partial x_k + \sum_j U_j I + CI]S[1 - (t/\alpha + x_1 + ... + x_n)/a]^{-1},$$

where $C = T/S$, while $I \in L_1$ is the constant unit operator. Seek a solution $U_j(t,x) = U(z)$, where $z = t/\alpha + x_1 + ... + x_n$. Then we get the equation:

$$\alpha^{-1}dU(z)/dz = (NndU(z)/dz + NUI + CI)S(1 - z/a)^{-1}.$$

Choose a number $\alpha > 0$ sufficiently large such that $1/\alpha - NnS(1-z/a)^{-1} \neq 0$ in a neighborhood of $z = 0$. Then the problem is:

$$(\alpha^{-1} - S(1-z/a)^{-1}Nn)dU(z)/dz = I(NU + C),$$

hence

$$(NU + C)^{-1}(dU(z)/dz) = Ig(z),$$

where $g(z) = (\alpha^{-1} - NnS(1-z/a)^{-1})^{-1}$ due to the alternativity of $\mathbf{K}$. The integration of the latter equation gives the z-differentiable (holomorphic) solution due to Theorem 3.2.14 [40] and Remark 2.18.1 [25, 26], since the mapping g is z-differentiable. Now estimate its expansion coefficients. For this consider an algebraic embedding θ of the complex field $\mathbf{C}$ into $\mathbf{K}$. The restriction of $U(z)$ on this copy of $\mathbf{C}$ has the form:

$$U(z) = [\exp((\int_0^z y(\xi)d\xi)N/C) - 1]C/N,$$

for each $z \in \theta(\mathbf{C})$, where $y(z) = (\alpha^{-1} - NnS(1-z/a)^{-1})^{-1}CS(1-z/a)^{-1}$. The function $S(1-z/a)^{-1}$ has the nonnegative expansion coefficients by powers z^k. Hence $y(\xi)$ also has the nonnegative expansion coefficients by ξ. Thus the indefinite integral

$$s(z) = N\int_0^z y(\xi)d\xi/C$$

also has nonnegative expansion coefficients by z and inevitably $\exp(s(z)) - 1 = s(z) + s^2(z)/2 + ...$ and thus $U(z)$ also have nonnegative expansion coefficients by $z \in \theta(\mathbf{C})$. But the algebra $\mathbf{K}$ is alternative and hence power associative, moreover, $\bigcup_\theta \theta(\mathbf{C}) = \mathbf{K}$. Thus expansion coefficients of $U(x_1 + x_2 + ... + x_n)$ by degrees of x_1,...,x_n are non-negative, consequently, $U(0, x_1, ..., x_n)$ is the majorant of zero. Therefore, functions $U_j(t, x_1, ..., x_n) = U(t/\alpha + x_1 + ... + x_n)$ compose the solution of problem II.

The problem I solved above can be reformulated in such a fashion, that there exists an inverse operator K^{-1} to the differential operator together with initial conditions which compose the operator equation $Ku = g$ in the space of holomorphic vector-functions u, g.

Consider now the case, when F_j are not $\mathbf{R}$-homogeneous $\mathbf{K}$-additive by u_j or $v_{j,s}$ for some j, s. Denote the operator corresponding to this problem by M and the problem is $Mu = g$ in the space of holomorphic vector-functions. Let X be a $\mathbf{K}$-vector space.

Then we call a subset W in X $\mathbf{K}$-convex if $(a_1v_1)b_1 + a_2(v_2b_2) \in W$ for each $a_1, a_2, b_1, b_2 \in B(\mathbf{K}, 0, 1)$ and each $v_1, v_2 \in W$. Since F_j are holomorphic by (t, x) and locally analytic by u_j, $v_{j,s}$ for each j, s, then we can take power series expansions by these variables and consider a sufficiently small $\mathbf{K}$-convex neighborhood of the point of zero initial conditions, where corrections of the second and higher order terms in u_j, $v_{j,s}$ are much smaller. This neighborhood W can be chosen such that $M = K + (M - K)$ with $\|M - K\|_W/\|K\|_W < 1/2$, where $\|M\|_W := \sup_{0 \neq u \in W} |Mu|/|Ku|$. Hence M is invertible on W such that the inverse operator exists $M^{-1} : M(W) \to W$.

The inverse of a holomorphic function is holomorphic (see also Theorem 2). That is functions defining M^{-1} are holomorphic by (x, u) also by t, when $t \in \mathbf{K}$, moreover, they are locally analytic by others variables. In view of the proof above we can construct a solution u of $Ku(t, x) = g$ for each (t, x) with $|(t, x)| < R$ for some $0 < R < \infty$ such that $u \in W/2$. Therefore, we get a solution of the problem $Mu = g$ also for each (t, x) in a sufficiently small neighborhood of 0. In view of the proof above this leads to the existence of a solution of the general problem.

3. Note. Instead of initial conditions at each t with $Re(t) = Re(t_0)$ it is possible for t in a $2^r - 1$-connected domain P in either the quaternion skew field $\mathbf{H}$ or in the octonion algebra $\mathbf{O}$ consider initial conditions for each $t \in \partial P$, when P in this section instead of U there satisfies conditions of Remark 2.2.1 below. The proof is quite analogous in this case. This follows also by applying the non-commutative analog of the Riemann mapping Theorem 2.43 [40] establishing $\mathcal{A}_r$-holomorphic diffeomorphism between different domains in $\mathcal{A}_r$ under definite rather mild conditions.

4. Notation. If $f : U \to \mathcal{A}_r$ is either z-differentiable or $\tilde{z}$-differentiable at $a \in U$ or on U, then we can write also $D_{\tilde{z}}$ instead of $\partial_{\tilde{z}}$ and D_z instead of ∂_z at $a \in U$ or on U respectively in situations, when it can not cause a confusion, where U is open in $\mathcal{A}_r^n$.

5. Proposition. *Let $g : U \to \mathcal{A}_r^m$, $r \geq 2$, and $f : W \to \mathcal{A}_r^n$ be two differentiable functions on U and W respectively such that $g(U) \subset W$, U is open in $\mathcal{A}_r^k$, W is open in $\mathcal{A}_r^m$, $k, n, m \in \mathbf{N}$, where f and g are simultaneously either $(z, \tilde{z})$, or z, or $\tilde{z}$-differentiable. Then the composite function $f \circ g(z) := f(g(z))$ is differentiable on U and*

$$(Df \circ g(z)).h = (Df(g)).((Dg(z)).h)$$

for each $z \in U$ and each $h \in \mathcal{A}_r^k$, and hence $f \circ g$ is of the same type of differentiability as f and g.

Proof. Theorems 2.11, 2.15, 2.16, 3.10 and Corollary 2.13 [25, 26] establish the equivalence of notions of $\mathcal{A}_r$-holomorphic, $\mathcal{A}_r$-integral holomorphic and $\mathcal{A}_r$ locally z-analytic classes of functions on domains satisfying definite conditions.

In view of these results it is sufficient to prove this Proposition on open domains U and W, where $g(U) \subset W$ with $U = g^{-1}(W)$, if others conditions are the same, since z-differentiability is equivalent with the local z-analyticity (that is in the z-representation). Indeed, the composition of two locally z-analytic functions $f \circ g$ with such domains is the locally analytic function.

Since g is differentiable, the function g is continuous and $g^{-1}(W)$ is open in $\mathcal{A}_r^k$. In view of [25, 26] if f and g are simultaneously either z-differentiable or $\tilde{z}$-differentiable, then either $\partial_{\tilde{z}} f = 0$ and $\partial_{\tilde{z}} g = 0$ or $\partial_z f = 0$ and $\partial_z g = 0$ correspondingly on their domains.

Consider the increment of the composite function

$$f \circ g(z+h) - f \circ g(z) = (Df(g))|_{g=g(z)}.(g(z+h) - g(z)) + \epsilon_f(\eta)|\eta|,$$

where $\eta = g(z+h) - g(z)$, $g(z+h) - g(z) = (Dg(z)).h + \epsilon_g(h)|h|$ (see §2.2 [25, 26]). Since the derivative Df is $\mathcal{A}_r^m$-additive and $\mathbf{R}$-homogeneous (and continuous) operator on $\mathcal{A}_r^m$, we have

$$f \circ g(z+h) - f \circ g(z) = (Df(g))|_{g=g(z)}.((Dg(z)).h) + \epsilon_{f \circ g}(h)|h|,$$

where

$$\epsilon_{f\circ g}(h)|h| := \epsilon_f((Dg(z)).h + \epsilon_g(h)|h|)|(Dg(z)).h + \epsilon_g(h)|h||$$
$$+[(Df(g))|_{g=g(z)}.(\epsilon_g(h))]|h|),$$
$$|(Dg(z)).h + \epsilon_g(h)|h|| \leq [\|Dg(z)\| + |\epsilon_g(h)|]|h|,$$

hence

$$|\epsilon_{f\circ g}(h)| \leq |\epsilon_f((Dg(z)).h + \epsilon_g(h)|h|)|[\|Dg(z)\| + |\epsilon_g(h)|] + \|(Df(g))|_{g=g(z)}\||\epsilon_g(h)|$$

and inevitably $\lim_{h\to 0}\epsilon_{f\circ g}(h) = 0$. Moreover, $\epsilon_{f\circ g}(h)$ is continuous in h, since ϵ_g and ϵ_f are continuous functions, Df and Dg are continuous operators. Evidently, if $\partial_{\tilde{z}} f = 0$ and $\partial_{\tilde{z}} g = 0$ on domains of f and g respectively, then $\partial_{\tilde{z}} f \circ g = 0$ on V, since $D = \partial_z + \partial_{\tilde{z}}$.

Suppose now that there are phrases corresponding to f_n and g_n denoted by μ_n and ν_n such that f_n and g_n uniformly converge to f and g respectively on each bounded canonical closed subset in W and U from a family $\mathcal{W}$ or $\mathcal{U}$ respectively, where $\mathcal{W}$ and $\mathcal{U}$ are coverings of W and U correspondingly. While $D_z\mu_n$ and $D_z\nu_n$ are fundamental sequences uniformly on each bounded canonical closed subset $P \in \mathcal{W}$ and $Q \in \mathcal{U}$ correspondingly relative to the operator norm (see Definitions in §2 [25, 26]). Then the sequence $\mu_n \circ \nu_n$ converges on each bounded canonical closed subset Q_1 such that $Q_1 \subset Q \cap g^{-1}(P)$, where $P \in \mathcal{W}$ and $Q \in \mathcal{U}$ in U, when n tends to the infinity. Moreover, $D_y\mu_n(y,\tilde{y}).(D_z\nu_n(z,\tilde{z}))|_{y=\nu_n(z,\tilde{z})}$ and $D_{\tilde{y}}\mu_n(y,\tilde{y}).(D_z\tilde{\nu}_n(z,\tilde{z}))|_{y=\nu_n(z,\tilde{z})}$ are the fundamental sequences of operators on each bounded canonical closed subset Q_1 so that $Q_1 \subset Q \cap g^{-1}(P)$, where $P \in \mathcal{W}$ and $Q \in \mathcal{U}$. The family $\mathcal{Q} := \{Q_1\}$ specified above evidently is the covering of V.

It remains to verify, that

$$D_z[\mu_n \circ \nu_n(z,\tilde{z})] = [(D_y\mu_n(y,\tilde{y}).(D_z\nu_n(z,\tilde{z})) + (D_{\tilde{y}}\mu_n(y,\tilde{y}).(D_z\tilde{\nu}_n(z,\tilde{z}))]|_{y=\nu_n(z,\tilde{z})}.$$

Since the derivation operator D_z by z is $\mathbf{R}$-homogeneous and $\mathcal{A}_r$-additive, it is sufficient to verify this in the case $D_z[\eta \circ \psi(z,\tilde{z})]$ locally in balls, where both series uniformly converge. Here the phrase is written as:

$$\eta = \eta(z,\tilde{z}) = \sum_k \{A_k, z, \tilde{z}\}_{q(k)} := \{a_{k,1} \overbrace{z_{l_1}^{k_1}} ... a_{k,p} \overbrace{z_{l_p}^{k_p}}\}_{q(k)}, \quad (1)$$

$k = (k_1, ..., k_p)$, $p = p(k) \in \mathbf{N}$, $0 \leq k_j \in \mathbf{Z}$ and $l_j \in \{1,2\}$ for each j, $a_{k,1}, ..., a_{k,p} \in \mathcal{A}_r$ are constants, $\overbrace{z_1} = z$, $\overbrace{z_2} = \tilde{z}$, z is the Cayley-Dickson variable, a vector q indicates on an order of the multiplication, $A_k = [a_{k,1}, ..., a_{k,p}]$, also

$$\psi = \psi(z,\tilde{z}) = \sum_m \{B_m, z, \tilde{z}\}_{q(m)}, \quad (2)$$

where $m = (m_1, ..., m_s)$, $B_m = [b_{m,1}, ..., b_{m,s}]$, $b_{m,j} \in \mathcal{A}_r$ is a constant for each m, j, also $s \in \mathbf{N}$, $0 \leq m_j \in \mathbf{N}$ for each j. We have the identities

$$D_z \overbrace{z_l} = \delta_{l,1}\mathbf{1}, \quad D_{\tilde{z}} \overbrace{z_l} = \delta_{l,2}\tilde{\mathbf{1}}, \quad (3)$$

where $\mathbf{1}h := h$ and $\tilde{\mathbf{1}}h := \tilde{h}$ for each $h \in \mathcal{A}_r$. Using shifts $z \mapsto z - \zeta$ we can consider that the series are decomposed around a zero point. If a series η is uniformly converging on a canonical closed subset Q_1 as above, then

$$D_z\eta = \sum_k D_z\{a_{k,1} \overbrace{z_{l_1}^{k_1}} ...a_{k,p} \overbrace{z_{l_p}^{k_p}} \}_{q(k)} \tag{4}$$

is also uniformly converging on Q_1 in accordance with our supposition about convergence of phrases and their derivatives.

In suitable Q_1 we deduce from Formula (4) that

$$D_z[\eta \circ \psi(z, \tilde{z})] = D_z \sum_k \{a_{k,1} \overbrace{\psi_{l_1}^{k_1}} ...a_{k,p} \overbrace{\psi_{l_p}^{k_p}} \}_{q(k)} \tag{5}$$

$$= \sum_k D_z\{a_{k,1} \overbrace{\psi_{l_1}^{k_1}} ...a_{k,p} \overbrace{\psi_{l_p}^{k_p}} \}_{q(k)}$$

$$= \sum_k [\{a_{k,1}(D_z \overbrace{\psi_{l_1}^{k_1}})...a_{k,p} \overbrace{\psi_{l_p}^{k_p}} \}_{q(k)} + ... + \{a_{k,1} \overbrace{\psi_{l_1}^{k_1}} ...a_{k,p}(D_z \overbrace{\psi_{l_p}^{k_p}})\}_{q(k)}]$$

$$= \sum_k \sum_{{}_1m,...,{}_pm; p=p(k)} [\{a_{k,1}(D_z(\overbrace{\{B_{{}_1m}, z, \tilde{z}\}_{q({}_1m)}})_{l_1}^{k_1})...a_{k,p}(\overbrace{\{B_{{}_pm}, z, \tilde{z}\}_{q({}_pm)}})_{l_p}^{k_p} \}_{q(k)}$$

$$+ ... + \{a_{k,1}(\overbrace{\{B_{{}_1m}, z, \tilde{z}\}_{q({}_1m)}})_{l_1}^{k_1} ...a_{k,p}(D_z(\overbrace{\{B_{{}_pm}, z, \tilde{z}\}_{q({}_pm)}})_{l_p}^{k_p})\}_{q(k)}.$$

On the other hand, we have

$$D_z\{B_m, z, \tilde{z}\}_{q(m)} = \{b_{m,1}(D_z \overbrace{z_{l_1}^{m_1}})...b_{m,s} \overbrace{z_{l_s}^{k_s}} \}_{q(m)} + ... + \{b_{m,1} \overbrace{z_{l_1}^{m_1}} ...b_{m,s}(D_z \overbrace{z_{l_s}^{k_s}})\}_{q(k)}$$

due to the Leibniz rule, hence

$$D_z \overbrace{\psi^p} = \sum_{{}_1m,...,{}_pm} [(...(D_z \overbrace{\{B_{{}_1m}, z, \tilde{z}\}_{q({}_1m)}})...) \overbrace{\{B_{{}_pm}, z, \tilde{z}\}_{q({}_pm)}} \tag{6}$$

$$+ ... + (...(\overbrace{\{B_{{}_1m}, z, \tilde{z}\}_{q({}_1m)}} \overbrace{\{B_{{}_2m}, z, \tilde{z}\}_{q({}_2m)}})...)(D_z \overbrace{\{B_{{}_pm}, z, \tilde{z}\}_{q({}_pm)}})]$$

$$= (D_y \overbrace{y^p}).(D_z\psi) + (D_{\tilde{y}} \overbrace{y^p}).(D_z\tilde{\psi})$$

due to Formulas (3). Thus Formulas (5, 6) imply that

$$D_z(\eta \circ \psi(z, \tilde{z})) = [(D_y\eta(y, \tilde{y})).(D_z\psi(z, \tilde{z})) + (D_{\tilde{y}}\eta(y, \tilde{y})).(D_z\tilde{\psi}(z, \tilde{z}))]|_{y=\psi(z,\tilde{z})}. \tag{7}$$

Quite analogously or using the conjugation one deduces that

$$D_{\tilde{z}}(\eta \circ \psi(z, \tilde{z})) = [(D_{\tilde{y}}\eta(y, \tilde{y})).(D_{\tilde{z}}\psi(z, \tilde{z})) + (D_y\eta(y, \tilde{y})).(D_{\tilde{z}}\tilde{\psi}(z, \tilde{z}))]|_{y=\psi(z,\tilde{z})}. \tag{8}$$

Particularly, we get

$$D_z(\eta \circ \psi(z)) = (D_y\eta(y)).(D_z\psi(z))|_{y=\psi(z)}, \text{ when } D_{\tilde{z}}\eta(z,\tilde{z}) = 0 \text{ and } D_{\tilde{z}}\psi(z,\tilde{z}) = 0 \quad (9)$$

and

$$D_{\tilde{z}}(\eta \circ \psi(\tilde{z})) = (D_{\tilde{y}}\eta(\tilde{y})).(D_{\tilde{z}}\psi(\tilde{z}))|_{y=\psi(\tilde{z})}, \text{ when } D_z\eta(z,\tilde{z}) = 0 \quad (10)$$

and $D_z\psi(z,\tilde{z}) = 0$.

Combining Formulas $(7, 8)$ and taking into account the proof given above and applying Proposition 2.3 [25, 26] we infer the chain rule, when both η and ψ are either the z or $\tilde{z}$ or $(z,\tilde{z})$-differentiable, that is in all three considered cases.

6. Proposition. *Let suppositions of Proposition 5 be satisfied for two ($\mathcal{A}_r$ super-) z-differentiable functions f and g, where U and $W \subset \mathcal{A}_r$. Then*

$$(d^n f(g(z))/dz^n).(h_1,...,h_n) = f^{(n)}(g(z)).(g'(z).h_1,...,g'(z).h_n)+$$

$$\sum_{k,j=1,...,n;\ k\neq j;\ j_1<...<j_{n-2}\in\{1,...,n\}\setminus\{k,j\}} f^{(n-1)}(g(z)).(g^{(2)}.(h_k,h_j),g'(z).h_{j_1},...,g'(z).h_{j_{n-2}})+$$

$$\sum_{2\leq k\leq n-2;\ p_1+...+p_l=n,\ p_1\geq p_2\geq ...\geq p_l\geq 1,\ l=n-k}$$

$$f^{(n-k)}(g(z)).(g^{(p_1)}.(h_{j(1,p_1)},...,h_{j(p_1,p_1)}),...,g^{(p_l)}.(h_{j(1,p_l)},...,h_{j(p_l,p_l)}) \quad (1)$$

$$+f'(g(z)).(g^{(n)}.(h_1,...,h_n))$$

for all Cayley-Dickson numbers $h_1,...,h_n \in \mathcal{A}_r$, $2 \leq r$, $n \geq 2$, where $j(s,p) \in \{1,...,n\}$ are all pairwise distinct natural numbers for different (s,p), $j(s,p_k) < j(s+1,p_k)$ for all s and k, $j(1,k) < j(1,k+1)$ if $p_k = p_{k+1}$.

Proof. In accordance with §5 it is sufficient to prove Formula (1) for phrases $\eta(z)$ and $\psi(z)$, since each z-differentiable function is infinite z-differentiable (see §2 [25, 26]). To derive Formula (1) the mathematical induction will be used below. For $n = 1$ Formula (1) follows from that of Proposition 5. In accordance with the Leibnitz rule over the Cayley-Dickson algebra $\mathcal{A}_r$, $2 \leq r$, [25, 26] the Formula

$$(d\{\eta_1(z)...\eta_n(z)\}_q/dz).h = \sum_{j=1}^{n}\{\eta_1...[(d\eta_j(z)/dz).h]...\eta_n\}_q \quad (2)$$

is accomplished, where $\eta_1,...,\eta_n$ are phrases of z-differentiable functions on the domain W open in $\mathcal{A}_r$. Mention that the differential operator $\eta^{(k)}(z).(h_1,...,h_k)$ is as usually symmetric relative to all transpositions of $h_1,...,h_n$ (see also [25, 26]). Therefore, we deduce from the Leibnitz rule, that

$$(d^2\eta(\psi(z))/dz^2).(h_1,h_2) = d[\eta'(\psi(z)).[\psi'(z).h_1]/dz].h_2 =$$

$$\eta^{(2)}(\psi(z)).([\psi'(z).h_1],[\psi'(z).h_2]) + \eta'(\psi(z)).[\psi^{(2)}(z).(h_1,h_2)] \quad (3)$$

and

$$[d^3\eta(\psi(z))/dz^3].(h_1,h_2,h_3) = d\{[d^2\eta(\psi(z))/dz^2].(h_1,h_2)\}.h_3$$

$$= \eta^{(3)}(\psi(z)).([\psi'(z).h_1],[\psi'(z).h_2],[\psi'(z).h_3]) + \eta^{(2)}(\psi(z)).([\psi^{(2)}(z).(h_1,h_3)],[\psi'(z).h_2])+$$

$$\eta^{(2)}(\psi(z)).([\psi^{(2)}(z).(h_2,h_3)],[\psi'(z).h_1]) + \eta^{(2)}(\psi(z)).([\psi^{(2)}(z).(h_1,h_2)],[\psi'(z).h_3])$$

$+$

$$\eta'(\psi(z)).[\psi^{(3)}(z).(h_1,h_2,h_3)]. \quad (4)$$

Suppose now that Formula (1) for $d^k\eta(\psi(z))/dz^k$ is proved for each $k = 1,...,n$. It is sufficient according to the continuation of the induction to demonstrate it for $k = n+1$. From $(1,2)$ we infer the equality:

$$[d^{n+1}\eta(\psi(z))/dz^{n+1}].(h_1,...,h_{n+1}) = \{d[d^n\eta(\psi(z))/dz^n].(h_1,...,h_n)/dz\}.h_{n+1} =$$

$$\eta^{(n+1)}(\psi(z)).(\psi'(z).h_1,...,\psi'(z).h_{n+1})+$$

$$\sum_{j=1,...,n;\ 1\le j_1<...<j_{n-1}\le n,\ j_k\ne j\forall k} \eta^{(n)}(\psi(z)).([\psi^{(2)}(z).(h_j,h_{n+1})],[\psi'(z).h_{j_1}],...,\psi'(z).h_{j_{n-1}})+$$

$$\sum_{k,j=1,...,n;\ k\ne j;\ j_1<...<j_{n-2}\in\{1,...,n\}\setminus\{k,j\}}$$

$$\eta^{(n)}(\psi(z)).(\psi^{(2)}.(h_k,h_j),\psi'(z).h_{j_1},...,\psi'(z).h_{j_{n-2}},\psi'(z).h_{n+1})+$$

$$\sum_{2\le k\le n-1;\ p_1+...+p_l=n,\ p_1\ge p_2\ge...\ge p_l\ge 1,\ l=n-k,p_{l+1}=1} \eta^{(n+1-k)}$$

$$(\psi(z)).(\psi^{(p_1)}.(h_{j(1,p_1)},...,h_{j(p_1,p_1)}),...,\psi^{(p_l)}.(h_{j(1,p_l)},...,h_{j(p_l,p_l)}),\psi'(z).h_{n+1})$$

$$\sum_{2\le k\le n-1;\ p_1+...+p_l=n,\ p_1\ge p_2\ge...\ge p_l\ge 1,\ l=n-k,\ s=1,...,l}$$

$$\eta^{(n+1-k)}(\psi(z)).(\psi^{(p_1)}.(h_{j(1,p_1)},...,h_{j(p_1,p_1)}),...,\psi^{(p_s+1)}.(h_{j(1,p_s)},...,h_{j(p_s,p_s)},h_{n+1}),...,$$

$$\psi^{(p_l)}.(h_{j(1,p_l)},...,h_{j(p_l,p_l)}))$$

$$+\eta'(\psi(z)).(\psi^{(n+1)}.(h_1,...,h_{n+1})) \quad (5)$$

for all Cayley-Dickson numbers $h_1,...,h_{n+1}$. This is the same expression as in Formula (1) for $n+1$ instead of n after gathering together terms $\eta^{(j)}$ with the same j.

7. Note. In Propositions 5 and 6 the specification of phrases corresponding to functions in the respective representation is essential as well as for calculations of their derivative operators.

In this article some elementary facts about analytic functions z^n, $z^{1/n}$, $\exp(z)$ and $Ln(z)$ of the Cayley-Dickson variables are used. They were considered in details in previous works [26, 25, 40]. Recall that the exponential function is defined by the power series $\exp(z) := 1 + \sum_{n=1}^{\infty} z^n/n!$ converging on the entire Cayley-Dickson algebra $\mathcal{A}_r$. It has the periodicity property $\exp(M(\phi+2\pi k)) = \exp(M\phi)$ for each purely imaginary Cayley-Dickson number M of the unit norm $|M| = 1$ for any real number $\phi \in \mathbf{R}$ and each integer number $k \in \mathbf{Z}$. The restriction of such exponential function on each complex plane $\mathbf{C}_M := \mathbf{R} \oplus M\mathbf{R}$ coincides with the traditional complex exponential function. Since the inverse function $Ln(z)$ of $z = \exp(x)$ is defined on every complex plane $\mathbf{C}_M \setminus \{0\}$ with the pricked zero point, the logarithmic function $Ln(z)$ is defined on $\mathcal{A}_r \setminus \{0\}$. This logarithmic function is certainly multi-valued. Consider the bunch of complex planes $\mathbf{C}_M$ intersecting by the real line $\mathbf{R}i_0$ as the geometric realization of $\mathcal{A}_r$. Certainly $\mathbf{C}_M = \mathbf{C}_{-M}$, so we take the set

$$\mathbf{S}_r^+ := \{M \in \mathcal{A}_r : |M| = 1, M = M_1 i_1 + ... + M_{2^r-1} i_{2^r-1}, \text{ either } M_1 > 0, \text{ or } M_1 = 0$$

and

$$M_2 > 0, \text{ or..., or } M_1 = ... = M_{2^r-2} = 0$$

and

$$M_{2^r-1} > 0\},$$

where $M_1, ..., M_{2^r-1} \in \mathbf{R}$, $2 \le r$. Then $\mathcal{A}_r = \bigcup_{M \in \mathbf{S}_r^+} \mathbf{C}_M$.

If A and B are two subsets in a complete uniform space X and $\theta : A \to B$ is a continuous bijective mapping, the equivalence relation $a \Upsilon b$ by definition means $b = \theta(a)$ for $a \in A$ and $b \in B$; or $b \Upsilon a$ means $a = \theta^{-1}(b)$. When θ is uniformly continuous, θ has a uniformly continuous extension $\theta : cl(A) \to cl(B)$, where $cl(A)$ denotes the closure of the set A in X (see Theorem 8.3.10 in [6]). Certainly, the mapping θ can be specified by its graph $\{(x_1, x_2) : x_2 = \theta(x_1), x_1 \in A\}$. We say, that A and B are glued (by θ), if $B = \theta(A)$ and the natural quotient mapping $\pi : X \to X/\Upsilon$ is given, where X/Υ denotes the quotient space (see §2.4 [6]). For the uniformly continuous θ this means that the gluing is extended from A onto $cl(A)$.

In the complex case to construct the Riemann surface of the logarithmic function one takes traditionally the complex plane $\mathbf{C}$ cut by the set $Q_1 := \{z = x + iy \in \mathbf{C} : x < 0\}$ and marking two respective points x_1 and x_2 of two edges $Q_{1,1}$ and $Q_{1,2}$ of the cut Q_1 arising from each given point $x < 0$, where $i = i_1$. Then one embeds $\mathbf{C}$ into either $\mathbf{C} \times \mathbf{R}$ or $\mathbf{C} \times \mathbf{R}i$ and bents the obtained surface slightly along the perpendicular axis e_3 to $\mathbf{C}$ by neighborhoods of two edges of the cut Q_1 and gets the new surface $\mathcal{C}$. Taking the countable infinite family of such surfaces $\mathcal{C}^j$ with the edges of cuts $Q_{1,1}^j$ and $Q_{1,2}^j$ and gluing by respective points of the cuts $Q_{1,2}^j$ with $Q_{1,1}^{j+1}$ for each j one gets the Riemann surface of the logarithmic function, where $j \in \mathbf{Z}$ (see, for example, [22]).

Analogous procedure to construct the Riemann surface is in the cases $r \ge 2$: one cuts $\mathcal{A}_r$ by Q_r and gets two edges $Q_{r,1}$ and $Q_{r,2}$ of the cut. This is described below.

If K and $M = KL$ are two purely imaginary Cayley-Dickson numbers with $|K| = |M| = |L| = 1$ so that they are orthogonal $K \perp M$, that is $Re(KM) = 0$, then $(K\mathbf{R}) \oplus (M\mathbf{R}) = K(\mathbf{R} \oplus L\mathbf{R})$ and L is also purely imaginary. Consider any path $\gamma : [0,1] \to K\mathbf{R} \oplus M\mathbf{R}$ winding one time around zero such that $\gamma(t) \ne 0$ for each t. Each $z \in K\mathbf{R} \oplus M\mathbf{R}$ can be written in the polar form $z = |z|Ke^{L\phi} = |z|\exp(\pi K e^{L\phi}/2)$, where $\phi = \phi(z) \in \mathbf{R}$, since $Ke^{L\phi} = K\cos(\phi) + (KL)\sin(\phi)$ due to Euler's formula, hence $|Ke^{L\phi(\gamma(t))}| = 1$ for each t (see Section 3 in [26, 25]). In particular, $\gamma(t) = |\gamma(t)|\exp(\pi K e^{L\phi(t)}/2)$. But $Ke^{L\phi(\gamma(0))} = Ke^{L\phi(\gamma(1))}$ such that the logarithm $Ln\,\gamma(t) = Ln\,|\gamma(t)| + \pi K e^{L\phi(\gamma(t))}/2$ does not change its branch, when the path $\gamma(t) \in (K\mathbf{R}) \oplus (M\mathbf{R}) \setminus \{0\}$ winds around zero. Due to the homotopy theorem (see [26, 25, 38]) this means that the logarithm $Ln\,\gamma(t)$ does not change its branch, when

$Re(\gamma(t)) = 0$ for each t for the path γ winding around zero with $|\gamma(t)| > 0$ for each t.

The first simple construction for $r \geq 2$ is the following. Take the set $Q_r := (-\infty, 0)\mathbf{S}_r^+ := \{z = tx : t \in (-\infty, 0), x \in \mathbf{S}_r^+\}$ and cut $\mathcal{A}_r$ by Q_r. The set Q_r is the union $Q_r = \bigcup_{j=1}^{2^r-1} \Omega_{j,r}$ of subsets $\Omega_{j,r} := \{z \in Q_r : z_0 = 0, ..., z_j = 0\}$ so that $\Omega_{j,r}$ is contained in the boundary of the preceding set $\Omega_{j,r} \subset \partial\Omega_{j-1,r}$ for each $j = 1, ..., 2^r - 1$, $dim\, \Omega_{j,r} = dim\, \Omega_{j-1,r} - 1$, moreover, $\mathbf{R}\mathbf{S}_r^+ = \mathcal{I}_r := \{z \in \mathcal{A}_r : Re(z) = 0\}$. Therefore, from each point $z \in Q_r$ two and only two different points z_1 and z_2 arise.

It is useful to embed $\mathcal{A}_r$ either into $\mathcal{A}_r \times \mathbf{R}^{2^r-1}$ or into $\mathcal{A}_r \times \mathcal{I}_r$, where $\mathcal{I}_r := \{z \in \mathcal{A}_r : Re(z) = 0\}$. Then one marks all pairs of respective points z_1 and z_2 arising from $z \in Q_r$ after cutting, slightly bents the cut copy of $\mathcal{A}_r \setminus \{0\}$ by $(2^r - 1)$ axes perpendicular to $\mathcal{A}_r$ by two neighborhoods of two edges of the cut Q_r. Thus one gets the 2^r dimensional surface $\mathcal{C}_r$ with two edges $Q_{r,1}$ and $Q_{r,2}$ of the cut. Taking the countable infinite family of such surfaces $\mathcal{C}_r^j$ with edges of the cuts $Q_{r,1}^j$ and $Q_{r,2}^j$, $j \in \mathbf{Z}$, and gluing respective points of edges $Q_{r,2}^j$ with $Q_{r,1}^{j+1}$ for each j one gets the Riemann surface $\mathcal{R}_r = \mathcal{R}_{r,Ln}$ of the logarithmic function $Ln(z) : \mathcal{A}_r \setminus \{0\} \to \mathcal{R}_r$. Thus the latter mapping is already univalent with the image in the Riemann surface (see in details [26, 25, 40]).

For convenience we attach numbers 1 and 2 to faces in such manner that the winding around zero in the complex plane $\mathbf{C}_M$ embedded into $\mathcal{R}_r$ counterclockwise means the transition through the cut from $Q_{r,2}^j$ to $Q_{r,1}^j$ for each M in the connected set $\mathbf{S}_r^+$.

For the function $z^{1/n}$ with $n \in \mathbf{N}$ its Riemann surface $\mathcal{R}_{r,z^{1/n}}$ is obtained from n copies of surfaces $\mathcal{C}_r^j$, $j = 1, ..., n$, by gluing the corresponding points of edges $Q_{r,2}^j$ with $Q_{r,1}^{j+1}$ for $j = 1, ..., n-1$ and of $Q_{r,2}^n$ with $Q_{r,1}^1$ (see also [22, 26, 25, 40]).

Another more complicated construction is described below. Now we take the set

$$Q_r := \bigcup_{j=1}^{2^r-1} P_j,$$

where $P_j := \{z \in \mathcal{A}_r : z = z_0 i_0 + ... + z_{2^r-1} i_{2^r-1};\ z_0 < 0 \text{ and } z_j = 0\}$, where $z_0, ..., z_{2^r-1} \in \mathbf{R}$, $2 \leq r$.

Let $z = z_0 + z'$ be the Cayley-Dickson number with the negative real part $z_0 < 0$ and the imaginary part $Im(z) = z'$, which can be written in the form $z' = |z'| \exp(\pi K e^{L\phi(z')}/2)$ and $z = |z| e^{P\psi}$, where K, L and P are purely imaginary of the unit norm, ϕ and $\psi \in \mathbf{R}$ are reals. This gives the relations $\cos(\psi) = z_0/|z|$ so that the parameter ψ is in the interval $\pi/2 + 2\pi k < \psi < 3\pi/2 + 2\pi k$ for some integer number k. This means that for a continuous path γ contained in the set Q_r the parameter $\psi(\gamma(t))$ is the continuous function of the real variable $t \in \mathbf{R}$ and remains in the same interval $(\pi/2 + 2\pi k, 3\pi/2 + 2\pi k)$, consequently, $Ln\gamma(t)$ preserves it branch along such path $\gamma(t)$. This shows that after the first cut along Q_r the obtained sets $Q_{r,1}$ and $Q_{r,2}$ need not be further cut. Thus the described reason simplifies the construction of the Riemann surface.

Each continuous path $\gamma : [0, 1] \to \mathcal{A}_r$ can be decomposed as the sum as well as the composition up to the homotopy satisfying the conditions of the homotopy theorem [26, 25, 38] of paths $\gamma_{k,l}$ in complex planes $(\mathbf{R}i_k) \oplus (\mathbf{R}i_l)$ for each $k < l \in \Lambda$ for the corresponding subset $\Lambda \subset \{0, 1, ..., 2^r - 1\}$. If $|\gamma(t)| > 0$ for each t we take $\gamma_{k,l}$ with $|\gamma_{k,l}(t)| > 0$ on

$[0,1]$ for all $k < l \in \Lambda$. When $\gamma[0,1]$ does not intersect Q_r one can choose $\gamma_{k,l}$ with images $\gamma_{k,l}[0,1]$ also non-intersecting with Q_r for all $k < l \in \Lambda$. Due to the homotopy theorem the logarithm $Ln\,\gamma(t)$ does not change its branch along such continuous path $\gamma(t)$, since this is the case for $Ln\,\gamma_{k,l}(t)$ for all $k < l \in \Lambda$.

For each $z \in P_j \setminus \bigcup_{m,m\neq j} P_m$ cutting by Q_r gives two points. If $z \in (P_k \cap P_j) \setminus \bigcup_{m,m\neq k,m\neq j} P_m$ with $k < j$ cutting gives four points which can be organized into respective pairs after cutting of P_k and then of P_j. This procedure gives pairs $(z_{1,1}; z_{2,2})$ and $(z_{1,2}; z_{2,1})$. For $z \in (P_{k_1} \cap ... \cap P_{k_l}) \setminus \bigcup_{m,m\neq k_1,...,m\neq k_l} P_m$ consider pairs of points appearing from the preceding point after cutting of P_{k_j}, $j = 1,...,l$ by induction, where $k_1 < ... < k_l$. One can do it by induction by all $l = 1,...,2^r - 1$ and all possible subsets $1 \le k_1 < k_2 < ... < k_l \le 2^r - 1$. It produces points denoted by $z_{a_1,...,a_l}$. Here $a_j = 1$ corresponds to the face of the cut with $z_j \le 0$, while $a_j = 2$ corresponds to the face of the cut with $z_j \ge 0$. Two points $z_{a_1,...,a_l}$ and $z_{b_1,...,b_l}$ form the pair of respective points, when $a_j + b_j = 3$ for each j, where $l \in \{1,...,2^r - 1\}$ is the index of such points. To each point $z \in (P_{k_1} \cap ... \cap P_{k_l}) \setminus \bigcup_{m,m\neq k_1,...,m\neq k_l} P_m$ of index $l \ge 1$ the point $M \in \mathbf{S}_r^+$ corresponds such that $M = M_1 i_1 + ... + M_{2^r-1} i_{2^r-1}$ and $z \in \mathbf{C}_M$, where $M_{k_1} = 0$,...,$M_{k_l} = 0$.

Then as above after cutting of $\mathcal{A}_r$ by Q_r one gets the 2^r dimensional surface $\mathcal{C}_r$ with two edges $Q_{r,1}$ and $Q_{r,2}$ of the cut. For the countable infinite family of the surfaces $\mathcal{C}_r^j$ with edges of the cuts $Q_{r,1}^j$ and $Q_{r,2}^j$, $j \in \mathbf{Z}$, one glues respective points of edges $Q_{r,2}^j$ with $Q_{r,1}^{j+1}$ for each j. Thus one gets the Riemann surface $\mathcal{R}_r = \mathcal{R}_{r,Ln}$ of the logarithmic function $Ln(z) : \mathcal{A}_r \setminus \{0\} \to \mathcal{R}_r$. Therefore, the latter mapping is univalent with the image in the Riemann surface (see in details [26, 25, 40]). For simplicity one can glue at first pairs of respective points of index $l = 1$ and extend gluing by continuity on points of index $l > 1$ (see above). The first construction for $r \ge 2$ operates with points of index one only. That is why it is simpler than the second procedure.

The reader can lightly see that geometrically $\mathcal{R}_{r,f}$ (for both types described above) is the bunch $\bigcup\{\mathcal{R}_{1,f,M} : M \in \mathbf{S}_r^+\}$ of complex Riemann surfaces $\mathcal{R}_{1,f,M}$ for the restrictions $f|_{\mathbf{C}_M}$ on the complex planes $\mathbf{C}_M$ of the function f, where either $f = Ln$ or $f = z^{1/n}$ among those considered here.

The main problem of the multi-dimensional geometry is in depicting its objects on the two-dimensional sheet of paper so one uses either projections or sections of the multi-dimensional surfaces. The bunch interpretation $\bigcup_{M\in\mathbf{S}_r^+} \mathcal{R}_{1,f,M}$ helps to imagine how the 2^r-dimensional Riemann surface $\mathcal{R}_{r,f}$ is organized for $f = Ln$ and $f = z^{1/n}$ with $n \in \mathbf{N}$.

2.2. PDE of Diverse Types over the Octonion Algebra

1. Remarks. For a subset U in either the quaternion skew field $\mathbf{H} = \mathcal{A}_2$ or in the octonion algebra $\mathbf{O} = \mathcal{A}_3$ we put $\pi_{\mathsf{s},\mathsf{p},\mathsf{t}}(U) := \{\mathsf{u} : z \in U, z = \sum_{\mathsf{v}\in\mathbf{b}} w_\mathsf{v}\mathsf{v}, \mathsf{u} = w_\mathsf{s}\mathsf{s} + w_\mathsf{p}\mathsf{p}\}$ for each $\mathsf{s} \neq \mathsf{p} \in \mathbf{b}$, where $\mathsf{t} := \sum_{\mathsf{v}\in\mathbf{b}\setminus\{\mathsf{s},\mathsf{p}\}} w_\mathsf{v}\mathsf{v} \in \mathcal{A}_{r,\mathsf{s},\mathsf{p}} := \{z \in \mathcal{A}_r : z = \sum_{\mathsf{v}\in\mathbf{b}} w_\mathsf{v}\mathsf{v}, w_\mathsf{s} = w_\mathsf{p} = 0, w_\mathsf{v} \in \mathbf{R}\ \forall \mathsf{v} \in \mathbf{b}\}$, where $\mathbf{b} := \{i_0, i_1, ..., i_{2^r-1}\}$ is the family of standard generators of the algebra $\mathcal{A}_r$. Geometrically the domain $\pi_{\mathsf{s},\mathsf{p},\mathsf{t}}(U)$ means the projection on the complex plane $\mathbf{C}_{\mathsf{s},\mathsf{p}}$ of the intersection U with the plane $\tilde{\pi}_{\mathsf{s},\mathsf{p},\mathsf{t}} \ni \mathsf{t}$, $\mathbf{C}_{\mathsf{s},\mathsf{p}} := \{a\mathsf{s} + b\mathsf{p} : a, b \in \mathbf{R}\}$, since $\mathsf{s}\mathsf{p}^* \in \hat{b} := \mathbf{b}\setminus\{1\}$. Recall that in §§2.5-7 [25] for each continuous function $f : U \to \mathcal{A}_r$ it was defined the operator $\hat{f}$ by each variable $z \in \mathcal{A}_r$. If a function f is z-differentiable by the Cayley-Dickson variable $z \in U \subset \mathcal{A}_r$, $2 \le r$, then $\hat{f}(z) = dg(z)/dz$, where $(dg(z)/dz).1 = f(z)$.

A Hausdorff topological space X is said to be n-connected for $n \geq 0$ if each continuous map $f : S^k \to X$ from the k-dimensional real unit sphere into X has a continuous extension over $\mathbf{R^{k+1}}$ for each $k \leq n$ (see also [55]). A 1-connected space is also said to be simply connected.

It is supposed further, that a domain U in $\mathcal{A}_r^m$ has the property that

$(D1)$ each projection $\mathbf{p}_j(U) =: U_j$ is $(2^r - 1)$-connected;

$(D2)$ $\pi_{\mathsf{s,p,t}}(U_j)$ is simply connected in $\mathbf{C}$ for each $k = 0, 1, ..., 2^{r-1}$, $\mathsf{s} = i_{2k}$, $\mathsf{p} = i_{2k+1}$, $\mathsf{t} \in \mathcal{A}_{r,\mathsf{s,p}}$ and $u \in \mathbf{C}_{\mathsf{s,p}}$, for which there exists $z = \mathsf{u} + \mathsf{t} \in U_j$, where $e_j = (0, ..., 0, 1, 0, ..., 0) \in \mathcal{A}_r^m$ is the vector with 1 on the j-th place, $\mathbf{p}_j(z) = {}^jz$ for each $z \in \mathcal{A}_r^m$, $z = \sum_{j=1}^m {}^jze_j$, ${}^jz \in \mathcal{A}_r$ for each $j = 1, ..., m$, $m \in \mathbf{N} := \{1, 2, 3, ...\}$. Frequently we take $m = 1$. Hence forward we consider a domain U satisfying Conditions $(D1, D2)$ if any other is not outlined.

The family of all $\mathcal{A}_r$ locally analytic functions $f(x)$ on U with values in $\mathcal{A}_r$ is denoted by $\mathcal{H}(U, \mathcal{A}_r)$. It is supposed that a locally analytic function $f(x)$ is written in the x-representation $\nu(x)$, also denoted by $\nu = \nu^f$. The latter is equivalent to the super-differentiability of f (see [26, 25, 36]). Each such f is supposed to be specified by its phrase ν.

For each super-differentiable function $f(x)$ its non-commutative line integral $\int_\gamma f(x)dx$ in U is defined along a rectifiable path γ in U. It is the integral of a differential form $\hat{f}(x).dx$, where

$(I1)$ $\hat{f}(x) = dg(x)/dx$,

$(I2)$ $[dg(x)/dx)].1 = f(x)$ for each $x \in U$.

A branch of the non-commutative line integral can be specified with the help of either the left or right algorithm (see [26, 25, 36]). We take further for definiteness the left algorithm if something another will not described. For $f \in \mathcal{H}(U, \mathcal{A}_r)$ and a rectifiable path $\gamma : [a, b] \to \mathcal{A}_r$ the integral $\int_\gamma f(x)dx$ depends only on an initial $\alpha = \gamma(a)$ and final $\beta = \gamma(b)$ points due to the non-commutative analog of the homotopy theorem in U, where $a < b \in \mathbf{R}$. When initial and final points or a path are not marked we denote the operation of the non-commutative line integration in the domain U simply by $\int f(x)dx$ analogously to the indefinite integral.

To rewrite a function from real variables z_j in the z-representation the following identities are used:

(1) $z_j = (-zi_j + i_j(2^r - 2)^{-1}\{-z + \sum_{k=1}^{2^r-1} i_k(zi_k^*)\})/2$ for each $j = 1, 2, ..., 2^r - 1$,

(2) $z_0 = (z + (2^r - 2)^{-1}\{-z + \sum_{k=1}^{2^r-1} i_k(zi_k^*)\})/2$, where $2 \leq r \in \mathbf{N}$, z is a Cayley-Dickson number decomposed as

(3) $z = z_0i_0 + ... + z_{2^r-1}i_{2^r-1} \in \mathcal{A}_r$, $z_j \in \mathbf{R}$ for each j, $i_k^* = \tilde{i}_k = -i_k$ for each $k > 0$, $i_0 = 1$.

2. The Simplest PDE of the First Order has the form:

$$[dy(x)/dx].h = f(x), \tag{1}$$

where $h = h(x) \in \mathcal{A}_r$, $r \geq 2$, $f, h \in \mathcal{H}(U, \mathcal{A}_r)$, $|h(x)| > 0$ for λ-almost all $x \in U$, an open domain U satisfies Conditions $1(D1, D2)$. Here λ denotes the Lebesgue measure on $\mathcal{A}_r$ induced from that of on the real shadow $\mathbf{R}^{2^r}$. Using a non-commutative line integral in U we get:

$$\int_\gamma ([dy(x)/dx].h)dx = \int_\gamma f(x)dx. \tag{2}$$

As usually

$$Re(\beta) = (\beta + \beta^*)/2 \tag{3}$$

denotes the real part of $\beta \in \mathcal{A}_r$, $\beta^* = \tilde{\beta}$ denotes the conjugated number β, so that its square norm is given by the equation

$$|\beta|^2 = \beta\beta^*. \tag{4}$$

Suppose that $h = h(t)$ is a (super-)differentiable function on a domain U, $h : U \to \mathcal{A}_r$. Take a marked point α in U.

Let us consider the differential equation

$$(d\omega(x)/dx).1 = h(\omega), \omega = \omega_h. \tag{5}$$

This equation is equivalent to

$$(1/h(\omega))(d\omega(x)/dx).1 = 1. \tag{5.1}$$

Integrating both sides of (5.1) by x we deduce that

$$p(\omega) := \int_\gamma [(1/h(\omega))(d\omega(t)/dt).1]dt = \int (1/h(\omega))d\omega = \int_\alpha^x dt = x - \alpha, \tag{5.2}$$

since $(d[\int (d\omega/dx).1dx]/dx).1 = (d\omega/dx).1$, where $\gamma(a) = \alpha$ and $\gamma(b) = x$, $\gamma : [a, b] \to U$ is a rectifiable path. Therefore,

$$\omega_h(x) = \omega(x) = p^{-1}(x - \alpha) + \phi_1(x^\diamond, \alpha^\diamond) \tag{6}$$

satisfies Equation (5), where p^{-1} denotes the inverse function. This function $\omega(x)$ is defined up to a summand $\phi_1(x^\diamond, \alpha^\diamond)$, where $\phi_1(x^\diamond, \alpha^\diamond)$ is any (super-)differentiable function of the arguments

$$x^\diamond := x^\diamond{}_1 := x - Re(x) =: Im(x) \tag{7}$$

and $\alpha^\diamond := \alpha^\diamond{}_1 := Im(\alpha)$ which are to be rewritten with the help of Formulas $1(1-3)$ above. To avoid any confusion of the notation derivatives are denoted by $y' = dy/dx$ or d^jy/dx^j or $y^{(j)}$, $j = 1, 2, ...$, when it is necessary. Take $\omega(x)$ satisfying the initial condition:

$$\omega(\alpha) = \alpha, \tag{8}$$

consequently,

$$p^{-1}(0) + \phi_1(\alpha^\diamond, \alpha^\diamond) = \alpha. \tag{9}$$

When α is fixed, the function ϕ_1 will be written as $\phi_1(x^\diamond)$. Now we can use the change of variables Theorem 3.11 [38] for non-commutative line integrals over Cayley-Dickson algebras. Therefore,

$$\int_\gamma ([dy(x)/dx].h)dx = \int_\psi [dy(\omega)/d\omega].d\omega = y(\psi(b)) - y(\psi(a)), \tag{10}$$

where ψ is a rectifiable path so that

$$\psi(t) = \omega(\gamma(t)) \tag{11}$$

for all $t \in [a, b]$, since by the chain rule

$$[dy(\omega(x))/dx].dx = [dy(\omega)/d\omega].[(d\omega(x)/dx).dx]|_{\omega=\omega(x)} = dy(\omega(x)) \tag{12}$$

(see Proposition 5 in §2). On the other hand, in view of Conditions $1(I1, I2)$ each non-commutative integration $\int_\gamma f(x)dx$ specifies a branch up to a function $\phi_f = \phi_f({\beta^\diamond}_h, {\alpha^\diamond}_h)$ or $\phi_f({\beta^\diamond}_h)$, when $\alpha = \gamma(a)$ is marked (fixed) and $\beta = \gamma(b)$ can vary, where

$${\beta^\diamond}_h = \beta - pr_{h(\beta)}\beta, \tag{13}$$

$pr_{h(\beta)}\beta$ denotes the projection of β on $h(\beta)$.

We have $\psi(b) = \omega(\beta)$ and $\psi(a) = \omega(\alpha)$. Suppose that $y_1(\omega(x))$ and $y_2(\omega(x)+\phi_1(x^\diamond))$ are two solutions of the differential equation (1), then their difference $y_1(\omega(\zeta))-y_2(\omega(x)+\phi_1(x^\diamond))$ is the function of $x^\diamond$ for $Re(x) = Re(\zeta) = \alpha_0$ with $\omega(\zeta) = \omega(x) + \phi_1(x^\diamond)$, since $[d\phi_1(x^\diamond)/dx].1 = 0$ and $[d\{y_1(\omega(\zeta)) - y_2(\omega(x) + \phi_1(x^\diamond))\}/dx].1 =$
$[dy_1(\xi)/d\xi].([d\omega(x)/dx].1)|_{\xi=\omega(\zeta)} - [dy_2(\xi)/d\xi].([d(\omega(x)+\phi_1(x^\diamond))/dx].1)|_{\xi=\omega(x)+\phi_1(x^\diamond)}$
$= (1-1)f(\omega(\zeta)) = 0$, since the x-differentiability of the function $\phi_1(x^\diamond, \alpha^\diamond)$ is provided by Formulas $1(1-3)$. Thus the formula

$$y(\omega(x)) = \int_\alpha^x f(\omega(x))dx + \phi_f(x^\diamond) \tag{14}$$

gives the general solution of the (super)-differential equation (1), since the domain U satisfies Conditions $1(D1, D2)$, while $\phi_f(x^\diamond) \in \mathcal{H}(U, \mathcal{A}_r)$ due to the x-representation $1(1-3)$, where $\omega = \omega_h$. The initial condition now is $\phi_f({\alpha^\diamond}_h, {\alpha^\diamond}_h) = f(\alpha)$. Its partial solution is defined, when the initial condition

$$y(x)|_{x_0=\alpha_0} = \eta(x^\diamond) \tag{15}$$

is given on the hyperplane

$$H_{\alpha_0} := \{x \in \mathcal{A}_r : x_0 = \alpha_0\}, \tag{16}$$

when $h(x)$ is not parallel to H_{α_0} at each point in H_{α_0}, where

$$x = x_0 i_0 + x_1 i_1 + ... + x_{2^r-1} i_{2^r-1} \tag{17}$$

is the decomposition of the Cayley-Dickson number $x \in \mathcal{A}_r$, $x_0, ..., x_{2^r-1} \in \mathbf{R}$. Here $\eta(x^\diamond)$ is a function on ∂U having an $\mathcal{H}(W, \mathcal{A}_r)$ extension, where W is an open neighborhood of ∂U.

This means that the Cauchy problem can be formulated as given by Equations $(1, 15)$ on the half-space

$$P_{\alpha_0} := \{x \in \mathcal{A}_r : \ Re(x) \geq \alpha_0\}. \tag{18}$$

Certainly, this implies $\omega(\alpha) = \alpha$ for each $\alpha \in H_{\alpha_0}$. That is $\phi_f(x^\diamond{}_h) = \eta(x^\diamond) - \int_\alpha^x f(\tau)d\tau$ for each $x \in H_{\alpha_0}$.

Therefore, the non-commutative or non-associative analog of the Cauchy problem is given by Equations $(1, 15)$. In the previous papers [40, 40] pseudo-conformal mappings of the quaternion and octonion variables were investigated. Using the non-commutative or non-associative analog of the Riemann mapping theorem it is possible to formulate the non-commutative or non-associative analog of the Cauchy problem on more general hypersurfaces ∂V so that $\partial V = \nu(H_{\alpha_0})$, where ν is a (super-)differentiable (can be pseudo-conformal for $2 \leq r \leq 3$) diffeomorphism from P_{α_0} onto V, V is the canonical closed subset in $\mathcal{A}_r$, $V = cl(Int(V))$. As usually $cl(A)$ denotes the closure of a set A in $\mathcal{A}_r$, while $Int(A)$ denotes the interior of A in $\mathcal{A}_r$ (see [40, 40]). For such $U = V$ the initial condition is

$$y(x)|_{\partial U} = \eta(x^\diamond)|_{\partial U}, \tag{19}$$

when $h(x)$ is not tangent to ∂U at each point in ∂U, where $\eta(x^\diamond)$ is a function on ∂U having an $\mathcal{H}(W, \mathcal{A}_r)$ extension, where W is an open neighborhood of ∂U, $x^\diamond$ denotes an arbitrary point in ∂U. This initial condition is natural, since $y(x)$ is the function of 2^r real variables $x_0, ..., x_{2^r-1}$ or one Cayley-Dickson variable x, where $r \geq 2$.

3. An R-linear PDE of the First Order.

Consider a PDE, which is **R**-linear and $\mathcal{A}_r$-additive relative to an unknown function f of the first order:

$$(dy(x)/dx).h + b(x)y(x) = Q(x) \tag{1}$$

on a domain U satisfying Conditions $1(D1, D2)$, where a function $b(x)$ is real-valued and a mapping $Q(x)$ is octonion valued, $x \in U$, $r = 2$ or $r = 3$, $h = h(x) \in \mathcal{A}_r$. Suppose that $b(x)$, $Q(x)$ and $h(x)$ are locally analytic functions by x written in the x-representation. Then

$$y^1(\omega(x)) = \exp\{-\int_\alpha^x b(\omega(t))dt\}\phi_1(x^\diamond) \tag{2}$$

is any solution of the homogeneous equation, where $\omega = \omega_h(x)$ (see also Formulas $2(5 - 7)$). Here $\phi_1(x^\diamond)$ is an octonion-valued function of $x^\diamond$ which can be specified by an initial condition $2(15)$ or more generally $2(19)$. When $\phi_1(x^\diamond)$ is arbitrary Formula (2) gives the general solution of the homogeneous differential equation. The second summand

$$y^2(\omega(x)) := \exp\{-\int b(\omega(x))dx\}[\int \exp\{\int b(\omega(x))dx\}Q(x)dx]$$

is the special solution of the inhomogeneous differential equation. Therefore

$$y(\omega(x)) = y^1(\omega(x)) + \exp\{-\int b(\omega(x))dx\}\int \exp\{\int b(\omega(x))dx\}Q(x)dx. \tag{3}$$

3.1. Example. The PDE

$$[(dy(x)/dx).h]\xi(x) + y(x) = g(x) \qquad (1)$$

over octonions in the domain $Re(x) \geq \alpha_0$, where $\alpha_0 > 0$, $\xi(x)$ is any analytic real-valued function written in the x-representation due to Formulas $1(1-3)$, has the general solution:

$$y(\omega_v(x)) = y^1(\omega_v(x)) + \exp(-(\omega_v(x) - \alpha))[\int_\alpha^x \exp(\omega_v(\tau) - \alpha)g(\tau)d\tau], \qquad (2)$$

where $y^1(\omega_v(x)) = [\exp(-(\omega_v(x) - \alpha))]\phi(x^\diamond)$, while $v(x) := h(x)\xi(x)$.

4. A PDE with Separated Variables can be written in the form:

$$f(y)[dy(x)/dx].h(x) + s(x) = 0, \qquad (1)$$

where $s, h \in \mathcal{H}(U, \mathcal{A}_r)$, $f \in \mathcal{H}(W, \mathcal{A}_v)$, $2 \leq r \leq v$, $h(x)$ is a non-zero function, where U and W are domains satisfying Conditions $2.2.1(D1, D2)$. We seek a solution y so that $y(U) \subset W$ for suitable U and W. When a mapping f and its phrase ν^f are known, then a function $g(y) = \int f(y)dy$ and its phrase ν^g can be found easily with the help of the left algorithm. If an analytic function f is with real expansion coefficients, this integration reduces to the ordinary integration. Any other branch of the integral $\int f(y)dy$ may differ from that of given by the left algorithm on a term like $\phi_f(y - Re(y)) \in \mathcal{H}$. This means that y is a zero of the corresponding equation

$$g(y) = b(x), \qquad (2)$$

where $b(x) = -\phi_s(x^\diamond) - \int_\alpha^x s(t)dt$. Thus

$$y = g^{-1}(b(x)), \qquad (3)$$

but generally the inverse function g^{-1} need not be univalent [40] analogously to the complex case. The function $\phi_s(x^\diamond)$ may be given with the help of the boundary condition $2(19)$.

5. Example of non-Linear PDE. Consider the PDE

$$f(y)[dy(x)/dx].h(x) + y^m s(x) = 0, \qquad (1)$$

where $m \in \mathbf{R}$, $s, h \in \mathcal{H}(U, \mathcal{A}_r)$, $f \in \mathcal{H}(W, \mathcal{A}_r)$, $h(x)$ is real-valued, $h(x) \in \mathbf{R} \setminus \{0\}$, as in §4, $2 \leq r \leq v \leq 3$. Due to the alternativity of the octonion algebra the equivalent equation is:

$$(y^{-m} f(y))[dy(x)/dx].h(x) + s(x) = 0, \qquad (2)$$

where as usually $[dy(x)/dx].h$ denotes the (super) differential of y along a vector (Cayley-Dickson number) h. The power function y^m can be written in the form $y^m = \exp(m\ Ln(y))$. Since the exponential function is analytic on $\mathcal{A}_r$ and the logarithmic function is locally analytic on $\mathcal{A}_r \setminus \{0\}$. Equation (2) is of the same type as 4(1). Thus results of §4 permit to integrate non-linear octonion (super)-differential equations. The evaluation of several useful integrals over the Cayley-Dickson algebras was given in

[36, 38].

6. Bernoulli's Type PDE. Let the differential equation

$$[dy(x)/dx].h(x) + y^k p(x) = y^m s(x) \tag{1}$$

be given over octonions, where $h(x)$ is as in §2, $k, m \in \mathbf{R}$, $m \neq 0$, $m \neq k$. Suppose that a function $p(x)$ is real-valued, $h(x)$, $p(x)$ and $s(x)$ are x-differentiable on a domain U satisfying Conditions 2.2.1$(D1, D2)$. We change variables as $v^l = y$ for the parameter $l \in \mathbf{R} \setminus \{0\}$. When $v(x)$ and $(dv(x)/dx).h(x)$ commute for each $x \in P_\alpha$, this gives the following:

$$lv^{l-1}\{[dv(x)/dx].h(x)\} + v^{lk}p(x) = v^{lm}s(x). \tag{2}$$

The alternativity of the octonion algebra leads to:

$$lv^{l-lm-1}[dv(x)/dx].h(x) + v^{lk-lm}p(x) = s(x). \tag{3}$$

When $l = 1/(k-m)$ it reduces to

$$\{1/(k-m)\}v^{-1+(1-m)/(k-m)}[dv(x)/dx].h(x) + vp(x) = s(x). \tag{4}$$

Therefore, for $k = 1$ Equation (1) is the octonion analog of the Bernoulli's equation and due to §3 we infer:

$$v(\omega(x)) = v^1(\omega(x)) +$$

$$[\int_\alpha^x (1-m)s(\tau)\exp\{\int_\alpha^\tau (1-m)p(\xi)d\xi\}d\tau]\exp\{(m-1)\int_\alpha^x p(t)dt\}, \tag{5}$$

where $\omega(x)$ is given by Equations $2(5.2, 6)$,

$$v^1(\omega(x)) = \exp\{\int_\alpha^x (1-m)p(t)dt\}\phi_1(x^\diamond).$$

Thus $y = v^{1/l}$ is the solution of Cauchy's problem $6(1), 2(15)$.

7. PDE of the n-th Order by dy/dx. Let

$$\sum_{k=0}^{n}\{A_k, ([dy(x)/dx].h)\} = 0 \tag{1}$$

be the differential equation of the n-th order by dy/dx, where $h = h(x,y) \in \mathcal{A}_r$, $2 \leq r$, $\{A_k, z\} := \{a_{k,1}za_{k,2}z...za_{k,k+1}\}_{q(2k+1)}$ for $z \in \mathcal{A}_r$, $a_{k,j} = a_{k,j}(x,y)$ are coefficients, $q(2k+1)$ are vectors indicating on an order of multiplications. For $r = 2$ due to the associativity of the quaternion skew field vectors $q(2k+1)$ can be omitted. Suppose that some solution $[dy(x)/dx].h = {}_\alpha f(x,y)$ of (1) is found. Then the latter PDE is the first order relative to the derivative dy/dx. Apart from the complex case over the Cayley-Dickson algebra the decomposition $\sum_{k=0}^n \{A_k, z\} = (z - {}_1f(x,y))...(z - {}_nf(x,y))$ may not be existing. If it is the case, then the initial PDE reduces to n PDE of the first order relative to the derivative dy/dx: $[dy(x)/dx].h = {}_lf(x,y)$, $l = 1,...,n$. One such equation over the

Cayley-Dickson algebra $\mathcal{A}_r$ gives the corresponding system of 2^r PDE over the real field $\mathbf{R}$.

8. Example. The PDE

$$x([dy(x)/dx].h)^2 - y(x)([dy(x)/dx].h) - x\{([dy(x)/dx].h)x^{-1}y(x)\} + x = 0, \quad (1)$$

where $h = h(x)$, over quaternions after the multiplication on x^{-1} from the left takes the form

$$([dy(x)/dx].h)^2 - (x^{-1}y(x))([dy(x)/dx].h) - ([dy(x)/dx].h)(x^{-1}y(x)) + 1 = 0, \quad (2)$$

where $h = h(x) \in \mathbf{H}$. The quadratic polynomial $z^2 + bz + zb + c$ over the Cayley-Dickson algebra $\mathcal{A}_r$ can be presented as $(z - \lambda_1)(z - \lambda_2)$, when $bz + zb = -\lambda_1 z - z\lambda_2$ and $\lambda_1\lambda_2 = c$. Thus

$$z^2 + bz + zb + c = (z + b)^2 + (c - b^2) = 0 \quad (3)$$

and the zeros of the quadratic polynomial are the following:

$$\lambda_1 = -b - (b^2 - c)^{1/2}, \lambda_2 = -b + (b^2 - c)^{1/2}. \quad (4)$$

Equation (2) can be considered over octonions as well. Applying the latter formula to (2) one gets:

$$[dy(x)/dx].h = x^{-1}y(x) \pm \{(x^{-1}y(x))^2 - 1\}^{1/2}. \quad (5)$$

Each Cayley-Dickson non-zero number $z \in \mathcal{A}_r$, $2 \le r$, can be written in the polar form $z = |z|e^{M\phi}$, where $\phi \in \mathbf{R}$, $M \in \mathcal{A}_r$ is a purely imaginary Cayley-Dickson number with $|M| = 1$ (see §3 in [25, 26]). Therefore, $z^{1/2} = \rho e^{K\psi}$, where $\rho > 0$, $K \in \mathcal{A}_r$ is a purely imaginary Cayley-Dickson number, $|K| = 1$, $\psi \in \mathbf{R}$, so that $\rho^2 = |z|$, $e^{2K\psi} = e^{M\phi}$ or $\cos(2\psi) + K\sin(2\psi) = \cos(\phi) + M\sin(\phi)$. That is, $\cos(2\psi) = \cos(\phi)$, $K\sin(2\psi) = M\sin(\phi)$. For $M\sin(\phi) \ne 0$ it gives two solutions: $K = M$, $\psi = \phi/2\ (mod\ 2\pi)$ or $\psi = \phi/2 + \pi\ (mod\ 2\pi)$. For $M\sin(\phi) = 0$, i.e. $\phi = 0\ (mod\ \pi)$, the Cayley-Dickson number K may be arbitrary purely imaginary of the unit norm, $\psi = \phi/2\ (mod\ \pi)$. For $\phi = 0\ (mod\ 2\pi)$ one gets two solutions ρ and $-\rho$, since $e^{K\pi} = -1$. For $\phi = \pi\ (mod\ 2\pi)$ this gives the imaginary unit sphere of solutions, since $e^{K\pi/2} = K, e^{3K\pi/2} = -K, K^2 = -1$.

9. Transition Operator. In the PDE

$$\partial\mathcal{E}/\partial t + (s, \nabla\mathcal{E}) + \alpha\mathcal{E} = \delta(t, x) \quad (1)$$

the operator on the right side is called the transition operator, where $0 \le t \in \mathbf{R}$, $x \in \mathbf{R}^3$, α is a real constant, $\mathcal{E}$ is a generalized function. We suppose that s is a constant vector of unit norm $|s| = 1$, $z = z_0 i_0 + ... + z_3 i_3 \in \mathbf{H}$, $z_0 = t$, $z_j = x_j$ for each $j = 1, 2, 3$, where $\nabla f := (\partial f/\partial x_1, \partial f/\partial x_2, \partial f/\partial x_3)$ denotes the gradient of the differentiable function f,

$(x, y) = x_1 y_1 + x_2 y_2 + x_3 y_3$ denotes the scalar product in the Euclidean space $\mathbf{R}^3$. Let us seek a solution in the form:

$$\mathcal{E}(t, x) = \theta(t)e^{-\alpha t}u(t, x), \quad (2)$$

where θ and u are generalized functions. Then Equation (1) simplifies to

$$\theta^{(1)}e^{-\alpha t}u + \theta e^{-\alpha t}\partial u/\partial t + \theta e^{-\alpha t}(s, \nabla u) = \delta \tag{3}$$

or

$$\theta^{(1)}e^{-\alpha t} = -\theta e^{-\alpha t}[(\partial u/\partial t)/u + (s, \nabla u)/u] + \delta/u, \tag{4}$$

when $u = u(t, x) \neq 0$. The left side of (4) is independent of x, consequently, the right side also is independent of x. This means that $[-(\partial u/\partial t) - (s, \nabla u) + \delta/(\theta e^{-\alpha t})]$ is independent of x. We put $h = i_0 + s_1 i_1 + s_2 i_2 + s_3 i_3$ and consider u as the function $u = u(z)$ of the quaternion variable $z \in \mathbf{H}$. Therefore,

$$\int_0^x [-(\partial u/\partial t) - (s, \nabla u) + \delta e^{\alpha t}/\theta]dz_0 = -u(\omega_h(x)) - \phi_u(Im\, \omega_h(x)) - \int [\delta(z)e^{\alpha t}/\theta]dz_0.$$

The equality $\omega_h(y) = x$ gives the solution $y_j = x_j - ts_j$ for each $j = 1, 2, 3$, consequently, $u(t, x) = u(x - ts)$.

One can take any sequence of z-differentiable functions g_n with compact supports tending to $\delta(z)$, when n tends to the infinity. That is

$$\lim_{n\to\infty} \int_{-\infty}^{\infty} g_n(t)s(t)dt = s(0) = (\delta, s)$$

for any continuous function s on $\mathbf{R}$. Substituting this expression for u into (3) on gets the simplified PDE for $z \neq 0$:

$$\theta^{(1)}e^{-\alpha t} = -\theta e^{-\alpha t}[-su^{(1)} + su^{(1)}] = 0. \tag{5}$$

It is seen that the latter equation has the solution $\theta(t) = 1$ for $t > 0$ and $\theta(t) = 0$ for $t < 0$. At the same time this function θ as the generalized function has the derivative $\theta^{(1)}(t) = \delta(t)$, consequently, (3) and (5) imply $u(x) = \delta(x)$ and inevitably $u = \delta(x - ts)$. Thus

$$\mathcal{E}(t, x) = \theta(t)e^{-\alpha t}\delta(x - ts) \tag{6}$$

and the generalized solution of the transition equation

$$\partial\mathcal{E}/\partial t + (s, \nabla\mathcal{E}) + \alpha\mathcal{E} = p$$

is the following $\mathcal{E} * p$, where $p = p(z)$ is a generalized function or particularly a z-differentiable function, $z \in \mathbf{H}$.

Integration of other diverse type PDE are described in [28].

3. Line Integration and First and Second Order PDE over Cayley-Dickson Algebras

3.1. Line Integration of Dirac Operators Over Cayley-Dickson Algebras

1. Factorization and Integration of Equations.

We consider a second order partial differential equation:

$$Af = g, \tag{1}$$

where

$$A = \sum_{l,m=1}^{k} \mathbf{a}_{l,m} \partial^2 / \partial \tau_l \partial \tau_m + \sum_{l=1}^{k} \alpha_l \partial / \partial \tau_l$$

is a partial differential operator of the second order. Let us suppose that a quadratic form

$$a(\tau) := \sum_{l,m} \mathbf{a}_{l,m} \tau_l \tau_m$$

is non-degenerate and is not always negative, because otherwise we can consider $-A$. Moreover, let a matrix of coefficients be real and symmetric $\mathbf{a}_{l,m}(\tau) = \mathbf{a}_{m,l}(\tau) \in \mathbf{R}$, $\alpha_l, \tau_l \in \mathbf{R}$ for each $l, m = 1, ..., k$. Then we reduce this form $a(\tau)$ by an invertible $\mathbf{R}$ linear operator $C = C(\tau)$ to the sum of squares. This means, that

$$A = \sum_{l=1}^{k} \mathbf{b}_l \partial^2 / \partial s_l^2 + \sum_{l=1}^{k} \beta_l \partial / \partial s_l, \tag{2}$$

where $\partial s_j / \partial \tau_l = C_{l,j}(\tau)$, $C = (C_{l,j})$, with real-valued functions $\mathbf{b}_l$ and β_l for each l. Here

$$\mathbf{b}_l \delta_{j,l} = \sum_{p,m} \mathbf{a}_{p,m} C_{p,j} C_{m,l} \text{ and}$$

$$\beta_j = \sum \mathbf{a}_{p,m} (\partial C_{p,j} / \partial \tau_m) + \sum_{v=1}^{k} \alpha_v C_{v,j}$$

for all $j, l = 1, ..., k$. In the case when coefficients of A are constant, using a multiplier of the type $\exp(\sum_l \epsilon_l s_l)$ it is possible to reduce this equation to the case so that if $\mathbf{b}_l \neq 0$, then $\beta_l = 0$ (see §3, Chapter 4 in [52]). Therefore, one can as usually simplify the operator with the help of such change of coordinates and consider that only β_1 may be non-zero if $\mathbf{b}_1 = 0$.

Thus one can choose an invertible real matrix $(c_{h,m})_{h,m=1,...,k}$ corresponding to $C = C(\tau)$ so that $\mathbf{b}_l \leq 0$ for $p + 1 \leq l \leq k$ and $\mathbf{b}_l \geq 0$ for $0 < l \leq p$, where $0 < p \leq k$, $q := k - p$. When $q = 0$ and $\beta_l = 0$ for each l the operator is elliptic, for $q = 0$ and $\beta_1 \neq 0$ the operator is parabolic, for $0 < p < k$ and $\beta_l = 0$ for each l the operator is hyperbolic. Sometimes the matrix C can be chosen constant on a domain, where the signature (p, q) of the quadratic form $a(\tau)$ is constant. We suppose that the sums $\sum_{l=1}^{p} \mathbf{b}_l^2(x) > 0$ and $\sum_{l=p+1}^{k} \mathbf{b}_l^2(x) > 0$ are positive λ-almost everywhere on a domain U, where λ is a measure induced by the Lebesgue measure on the real shadow of the Cayley-Dickson algebra. Generally a natural number $k - p = q(x)$ may either be constant or change while crossing the surface $\{x \in U : \sum_{l=1}^{k} \mathbf{b}_l^2(x) = 0\}$, when the domain U satisfies Conditions 2.2.1$(D1, D2)$.

We consider elliptic and hyperbolic partial differential operators reduced to the sum of squares

$$A = [\sum_{l=0}^{k} \mathbf{b}_l(x) \partial^2 / \partial x_l^2], \tag{3}$$

where $\mathbf{b}_l(x) \in \mathbf{R}$ for all $x = x_0 i_0 + ... + x_{2^r-1} i_{2^r-1}$ in the open domain $U \subset \mathcal{A}_r$ satisfying Conditions 2.2.1$(D1, D2)$ in the Cayley-Dickson algebra $\mathcal{A}_r$, $1 \le k \le 2^r - 1$, $2 \le r \le 3$. Practically a coefficient $\mathbf{b}_l$ can depend only on $x_0, ..., x_k$ remaining z-differentiable in definite z-representations due to Formulas 2.2.1$(1-3)$ for each l.

More generally we can consider partial differential operators of the form

$$A = c_1 B_1 + ... + c_m B_m, \tag{4}$$

where $c_j B_j f = c_j (B_j f)$, while each

$$B_j = \sum_{k=m_1+...+m_{j-1}+1}^{m_1+...+m_j} \mathbf{b}_k(x) \partial^2 / \partial x_k^2 \tag{4'}$$

is an elliptic partial differential operator of the second order by variables

$$x_{m_1+...+m_{j-1}+1}, ..., x_{m_1+...+m_j};\ c_j \in \mathcal{A}_r$$

with $Re(c_j) \ge 0$ for each $1 \le j \le l$, $Re(c_j) < 0$ for every $j > l$, with $|c_j| = 1$ for each $j = 1, ...m$, where $1 \le r$, $1 \le l < m$, $m_0 = 0$.

We remind, that Dirac had used complexified bi-quaternions to solve Klein-Gordon's hyperbolic partial differential equation with constant coefficients appearing in spin problems. That is, he had decomposed d'Alembert's operator $\partial^2/\partial t^2 - \nabla^2$ as the product $\mathbf{i}^* \sigma \mathbf{i} \sigma$ over the complexified bi-quaternion algebra $\mathbf{H}_\mathbf{C}$ with the first order differential operator σ.

If follow this approach one takes the complexified Cayley-Dickson algebra

$$(\mathcal{A}_r)_\mathbf{C} = \mathcal{A}_r \oplus \mathcal{A}_r \mathbf{i}, \tag{5}$$

where $\mathbf{i}$ is taken to be commuting with i_j for each $j = 0, ..., 2^r - 1$. Now the algebra $(\mathcal{A}_r)_\mathbf{C}$ is already not the division algebra even for $2 \le r \le 3$. That is two non zero elements with zero product occur in it. Then each element $z = (z_1, 0)$ in $(\mathcal{A}_r)_\mathbf{C}$ can be written in the 2×2 matrix form $\begin{pmatrix} z_1 & 0 \\ 0 & z_1 \end{pmatrix}$ and $z = (0, z_2)$ can be written in the form $\begin{pmatrix} 0 & z_2 \\ -z_2 & 0 \end{pmatrix}$, where entries $z_1, z_2 \in \mathcal{A}_r$ are Cayley-Dickson numbers, $\mathbf{i} = \begin{pmatrix} 0 & 1 \\ -1 & 0 \end{pmatrix}$.

Let each coefficient c_j be written in the polar form

$$c_j = \exp(i_{\kappa(j)} \gamma_j) \tag{6}$$

with $0 \le |\gamma_j| \le \pi$, $j = 1, ..., m$, $1 \le r$, $1 \le \kappa(j) \le \kappa(j+1)$ for each j. Put $p = p_1 + ... + p_m$, where $p_j = 0$ for either $\gamma_j = 0$ or $\kappa(j) = \kappa(j-1)$, while $p_j = 1$ for $\gamma_j \ne 0$ and $\kappa(j) \ne \kappa(j-1)$. Up to an isomorphism we take the Cayley-Dickson algebra $\mathcal{A}_v$ with $v \ge r$ satisfying inequalities $2^{v-1} < 2^p(m+1) \le 2^v$. Further we make the complexification $(\mathcal{A}_v)_\mathbf{C}$ of the Cayley-Dickson algebra $\mathcal{A}_v$.

Take two non-negative integer numbers $0 \le r$ and v with $r \le v \in \mathbf{Z}$. We consider the quotient algebra over the real field $\mathcal{A}_v / \mathcal{A}_r =: \mathcal{A}_{r,v}$. For $r = v$ this algebra is isomorphic with the real field $\mathbf{R}$. For $r < v$ the algebra $\mathcal{A}_{r,v}$ is isomorphic with $\bigoplus_{k=0}^{2^{v-r}-1} \mathbf{R} i_{2^r k}$. The latter algebra is produced by generators $\{i_{2^r \beta} : \beta = 2^\gamma - 1;\ \gamma = 0, 1, ..., v - r\}$ and their finite ordered products, that gives the generators set $\{i_{2^r k} : k = 0, ..., 2^{v-r} - 1\}$, where generators satisfying the numbering rule $i_j i_{2^s} = i_{j+2^s}$ for each $1 \le s$, $j = 0, 1, ..., 2^s -$

1 can be taken up to an isomorphism of the Cayley-Dickson algebra $\mathcal{A}_{s+1}$. Therefore, the algebra $\mathcal{A}_{r,v}$ is isomorphic with the Cayley-Dickson algebra $\mathcal{A}_{v-r}$, since the doubling procedure can be started from another suitable purely imaginary Cayley-Dickson numbers such as generators [2, 18]. But we consider in $\mathcal{A}_{r,v}$ its specific generator basis $\{i_{2^r k} : k = 0, ..., 2^{v-r} - 1\}$.

For each Cayley-Dickson numbers $x, y \in \mathcal{A}_r$ we define the real-valued scalar product
(RS) $(x, y) = (x, y)_r := Re(x\tilde{y})$,
where $\tilde{z} = z^*$ denotes the conjugated number, while $Re(y) := (y + y^*)/2$ denotes the real part of y.

The real scalar product $(.,.)_r$ in $\mathcal{A}_r$ we extend on the algebra $\mathcal{A}_{r,v}$ as

$$< x, y >_{r,v} = x\tilde{y} = \sum_{j,k=0}^{2^{v-r}-1} x_{2^r j} y_{2^r k} i_{2^r j} i^*_{2^r k} \qquad (SP)$$

for each $x, y \in \mathcal{A}_{r,v}$, $x = \sum_{j=0}^{2^{v-r}-1} x_{2^r j} i_{2^r j}$, $x_{2^r j} \in \mathbf{R}$ for each $j = 0, ..., 2^{v-r} - 1$. Particularly, one gets $< x, y >_{0,v} = < x, y >_v$. In the case of the complexified algebra $(\mathcal{A}_{r,v})_{\mathbf{C}}$ the scalar product is:

$$< (a, b), (c, d) >_{r,v} = < (a, b), (c, d) > = (< a, c > - < b, d >, < a, d > + < b, c >), \quad (SPC)$$

for all (a, b) and $(c, d) \in (\mathcal{A}_{r,v})_{\mathbf{C}}$.

We recall a doubling procedure for a construction of the Cayley-Dickson algebra $\mathcal{A}_{r+1}$ from $\mathcal{A}_r$. Each Cayley-Dickson number $z \in \mathcal{A}_{r+1}$ is written in the form $z = \xi + \eta \mathbf{l}$, where $\mathbf{l}^2 = -1$, $\mathbf{l} \notin \mathcal{A}_r$, $\xi, \eta \in \mathcal{A}_r$. The addition of such numbers is componentwise. The conjugate of any Cayley-Dickson number z is given by the formula:

$$z^* := \xi^* - \eta \mathbf{l}. \qquad (M1)$$

The multiplication in $\mathcal{A}_{r+1}$ is defined by the following equation:

$$(\xi + \eta \mathbf{l})(\gamma + \delta \mathbf{l}) = (\xi\gamma - \tilde{\delta}\eta) + (\delta\xi + \eta\tilde{\gamma})\mathbf{l} \qquad (M2)$$

for each $\xi, \eta, \gamma, \delta \in \mathcal{A}_r$, $z := \xi + \eta \mathbf{l} \in \mathcal{A}_{r+1}$, $\zeta := \gamma + \delta \mathbf{l} \in \mathcal{A}_{r+1}$.

Using Formula $(M2)$ we get: $(b i_{2^r k})(i_{2^r k} b)^* = (b i_{2^r k})(b^* i^*_{2^r k}) = b^2 = (b^2 i_{2^r k}) i^*_{2^r k} = i_{2^r k}(i_{2^r k} b^2)^*$ for each $k \geq 1$ and $b \in \mathcal{A}_r$, since $i^*_j = -i_j$ for each $j \geq 1$. Another useful identity is the following: $(i_s i_{2^r j}) i^*_{2^r k} = -(i_s i_{2^r k}) i^*_{2^r j}$ for each $0 \leq s \leq 2^r - 1$ and $k \neq j$ with $k \geq 1$ and $j \geq 1$, since $(i_s i_{2^r j}) i_{2^r k} = (i_s i^*_{2^r k}) i_{2^r j}$. Certainly also the equality $(i_s i_0) i^*_j + (i_s i_j) i^*_0 = 0$ holds for each $j \geq 1$ and $1 \leq s \leq 2^r - 1$, since $i_0 = 1$. Therefore, Formulas (SP) and $(4')$ together with the latter identities imply:

$$< cBy, y >_v = \sum_{j=0}^{2^{v-r}-1} c < By_{2^r j}, y_{2^r j} >_{r,v} \qquad (6)$$

for each $c \in \mathcal{A}_r$ and a twice differentiable function y with values in $\mathcal{A}_{r,v}$.

Relative to the complex scalar product given by Equality (SPC) we decompose the operator A (see $(4, 4')$ above) in the form

$$A = (\mathbf{i}\sigma)(\mathbf{i}\sigma_1) + Q = -\sigma\sigma_1 + Q, \tag{7}$$

where σ, σ_1 and Q are partial differential operators of the first order, each complex number $\alpha \in \mathbf{C}$ is presented as a real 2×2 matrix. Particularly, $\mathbf{i} = \begin{pmatrix} 0 & 1 \\ -1 & 0 \end{pmatrix}$, $\mathbf{i}^* = \begin{pmatrix} 0 & -1 \\ 1 & 0 \end{pmatrix}$. Each subalgebra $\mathrm{g}_{j,k}$ constructed from two generators $i_j \neq i_k$ is associative, consequently, $(wi_k)(w^*i_k^*) = w^2$ and $w((wi_k)i_k) = -w^2$ for each $w = w_0 + w_j i_j$ with $w_0, w_j \in \mathbf{R}$. Therefore, we can take

$$\sigma f(z) = \sum_{j=1}^{m} \sum_{k=m_1+...+m_{j-1}+1}^{m_1+...+m_j} \mathbf{a}_k(z)(\partial f/\partial z_{2^r k})[w_j^* i_{2^r k}^*] \tag{8}$$

and

$$\sigma_1 f(z) = \sum_{j=1}^{m} \sum_{k=m_1+...+m_{j-1}+1}^{m_1+...+m_j} \mathbf{a}_k(z)(\partial f/\partial z_{2^r k})[w_j i_{2^r k}^*] \tag{9}$$

on the space of all $\mathcal{A}_r$-valued (super-)differentiable functions f for $r \leq 2$ or real-valued functions f for $3 \leq r$ of the $\mathcal{A}_{r,v}$ variable, since $(\mathbf{i}i_k)^2 = \mathbf{i}^2 i_k^2 = (-1)^2 = 1$ for each $k \geq 1$, where $w_j^2 = c_j$ for all j and $\mathbf{a}_k^2(x) = \mathbf{b}_k(x)$ for each k and x, $w_j \in \mathcal{A}_r$, $\mathbf{a}_k(x) \in \mathbf{R}$ for all k and x, $i_{2^r k} \in \mathcal{A}_{r,v}$, $z_{2^r k} = x_k$, $z = \sum_k z_{2^r k} i_{2^r k} \in \mathcal{A}_{r,v}$, $\partial f(z)/\partial z_{2^r k} = (df(z)/dz).i_{2^r k}$. For $b = \partial^2 f/\partial z_{2^r k}^2$ and $\mathbf{l} = i_{2^r k}$ and $w \in \mathcal{A}_r$ one has the identities: $(b(w\mathbf{l}))(w^*\mathbf{l}) = ((wb)\mathbf{l})(w^*\mathbf{l}) = -w(wb) = -w^2 b$ and $(((b\mathbf{l})w^*)\mathbf{l})w = (((bw)\mathbf{l})\mathbf{l})w = -(bw)w = -bw^2$ in the considered here cases. The operator Q is given by the equality:

$$Qf(z) - \sum_{j_1,j_2=1}^{m} \sum_{k_1=m_1+...+m_{j_1-1}+1}^{m_1+...+m_{j_1}} \sum_{k_2=m_1+...+m_{j_2-1}+1}^{m_1+...+m_{j_2}}$$

$$\mathbf{a}_{k_1}(z)\{(\partial \mathbf{a}_{k_2}(z)/\partial z_{2^r k_1})(\partial f/\partial z_{2^r k_2})[w_{j_2} i_{2^r k_2}^*]\}[w_{j_1}^* i_{2^r k_1}^*], \tag{10}$$

since $\mathbf{i}i_k = i_k\mathbf{i}$ in the complexified Cayley-Dickson algebra $(\mathcal{A}_v)_{\mathbf{C}}$ for each k. The latter equality (10) shows, that the differential operator Q is non-zero, when $\mathbf{a}_k(z)$ are non-constant coefficients.

If use $i_0 = 1$ and $\partial/\partial z_0$ also one can write out d'Alembert's operator in our notation

$$\partial^2/\partial z_0^2 - \sum_{j=1}^{3} \partial^2/\partial z_j^2 =$$

$$= (\mathbf{i}^*\partial/\partial z_0 + i_1\partial/\partial z_1 + i_2\partial/\partial z_2 + i_3\partial/\partial z_3)(\mathbf{i}\partial/\partial z_0 + i_1\partial/\partial z_1 + i_2\partial/\partial z_2 + i_3\partial/\partial z_3). \tag{11}$$

We recall, that the Cayley-Dickson algebra $\mathcal{A}_r$ is power associative, that is $z^k z^l = z^{k+l}$ for all natural numbers k and l. But the complexified Cayley-Dickson algebra $(\mathcal{A}_r)_{\mathbf{C}}$ is not power associative for $r \geq 3$, since the Cayley-Dickson algebra $\mathcal{A}_r$ is not associative for $r \geq 3$. Therefore, we do not widely use the complexified Cayley-Dickson algebras, but we utilize the Cayley-Dickson algebras $\mathcal{A}_v$ over the real field $\mathbf{R}$, when something other will not be specified.

With these decomposition of operators given by Equations $(7-9, 11)$ the differential equation (1) can be integrated with the help of the non-commutative line integration. We consider at first the partial differential equation

$$\Upsilon f = g \tag{12}$$

on an open domain U in $\mathcal{A}_v$, where

$$\Upsilon f = \sum_{j=0}^{2^v-1} (\partial f/\partial z_j)[i_j^* \psi_j(z)], \tag{13}$$

f and g and $\psi_j(z)$ are $\mathcal{A}_v$-valued functions on the domain U satisfying Conditions $2.2.1(D1, D2)$, where $g,\ \psi_j \in C^0(U, \mathcal{A}_v)$ for each j, particularly they may be $\mathcal{A}_v$ (super-) differentiable functions.

2. Line Anti-derivatives over Cayley-Dickson Algebras.

Theorem. *Let a first order partial differential operator Υ be given by Equation 1(13) with real-valued continuous functions $\psi_j(z) \in C^0(U, \mathcal{A}_v)$ for each j such that $\psi_j(z) \neq 0$ for each $z \in U$ and each $j = 0, ..., n$, where a domain U satisfies Conditions 2.2.1$(D1, D2)$, ${}_0z$ is a marked point in U, $1 < n < 2^v$, $2 \leq v$. Then a line integral $\mathcal{I}_\Upsilon : C^0(U, \mathcal{A}_v) \to C^1(U, \mathcal{A}_v)$, $\mathcal{I}_\Upsilon f(z) := {}_\Upsilon \int_{{}_0z}^z f(y)dy$ on $C^0(U, \mathcal{A}_v)$ exists so that*

$$\Upsilon \mathcal{I}_\Upsilon f(z) = f(z) \tag{1}$$

for each $z \in U$; this anti-derivative is **R***-linear (or* **H***-left-linear when $v = 2$):*

$$\mathcal{I}_\Upsilon[af(z) + bg(z)] = a\mathcal{I}_\Upsilon f(z) + b\mathcal{I}_\Upsilon g(z) \tag{2}$$

for any real constants $a, b \in \mathbf{R}$ (or $a, b \in \mathbf{H}$ for $v = 2$) and continuous functions $f, g \in C^0(U, \mathcal{A}_v)$. If there is a second anti-derivative $\mathcal{I}_{\Upsilon,2} f(z)$, then $\mathcal{I}_\Upsilon f(z) - \mathcal{I}_{\Upsilon,2} f(z)$ belongs to the kernel $ker(\Upsilon)$ of the operator Υ.

The proof is given in §2.4 [33].

3. Example. If ψ_j depends only on z_j for each j, there exists a C^1 differentiable change of variables $\zeta = \zeta(z)$ so that $\partial g(\zeta)/\partial \zeta_j = (\partial g(z)/\partial z_j)\psi_j(z)$ for each differentiable function $g : U \to \mathcal{A}_v$ by real variables $z_0, ..., z_{2^v-1}$ on U, where

$$(\partial z_k/\partial \zeta_j) = \delta_{j,k}\psi_k(z) \tag{1}$$

for all j and k, $\delta_{j,j} = 1$, while $\delta_{j,k} = 0$ for each $j \neq k$. We take new functions ${}_jg$ satisfying the equation:

$${}_jg(i_j^* z) = g(z) \tag{2}$$

for each $z \in U$ and all j. We also put

$$\eta_j(z) = i_j^* z. \tag{3}$$

The multiplication of generators implies that $i_j^*(i_jz) = z$ for all $j = 0, ..., 2^v - 1$ and $z \in \mathcal{A}_v$. Therefore, from Equations $(1, 2)$ we deduce that

$$(dg(z)/dz).i_j = (d_jg(\eta_j)/d\eta_j).[(d\eta_j/dz).i_j] = (d_jg(\eta_j)/d\eta_j).1 = (d_kg(\eta_k)/d\eta_k).[i_k^*i_j], \quad (4)$$

since $(d\eta_j/dz).i_j = i_j^*i_j = 1$ for each j. Then we take the integral

$$(5) \quad {}_{\Upsilon}\int_{0z}^{z} g(y)dy := n^{-1}\sum_{j=1}^{n}\int_{0z}^{z} {}_jg(\eta_j(y))i_jd\eta_j(y),$$

since $\int_{0z}^{z} {}_jg(\eta_j(y))i_jd\eta_j(y) = (\int_{0z}^{z} {}_jg(\eta_j(y))d\eta_j(y))i_j$.

Mention that generally $\Upsilon(f(z)b)$ may be not equal to $(\Upsilon f(z))b$ for a constant $b \in \mathcal{A}_v \setminus \mathbf{R}$ and a function $f \in C^1(U, \mathcal{A}_v)$ with $v \geq 2$, since the Cayley-Dickson algebra is non-commutative.

This theorem can be generalized in the following manner encompassing wider class of partial differential operators of the first order over the Cayley-Dickson algebras.

4. Theorem. *Suppose that a first order partial differential operator Υ is given by the formula*

$$\Upsilon f = \sum_{j=0}^{n}(\partial f/\partial z_j)\phi_j^*(z), \quad (1)$$

where $\phi_j(z) \neq \{0\}$ for each $z \in U$ and $\phi_j(z) \in C^0(U, \mathcal{A}_v)$ for each $j = 0, ..., n$ such that $Re(\phi_j(z)\phi_k^*(z)) = 0$ for each $z \in U$ and each $0 \leq j \neq k \leq n$, where a domain U satisfies Conditions 2.2.1$(D1, D2)$, ${}_0z$ is a marked point in U, $1 < n < 2^v$, $2 \leq v$. Suppose also that a system $\{\phi_0(z), ..., \phi_n(z)\}$ is for $n = 2^v - 1$, or can be completed by Cayley-Dickson numbers $\phi_{n+1}(z), ..., \phi_{2^v-1}(z)$, such that (α) $alg_{\mathbf{R}}\{\phi_j(z), \phi_k(z), \phi_l(z)\}$ is alternative for all $0 \leq j, k, l \leq 2^v - 1$ and (β) $alg_{\mathbf{R}}\{\phi_0(z), ..., \phi_{2^v-1}(z)\} = \mathcal{A}_v$ for each $z \in U$. Then a line integral $\mathcal{I}_\Upsilon : C^0(U, \mathcal{A}_v) \to C^1(U, \mathcal{A}_v)$, $\mathcal{I}_\Upsilon f(z) := {}_\Upsilon\int_{0z}^{z} f(y)dy$ on $C^0(U, \mathcal{A}_v)$ exists so that

$$\Upsilon\mathcal{I}_\Upsilon f(z) = f(z) \quad (2)$$

for each $z \in U$; this anti-derivative is $\mathbf{R}$-linear (or $\mathbf{H}$-left-linear when $v = 2$). If there is a second anti-derivative $\mathcal{I}_{\Upsilon,2}f(z)$, then $\mathcal{I}_\Upsilon f(z) - \mathcal{I}_{\Upsilon,2}f(z)$ belongs to the kernel $ker(\Upsilon)$ of the operator Υ.

The proof of this theorem is given in §2.5 [33].

3.2. Decompositions of PDE Over Cayley-Dickson Algebras

Henceforward, notations and definitions of §2.2.1 and §3.1 are used.

1. Transformation of the First Order Partial Differential Operator over the Cayley-Dickson Algebras.

To simplify the operator Υ and its particular variant σ one can use a change of variables. We consider this operator in the form:

$$\Upsilon f = \sum_{j=0}^{2^v-1}(\partial f/\partial z_j)\eta_j(z), \quad (1)$$

with either $\eta_j(z) = i_j^*\psi_j(z)$ or $\eta_j(z) = \phi_j^*(z) \in \mathcal{A}_v$ for each j (see Theorems 2.4.1 and 2.5.2 in [33]). For it we seek the change of variables $x = x(z)$ so that

$$\sum_{j=0}^{2^v-1} (\partial x_l/\partial z_j)\omega_j(z) = t_l, \tag{2}$$

where $t_l \in \mathcal{A}_v$ is a constant for each l, for η_j not being identically zero, while ω_j is chosen arbitrarily also z-differentiable so that the resulting matrix Ω will not be degenerate, i.e. its rows are real-independent as vectors (see below). Certainly $(\partial x_l/\partial z_j) \in \mathbf{R}$ are real partial derivatives, since x_l and z_j are real coordinates. We suppose that the functions $\eta_j(z)$ are linearly independent over the real field for each z in the domain U. Using the standard basis of generators $\{i_j : j = 0, ..., 2^v - 1\}$ of the Cayley-Dickson algebra $\mathcal{A}_v$ and the decompositions $\omega_j = \sum_k \omega_{j,k} i_k$ and $t_j = \sum_k t_{j,k} i_k$ with real elements $\omega_{j,k}$ and $t_{j,k}$ for all j and k we rewrite System (2) in the matrix form:

$$(\partial x_l/\partial z_j)_{l,j=0,...,2^v-1}\Omega = T, \tag{3}$$

where $\Omega = (\omega_{j,k})_{j,k=0,...,2^v-1}$, $T = (t_{j,k})_{j,k=0,...,2^v-1}$. Suppose that the functions $\omega_j(z)$ are arranged into the family $\{\omega_j : j = 0, ..., 2^v - 1\}$ as above and are such that the matrix $\Omega(z)$ is non-degenerate for all z in the domain U. For example, this is always the case, when $|\omega_j(z)| > 0$ and $Re[\omega_j(z)\omega_k(z)^*] = 0$ for each $j \neq k$ for each $z \in U$. Here particularly $\omega_j(z) = \eta_j(z)$ can also be taken for all $j = 0, ..., 2^v - 1$ and $z \in U$. Therefore, Equality (3) becomes equivalent to

$$(\partial x_l/\partial z_j)_{l,j=0,...,2^v-1} = T\Omega^{-1}. \tag{4}$$

We take the real matrix T of the same rank as the real matrix $(\omega_{j,k})_{j,k=0,...,2^v-1}$. Thus (4) is the linear system of partial differential equations of the first order over the real field. It can be solved by the standard methods [45].

We remind how each linear partial differential equation (3) or (4) can be resolved. Write it in the form:

$$X_1(x_1, ..., x_n, u)\partial u/\partial x_1 + ... + X_n(x_1, ..., x_n, u)\partial u/\partial x_n = R(x_1, ..., x_n, u) \tag{5}$$

with u and $x_1, ..., x_n$ here instead of x_l and $z_0, ..., z_{2^v-1}$ in (3) seeking simultaneously suitable R corresponding to $t_{l,k}$. A function $u = u(x_1, ..., x_n)$ defined and continuous with its partial derivatives $\partial u/\partial x_1$, ...,$\partial u/\partial x_n$ in some domain V of variables $x_1, ..., x_n$ in $\mathbf{R}^n$ making (5) the identity is called a solution of this linear equation. If $R = 0$ identically, then the equation is called homogeneous. A solution $u = const$ of the homogeneous equation

$$X_1(x_1, ..., x_n, u)\partial u/\partial x_1 + ... + X_n(x_1, ..., x_n, u)\partial u/\partial x_n = 0 \tag{6}$$

is called trivial. Then one composes the equations:

$$dx_1/X_1(x) = dx_2/X_2(x) = ... = dx_n/X_n(x), \tag{7}$$

where $x = (x_1, ..., x_n)$. This system is called the system of ordinary differential equations in the symmetric form corresponding to the homogeneous linear equation in partial derivatives. It is supposed that the coefficients $X_1, ..., X_n$ are defined and continuous together with their first order partial derivatives by $x_1, ..., x_n$ and that $X_1, ..., X_n$ are not simultaneously zero in a neighborhood of some point x^0. Such point x^0 is called non singular. For example when the function X_n is non-zero System (7) can be written as:

$$dx_1/dx_n = X_1/X_n, ..., dx_{n-1}/dx_n = X_{n-1}/X_n. \qquad (8)$$

This system satisfies conditions of the theorem about an existence of integrals of the normal system. A system of n differential equations

$$dy_k/dx = f_k(x, y_1, ..., y_n), k = 1, ..., n, \qquad (9)$$

is called normal of the n-th order. It is called linear if all functions f_k depend linearly on $y_1, ..., y_n$. Any family of functions $y_1, ..., y_n$ satisfying (9) in some interval (a, b) is called its solution. A function $g(x, y_1, ..., y_n)$ different from a constant identically and differentiable in a domain D and such that its partial derivatives $\partial g/\partial y_1$,...,$\partial/\partial y_n$ are not simultaneously zero in D is called an integral of System (9) in D if the complete differential $dg = (\partial g/\partial x)dx + (\partial g/\partial y_1)dy_1 + ... + (\partial g/\partial y_n)dy_n$ becomes identically zero, when the differentials dy_k are substituted on their values from (9), that is $(\partial g(x, y)/\partial x) + (\partial g/\partial y_1)f_1(x, y) + ... + (\partial g(x, y)/\partial y_n)f_n(x, y) = 0$ for each $(x, y) \in D$, where $y = (y_1, ..., y_n)$. The equality $g(x, y) = const$ is called the first integral of System (9).

It is supposed that each function $f_k(x, y)$ is continuous on D and satisfies the Lipschitz conditions by variables $y_1, ..., y_n$:

$$|f_k(x, y) - f_k(x, z)| \le C_k|y - z| \qquad (L)$$

for all (x, y) and $(x, z) \in D$, where C_k are constants. Then System (9) has exactly n independent integrals in some neighborhood D^0 of a marked point (x^0, y^0) in D, when the Jacobian $\partial(g_1, ..., g_n)/\partial(y_1, ..., y_n)$ is not zero on D^0 (see Section 5.3.3 [45]).

In accordance with Theorem 12.1,2 [45] each integral of System (7) is a non-trivial solution of Equation (6) and vice versa each non-trivial solution of Equation (6) is an integral of (7). If $g_1(x_1, ..., x_n)$,...,$g_{n-1}(x_1, ..., x_n)$ are independent integrals of (7), then the function

$$u = \Phi(g_1, ..., g_{n-1}), \qquad (10)$$

where Φ is an arbitrary function continuously differentiable by $g_1, ..., g_{n-1}$, is the solution of (6). Formula (10) is called the general solution of Equation (6).

To the non-homogeneous Equation (5) the system

$$dx_1/X_1 = ... = dx_n/X_n = du/R \qquad (11)$$

is posed. System (11) gives n independent integrals $g_1, ..., g_n$ and the general solution

$$\Phi(g_1(x_1, ..., x_n, u), ..., g_n(x_1, ..., x_n, u)) = 0 \qquad (12)$$

of (5), where Φ is any continuously differentiable function by $g_1, ..., g_n$. If the latter equation is possible to resolve relative to u this gives the solution of (5) in the explicit form $u = u(x_1, ..., x_n)$ which generally depends on Φ and $g_1, ..., g_n$. Therefore, Formula (12) for different R and u and X_j corresponding to $t_{l,k}$ and x_l and $\omega_{j,k}$ respectively can be satisfied in (3) or (4), the variables x_j are used in (12) instead of z_j in $(3,4)$, where $k = 0, ..., 2^v - 1$.

Thus after the change of the variables the operator Υ takes the form:

$$\Upsilon f = \sum_{j=0}^{2^v-1} (\partial f/\partial x_j)t_j \tag{13}$$

with constants $t_j \in \mathcal{A}_v$. Undoubtedly, also the operator Υ with $j = 0, ..., n$, $2^{v-1} \le n \le 2^v - 1$ instead of $2^v - 1$ can also be reduced to the form $\Upsilon f = \sum_{j=0}^{n} (\partial f/\partial x_j)t_j$, when the rank is $rank(\omega_{j,k}) = n + 1$ in a basis of generators $N_0, ..., N_n$, where $N_0,...,N_{2^v-1}$ is a generator basis of the Cayley-Dickson algebra $\mathcal{A}_v$. Particularly, if the rank is $rank(\omega_{j,k}) = m \le 2^v$ and T has the unit upper left $m \times m$ block and zeros outside it, then $t_j = N_j$ for each $j = 0, ..., m - 1$ can be chosen.

One can mention that direct algorithms of Theorems 2.4.1 and 2.5.2 [33] may be simpler for finding the anti-derivative operator $\mathcal{I}_\Upsilon$, than this preliminary transformation of the partial differential operator Υ to the standard form (13).

2. Definitions.

Let X and Y be two $\mathbf{R}$ linear normed spaces which are also left and right $\mathcal{A}_r$ modules, where $1 \le r$. Let Y be complete relative to its norm. We put $X^{\otimes k} := X \otimes_{\mathbf{R}} ... \otimes_{\mathbf{R}} X$ is the k times ordered tensor product over $\mathbf{R}$ of X. By $L_{q,k}(X^{\otimes k}, Y)$ we denote a family of all continuous k times $\mathbf{R}$ poly-linear and $\mathcal{A}_r$ additive operators from $X^{\otimes k}$ into Y. Then $L_{q,k}(X^{\otimes k}, Y)$ is also a normed $\mathbf{R}$ linear and left and right $\mathcal{A}_r$ module complete relative to its norm. In particular, $L_{q,1}(X, Y)$ is denoted also by $L_q(X, Y)$.

We present X as the direct sum $X = X_0 i_0 \oplus ... \oplus X_{2^r-1} i_{2^r-1}$, where $X_0,...,X_{2^r-1}$ are pairwise isomorphic real normed spaces. If $A \in L_q(X, Y)$ and $A(xb) = (Ax)b$ or $A(bx) = b(Ax)$ for each $x \in X_0$ and $b \in \mathcal{A}_r$, then an operator A we call right or left $\mathcal{A}_r$-linear respectively.

An $\mathbf{R}$ linear space of left (or right) k times $\mathcal{A}_r$ poly-linear operators is denoted by $L_{l,k}(X^{\otimes k}, Y)$ (or $L_{r,k}(X^{\otimes k}, Y)$ respectively).

As usually a support of a function $g : S \to \mathcal{A}_r$ on a topological space S is by the definition $supp(g) = cl\{t \in S : g(t) \ne 0\}$, where the closure is taken in S.

We consider a space of test function $\mathcal{D} := \mathcal{D}(\mathbf{R}^n, Y)$ consisting of all infinite differentiable functions $f : \mathbf{R}^n \to Y$ on $\mathbf{R}^n$ with compact supports. A sequence of functions $f_n \in \mathcal{D}$ tends to zero, if all f_n are zero outside some compact subset K in the Euclidean space $\mathbf{R}^n$, while on it for each $k = 0, 1, 2, ...$ the sequence $\{f_n^{(k)} : n \in \mathbf{N}\}$ converges to zero uniformly. Here as usually $f^{(k)}(t)$ denotes the k-th derivative of f, which is a k times $\mathbf{R}$ poly-linear symmetric operator from $(\mathbf{R}^n)^{\otimes k}$ to Y, that is $f^{(k)}(t).(h_1, ..., h_k) = f^{(k)}(t).(h_{\sigma(1)}, ..., h_{\sigma(k)}) \in Y$ for each $h_1, ..., h_k \in \mathbf{R}^n$ and every transposition $\sigma : \{1, ..., k\} \to \{1, ..., k\}$, σ is an element of the symmetric group S_k, $t \in \mathbf{R}^n$. For convenience one puts $f^{(0)} = f$. In particular, $f^{(k)}(t).(e_{j_1}, ..., e_{j_k}) = \partial^k f(t)/\partial t_{j_1}...\partial t_{j_k}$ for all $1 \le j_1, ..., j_k \le n$, where $e_j = (0, ..., 0, 1, 0, ..., 0) \in \mathbf{R}^n$ with 1 on the j-th place.

Such convergence in $\mathcal{D}$ defines closed subsets in this space $\mathcal{D}$, their complements by the definition are open, that gives the topology on $\mathcal{D}$. The space $\mathcal{D}$ is $\mathbf{R}$ linear and right and left $\mathcal{A}_r$ module.

By a generalized function of class $\mathcal{D}' := [\mathcal{D}(\mathbf{R}^n, Y)]'$ is called a continuous $\mathbf{R}$-linear $\mathcal{A}_r$-additive function $g : \mathcal{D} \to \mathcal{A}_r$. The set of all such functionals is denoted by $\mathcal{D}'$. That is, g is continuous, if for each sequence $f_n \in \mathcal{D}$, converging to zero, a sequence of numbers $g(f_n) =: [g, f_n) \in \mathcal{A}_r$ converges to zero for n tending to the infinity.

A generalized function g is zero on an open subset V in $\mathbf{R}^n$, if $[g, f) = 0$ for each $f \in \mathcal{D}$ equal to zero outside V. By a support of a generalized function g is called the family, denoted by $supp(g)$, of all points $t \in \mathbf{R}^n$ such that in each neighborhood of each point $t \in supp(g)$ the functional g is different from zero. The addition of generalized functions g, h is given by the formula:

$$[g + h, f) := [g, f) + [h, f). \tag{1}$$

The multiplication $g \in \mathcal{D}'$ on an infinite differentiable function w is given by the equality:

$$[gw, f) = [g, wf) \tag{2}$$

either for $w : \mathbf{R}^n \to \mathcal{A}_r$ and each test function $f \in \mathcal{D}$ with a real image $f(\mathbf{R}^n) \subset \mathbf{R}$, where $\mathbf{R}$ is embedded into Y; or $w : \mathbf{R}^n \to \mathbf{R}$ and $f : \mathbf{R}^n \to Y$.

A generalized function g' prescribed by the equation:

$$[g', f) := -[g, f') \tag{3}$$

is called a derivative g' of a generalized function g, where $f' \in \mathcal{D}(\mathbf{R}^n, L_q(\mathbf{R}^n, Y))$, $g' \in [\mathcal{D}(\mathbf{R}^n, L_q(\mathbf{R}^n, Y))]'$.

Another space $\mathcal{B} := \mathcal{B}(\mathbf{R}^n, Y)$ of test functions consists of all infinite differentiable functions $f : \mathbf{R}^n \to Y$ such that the limit $\lim_{|t| \to +\infty} |t|^m f^{(j)}(t) = 0$ exists for each $m = 0, 1, 2, ..., j = 0, 1, 2, ...$ A sequence $f_n \in \mathcal{B}$ is called converging to zero, if the sequence $|t|^m f_n^{(j)}(t)$ converges to zero uniformly on $\mathbf{R}^n \setminus B(\mathbf{R}^n, 0, R)$ for each $m, j = 0, 1, 2, ...$ and each $0 < R < +\infty$, where $B(Z, z, R) := \{y \in Z : \rho(y, z) \le R\}$ denotes a ball with center at z of radius R in a metric space Z with a metric ρ. The family of all $\mathbf{R}$-linear and $\mathcal{A}_r$-additive functionals on $\mathcal{B}$ is denoted by $\mathcal{B}'$.

In particular we can take $X = \mathcal{A}_r^\alpha$, $Y = \mathcal{A}_r^\beta$ with $1 \le \alpha, \beta \in \mathbf{Z}$. Analogously spaces $\mathcal{D}(U, Y)$, $[\mathcal{D}(U, Y)]'$, $\mathcal{B}(U, Y)$ and $[\mathcal{B}(U, Y)]'$ are defined for domains U in $\mathbf{R}^n$. For definiteness we write $\mathcal{B}(U, Y) = \{f|_U : f \in \mathcal{B}(\mathbf{R}^n, Y)\}$ and $\mathcal{D}(U, Y) = \{f|_U : f \in \mathcal{D}(\mathbf{R}^n, Y)\}$.

A function $g : U \to \mathcal{A}_v$ is called locally integrable, if it is absolutely integrable on each bounded λ measurable sub-domain V in U, i.e.

$$\int_V |g(z)| \lambda(dz) < \infty,$$

where λ denotes the Lebesgue measure on U.

A generalized function f is called regular if locally integrable functions ${}_{j,k}f^1, {}_l f^2 : U \to \mathcal{A}_v$ exist such that

$$[f, \omega) = \int_U \{\sum_{j,k,l} {}_{j,k}f^1(z)_k\omega(z)_j f^2(z)\}_{q(3)} \lambda(dz),$$

for each test function $\omega \in \mathcal{B}(U,Y)$ or $\omega \in \mathcal{D}(U,Y)$ correspondingly, where $\omega = ({}_1\omega, ..., {}_\beta\omega)$, ${}_k\omega(z) \in \mathcal{A}_v$ for each $z \in U$ and all k, $q(3)$ is a vector indicating on an order of the multiplication in the curled brackets and it may depend on the indices $j, l = 1, ..., \alpha$, $k = 1, ..., \beta$.

We supply the space $\mathcal{B}(\mathbf{R}^n, Y)$ with the countable family of semi-norms

$$p_{\alpha,k}(f) := \sup_{x \in \mathbf{R}^n} |(1 + |x|)^k \partial^\alpha f(x)| \tag{4}$$

inducing its topology, where $k = 0, 1, 2, ...$; $\alpha = (\alpha_1, ..., \alpha_n)$, $0 \le \alpha_j \in \mathbf{Z}$. On this space we take the space $\mathcal{B}'(\mathbf{R}^n, Y)_l$ of all Y valued continuous generalized functions (functionals) of the form

$$f = f_0 i_0 + ... + f_{2^v-1} i_{2^v-1} \tag{5}$$

and $g = g_0 i_0 + ... + g_{2^v-1} i_{2^v-1}$, where f_j and $g_j \in \mathcal{B}'(\mathbf{R}^n, Y)$, with restrictions on $\mathcal{B}(\mathbf{R}^n, \mathbf{R})$ being real- or $\mathbf{C_i} = \mathbf{R} \oplus \mathbf{iR}$- valued generalized functions $f_0, ..., f_{2^v-1}$, $g_0, ..., g_{2^v-1}$ respectively. Let $\phi = \phi_0 i_0 + ... + \phi_{2^v-1} i_{2^v-1}$ with $\phi_0, ..., \phi_{2^v-1} \in \mathcal{B}(\mathbf{R}^n, \mathbf{R})$, then

$$[f, \phi) = \sum_{k,j=0}^{2^v-1} [f_j, \phi_k) i_k i_j. \tag{6}$$

Let their convolution be defined in accordance with the formula:

$$[f * g, \phi) = \sum_{j,k=0}^{2^v-1} ([f_j * g_k, \phi) i_j) i_k \tag{7}$$

for each $\phi \in \mathcal{B}(\mathbf{R}^n, Y)$. Particularly,

$$(f * g)(x) = f(x - y) * g(y) = f(y) * g(x - y) \tag{8}$$

for all $x, y \in \mathbf{R}^n$ due to (7), since the latter equality is satisfied for each pair f_j and g_k.

3. The Decomposition Theorem of Partial Differential Operators over the Cayley-Dickson Algebras.

We consider a partial differential operator of order u:

$$Af(x) = \sum_{|\alpha| \le u} \mathbf{a}_\alpha(x) \partial^\alpha f(x), \tag{1}$$

where $\partial^\alpha f = \partial^{|\alpha|} f(x) / \partial x_0^{\alpha_0} ... \partial x_n^{\alpha_n}$, $x = x_0 i_0 + ... x_n i_n$, $x_j \in \mathbf{R}$ for each j, $1 \le n = 2^r - 1$, $\alpha = (\alpha_0, ..., \alpha_n)$, $|\alpha| = \alpha_0 + ... + \alpha_n$, $0 \le \alpha_j \in \mathbf{Z}$. By the definition this means that the principal symbol

$$A_0 := \sum_{|\alpha| = u} \mathbf{a}_\alpha(x) \partial^\alpha \tag{2}$$

has α so that $|\alpha| = u$ and $\mathbf{a}_\alpha(x) \in \mathcal{A}_r$ is not identically zero on a domain U in $\mathcal{A}_r$. As usually $C^k(U, \mathcal{A}_r)$ denotes the space of k times continuously differentiable functions by all real variables $x_0, ..., x_n$ on U with values in $\mathcal{A}_r$, while the x-differentiability corresponds to the super-differentiability by the Cayley-Dickson variable x.

Speaking about locally constant or locally differentiable coefficients we shall undermine that a domain U is the union of sub-domains U^j satisfying conditions $2.6(D1, i - vii)$ and $U^j \cap U^k = \partial U^j \cap \partial U^k$ for each $j \neq k$. All coefficients $\mathbf{a}_\alpha$ are either constant or differentiable of the same class on each $Int(U^j)$ with the continuous extensions on U^j. More generally it is up to a C^u or x-differentiable diffeomorphism of U respectively.

If an operator A is of the odd order $u = 2s - 1$, then an operator E of the even order $u + 1 = 2s$ by variables (t, x) exists so that

$$Eg(t,x)|_{t=0} = Ag(0,x) \qquad (3)$$

for any $g \in C^{u+1}([c,d] \times U, \mathcal{A}_r)$, where $t \in [c,d] \subset \mathbf{R}$, $c \leq 0 < d$, for example, $Eg(t,x) = \partial(tAg(t,x))/\partial t$.

Therefore, it remains the case of the operator A of the even order $u = 2s$. Take $z = z_0 i_0 + ... + z_{2^v-1} i_{2^v-1} \in \mathcal{A}_v$, $z_j \in \mathbf{R}$. Operators depending on a less set $z_{l_1}, ..., z_{l_n}$ of variables can be considered as restrictions of operators by all variables on spaces of functions constant by variables z_s with $s \notin \{l_1, ..., l_n\}$.

Theorem. *Let $A = A_u$ be a partial differential operator of an even order $u = 2s$ with locally constant or variable (locally) $C^{s'}$ or x-differentiable on U coefficients $\mathbf{a}_\alpha(x) \in \mathcal{A}_r$ such that it has the form*

$$Af = c_{u,1}(B_{u,1}f) + ... + c_{u,k}(B_{u,k}f), \qquad (4)$$

where each

$$B_{u,p} = B_{u,p,0} + Q_{u-1,p} \qquad (5)$$

is a partial differential operator by variables

$$x_{m_{u,1}+...+m_{u,p-1}+1}, ..., x_{m_{u,1}+...+m_{u,p}}$$

and of the order u, $m_{u,0} = 0$, $c_{u,k}(x) \in \mathcal{A}_r$ for each k, its principal part

$$B_{u,p,0} = \sum_{|\alpha|=s} \mathbf{a}_{p,2\alpha}(x)\partial^{2\alpha} \qquad (6)$$

is elliptic with real coefficients $\mathbf{a}_{p,2\alpha}(x) \geq 0$, either $0 \leq r \leq 3$ and $f \in C^u(U, \mathcal{A}_r)$, or $r \geq 4$ and $f \in C^u(U, \mathbf{R})$. Then three partial differential operators Υ^s and Υ_1^s and Q of orders s and p with $p \leq u - 1$ with locally constant or variable (locally) $C^{s'}$ or x-differentiable correspondingly on U coefficients with values in $\mathcal{A}_v$ exist and coefficients of the third operator Q may be generalized functions, when coefficients of A are discontinuous locally constant or $C^{s'}$ discontinuous on the entire U or when $s' < s$, $r \leq v$, such that

$$Af = \Upsilon^s(\Upsilon_1^s f) + Qf. \qquad (7)$$

4. Corollary. *Let suppositions of Theorem 3 be satisfied. Then a change of variables locally affine or variable C^1 or x-differentiable on U correspondingly on U exists so that the principal part $A_{2,0}$ of A_2 becomes with constant coefficients, when $\mathbf{a}_{p,2\alpha} > 0$ for each* *p, α and x.*

5. Corollary. *If two operators $E = A_{2s}$ and $A = A_{2s-1}$ are related by Equation 3(3), and A_{2s} is presented in accordance with Formulas 3(4, 5), then three operators Υ^s, Υ^{s-1} and Q of orders s, $s-1$ and $2s-2$ exist so that*

$$A_{2s-1} = \Upsilon^s \Upsilon^{s-1} + Q. \tag{1}$$

The proof of these theorem and corollaries are contained in [42].

6. Partial Differential Operators with Generalized Coefficients.
Let an operator Q be given by Formula (i) $[Af, \omega^{\otimes(u+1)}) = [\Upsilon^s(\Upsilon^s_1 f) + Qf, \omega^{\otimes(u+1)})$ for each real-valued test function ω on a domain U. Initially it is considered as a domain in the Cayley-Dickson algebra $\mathcal{A}_v$. But in the case when Q and f depend on smaller number of real coordinates $z_0, ..., z_{n-1}$ we can take the real shadow of U and its sub-domain V of variables $(z_0, ..., z_{n-1})$, where z_k are marked for example being zero for all $n \le k \le 2^v - 1$. Thus we take a domain V which is a canonical closed subset in the Euclidean space $\mathbf{R}^n$, $2^{v-1} \le n \le 2^v - 1$, $v \ge 2$.

A canonical closed subset P of the Euclidean space $X = \mathbf{R}^n$ is called a quadrant if it can be given by the condition $P := \{x \in X : q_j(x) \ge 0\}$, where $(q_j : j \in \Lambda_P)$ are linearly independent elements of the topologically adjoint space X^*. Here $\Lambda_P \subset \mathbf{N}$ (with $card(\Lambda_P) = k \le n$) and k is called the index of P. If $x \in P$ and exactly j of the q_i's satisfy $q_i(x) = 0$ then x is called a corner of index j.

That is P is affine diffeomorphic with $P^n = \prod_{j=1}^n [\mathsf{a}_j, b_j]$, where $-\infty \le \mathsf{a}_j < b_j \le \infty$, $[\mathsf{a}_j, b_j] := \{x \in \mathbf{R} : \mathsf{a}_j \le x \le b_j\}$ denotes the segment in $\mathbf{R}$. This means that there exists a vector $p \in \mathbf{R}^n$ and a linear invertible mapping C on $\mathbf{R}^n$ so that $C(P) - p = P^n$. We put $t^{j,1} := (t_1, ..., t_j, ..., t_n : t_j = \mathsf{a}_j)$, $t^{j,2} := (t_1, ..., t_j, ..., t_n : t_j = b_j)$. Consider $t = (t_1, ..., t_n) \in P^n$.

This permits to define a manifold M with corners. It is a metric separable space modelled on $X = \mathbf{R}^n$ and is supposed to be of class C^s, $1 \le s$. Charts on M are denoted (U_l, u_l, P_l), that is, $u_l : U_l \to u_l(U_l) \subset P_l$ is a C^s-diffeomorphism for each l, U_l is open in M, $u_l \circ {u_j}^{-1}$ is of C^s class of smoothness from the domain $u_j(U_l \cap U_j) \ne \emptyset$ onto $u_l(U_l \cap U_j)$, that is, $u_j \circ u_l^{-1}$ and $u_l \circ u_j^{-1}$ are bijective, $\bigcup_j U_j = M$.

A point $x \in M$ is called a corner of index j if there exists a chart (U, u, P) of M with $x \in U$ and $u(x)$ is of index $ind_M(x) = j$ in $u(U) \subset P$. A set of all corners of index $j \ge 1$ is called a border ∂M of M, x is called an inner point of M if $ind_M(x) = 0$, so $\partial M = \bigcup_{j \ge 1} \partial^j M$, where $\partial^j M := \{x \in M : ind_M(x) = j\}$ (see also [46]). We consider that

$(D1)$ V is a canonical closed subset in the Euclidean space $\mathbf{R}^n$, that is $V = cl(Int(V))$, where $Int(V)$ denotes the interior of V and $cl(V)$ denotes the closure of V.

Particularly, the entire space $\mathbf{R}^n$ may also be taken.

Let a manifold W be satisfying the following conditions $(i - v)$.

(i). The manifold W is continuous and piecewise C^α, where C^l denotes the family of l times continuously differentiable functions. This means by the definition that W as the manifold is of class $C^0 \cap C^\alpha_{loc}$. That is W is of class C^α on open subsets $W_{0,j}$ in W and $W \setminus (\bigcup_j W_{0,j})$ has a codimension not less than one in W.

(ii). $W = \bigcup_{j=0}^m W_j$, where $W_0 = \bigcup_k W_{0,k}$, $W_j \cap W_k = \emptyset$ for each $k \neq j$, $m = dim_{\mathbf{R}} W$, $dim_{\mathbf{R}} W_j = m - j$, $W_{j+1} \subset \partial W_j$.

(iii). Each W_j with $j = 0, ..., m-1$ is an oriented C^α-manifold, W_j is open in $\bigcup_{k=j}^m W_k$. An orientation of W_{j+1} is consistent with that of ∂W_j for each $j = 0, 1, ..., m-2$. For $j > 0$ the set W_j is allowed to be void or non-void.

(iv). A sequence W^k of C^α orientable manifolds embedded into $\mathbf{R}^n$, $\alpha \geq 1$, exists such that W^k uniformly converges to W on each compact subset in $\mathbf{R}^n$ relative to the metric $dist$.

For two subsets B and E in a metric space X with a metric ρ we put

$$dist(B, E) := \max\{\sup_{b \in B} dist(\{b\}, E), \sup_{e \in E} dist(B, \{e\})\}, \tag{1}$$

where

$$dist(\{b\}, E) := \inf_{e \in E} \rho(b, e), dist(B, \{e\}) := \inf_{b \in B} \rho(b, e), b \in B, e \in E.$$

Generally, $dim_{\mathbf{R}} W = m \leq n$. Let $(e_1^k(x), ..., e_m^k(x))$ be a basis in the tangent space $T_x W^k$ at $x \in W^k$ consistent with the orientation of W^k, $k \in \mathbf{N}$.

We suppose that the sequence of orientation frames $(e_1^k(x_k), ..., e_m^k(x_k))$ of W^k at x_k converges to $(e_1(x), ..., e_m(x))$ for each $x \in W_0$, where $\lim_k x_k = x \in W_0$, while $e_1(x)$,...,$e_m(x)$ are linearly independent vectors in $\mathbf{R}^n$.

(v). Let a sequence of Riemann volume elements λ_k on W^k (see §XIII.2 [58]) induce a limit volume element λ on W, that is, $\lambda(B \cap W) = \lim_{k \to \infty}(B \cap W^k)$ for each compact canonical closed subset B in $\mathbf{R}^n$, consequently, $\lambda(W \setminus W_0) = 0$.

(vi). We shall consider surface integrals of the second kind, i.e. by the oriented surface W (see (iv)), where each W_j, $j = 0, ..., m-1$ is oriented (see also §XIII.2.5 [58]).

Suppose that a boundary ∂U of U satisfies Conditions $(i - v)$ and

(vii) let the orientations of ∂U^k and U^k be consistent for each $k \in \mathbf{N}$ (see Proposition 2 and Definition 3 [58]).

Particularly, the Riemann volume element λ_k on ∂U^k is consistent with the Lebesgue measure on U^k induced from $\mathbf{R}^n$ for each k. This induces the measure λ on ∂U as in (v). This consideration certainly encompasses the case of a domain U with a C^α boundary ∂U as well.

Suppose that U_1,...,U_l are domains in $\mathbf{R}^n$ satisfying conditions $(D1, i - vii)$ and such that $U_j \cap U_k = \partial U_j \cap \partial U_k$ for each $j \neq k$, $U = \bigcup_{j=1}^l U_j$. Consider a function $g : U \to \mathcal{A}_v$ such that each its restriction $g|_{U_j}$ is of class C^s, but g on the entire domain U may be discontinuous or not C^k, where $0 \leq k \leq s$, $1 \leq s$. If $x \in \partial U_j \cap \partial U_k$ for some $j \neq k$, $x \in Int(U)$, such that x is of index $m \geq 1$ in U_j (and in U_k also). Then there exists canonical C^α local coordinates $(y_1, ..., y_n)$ in a neighborhood V_x of x in U such that $S := V_x \cap \partial^m U_j = \{y : y \in V_x; \ y_1 = 0, ..., y_m = 0\}$. Using locally finite coverings of the locally compact topological space $\partial U_j \cap \partial U_k$ without loss of generality we suppose that C^α functions $P_1(z), ..., P_m(z)$ on $\mathbf{R}^n$ exist with $S = \{z : z \in \mathbf{R}^n; P_1(z) = 0, ..., P_m(z) = 0\}$. Therefore, on the surface S the delta-function $\delta(P_1, ..., P_m)$ exists, for $m = 1$ denoting them $P = P_1$ and $\delta(P)$ respectively (see §III.1 [9]). It is possible to choose $y_j = P_j$ for $j = 1, ..., m$. Using generalized functions with definite supports, for example compact supports, it is possible without loss of generality consider that $y_1, ..., y_n \in \mathbf{R}$ are real variables. Let $\theta(P_j)$ be the characteristic function of the domain $\mathcal{P}_j := \{z : P_j(z) \geq 0\}$, $\theta(P_j) := 1$ for $P_j \geq 0$ and $\theta(P_j) = 0$ for $P_j < 0$. Then the generalized function $\theta(P_1, ..., P_m) := \theta(P_1)...\theta(P_m)$ can be considered as the direct product of generalized functions $\theta(y_1)$,...,$\theta(y_m)$, $1(y_{m+1}, ..., y_n) \equiv 1$, since variables $y_1, ..., y_n$ are independent. Then in the class of generalized functions the following formulas are valid:

$$\partial\theta(P_j)/\partial z_k = \delta(P_j)(\partial P_j/\partial z_k) \tag{2}$$

for each $k = 1, ..., n$, consequently,

$$grad[\theta(P_1, ..., P_m)] = \sum_{j=1}^{m} [\theta(P_1)...\theta(P_{j-1})\delta(P_j)(grad(P_j))\theta(P_{j+1})...\theta(P_m)], \tag{3}$$

where, $grad\ g(z) := (\partial g(z)/\partial z_1, ..., \partial g(z)/\partial z_n)$ (see Formulas III.1.3$(1, 7, 7', 9)$ and III.1.9(6) [9]).

Let for the domain U in the Euclidean space $\mathbf{R^n}$ the set of internal surfaces $cl_U[Int_{\mathbf{R}^n}(U) \cap \bigcup_{j \neq k}(\partial U_j \cap \partial U_k)]$ in U on which a function $g : U \to \mathcal{A}_v$ or its derivatives may be discontinuous is presented as the disjoint union of surfaces Γ_j, where each surface Γ^j is the boundary of the sub-domain

$$\mathcal{P}^j := \{P_{j,1}(z) \geq 0, ..., P_{j,m_j}(z) \geq 0\}, \Gamma^j = \partial\mathcal{P}^j = \bigcup_{k=1}^{m_j} \partial^k \mathcal{P}^j,]eqno(4)$$

$m_j \in \mathbf{N}$ for each j, $cl_X(V)$ denotes the closure of a subset V in a topological space X, $Int_X(V)$ denotes the interior of V in X. By its construction $\{\mathcal{P}^j : j\}$ is the covering of U which is the refinement of the covering $\{U_k : k\}$, i.e. for each $\mathcal{P}^j$ a number k exists so that $\mathcal{P}^j \subset U_k$ and $\partial\mathcal{P}^j \subset \partial U_k$ and $\bigcup_j \mathcal{P}^j = \bigcup_k U_k = U$. We put

$$h_j(z(x)) = h(x)|_{x \in \Gamma^j} := \lim_{y_{j,1}\downarrow 0, ..., y_{j,n}\downarrow 0} g(z(x + y)) - \lim_{y_{j,1}\downarrow 0, ..., y_{j,n}\downarrow 0} g(z(x - y)) \tag{5}$$

in accordance with the supposition made above that g can have only discontinuous of the first kind, i.e. the latter two limits exist on each Γ^j, where x and y are written in coordinates in $\mathcal{P}^j$, $z = z(x)$ denotes the same point in the global coordinates z of the Euclidean space $\mathbf{R}^n$. We take new continuous function

$$g^1(z) = g(z) - \sum_j h_j(z)\theta(P_{j,1}(z), ..., P_{j,m_j}(z)). \tag{6}$$

Let the partial derivatives and the gradient of the function g^1 be calculated piecewise one each U_k. Since g^1 is the continuous function, it is the regular generalized function by the definition, consequently, its partial derivatives exist as the generalized functions. If $g^1|_{U_j} \in C^1(U_j, \mathcal{A}_v)$, then $\partial g^1(z)/\partial z_k$ is the continuous function on U_j, i.e. in this case $\partial g^1(z)\chi_{U_j}(z)/\partial z_k$ is the regular generalized function on U_j for each k, where $\chi_G(z)$ denotes the characteristic function of a subset G in $\mathcal{A}_v$, $\chi_G(z) = 1$ for each $z \in G$, while $\chi(z) = 0$ for $z \in \mathcal{A}_v \setminus G$. Therefore, $g^1(z) = \sum_j g^1(z)\chi_{U_j \setminus \bigcup_{k<j} U_k}(z)$, where $U_0 = \emptyset$, $j, k \in \mathbf{N}$.

On the other hand, the function $g(z)$ is locally continuous on U, consequently, it defines the regular generalized function on the space $\mathcal{D}(U, \mathcal{A}_v)$ of test functions as

$$[g, \omega) := \int_U g(z)\omega(z)\lambda(dz),$$

where λ is the Lebesgue measure on U induced by the Lebesgue measure on the real shadow $\mathbf{R}^{2^v}$ of the Cayley-Dickson algebra $\mathcal{A}_v$, $\omega \in \mathcal{D}(U, \mathcal{A}_v)$. Thus partial derivatives of g exist as generalized functions.

In accordance with Formulas $(2, 3, 6)$ we infer that the gradient of the function $g(z)$ on the domain U is the following:

$$grad\ g(z) = grad\ g^1(z) + \sum_j h_j(z) grad\ \theta(P_{j,1}, ..., P_{j,m_j}). \tag{7}$$

Thus Formulas $(3, 7)$ permit calculations of coefficients of the partial differential operator Q given by Formula $3(7)$.

7. Line Generalized Functions.

Let U be a domain satisfying Conditions $2.2.1(D1, D2)$ and $6(D1, i - vii)$. We embed the Euclidean space $\mathbf{R}^n$ into the Cayley-Dickson algebra $\mathcal{A}_v$, $2^{v-1} \le n \le 2^v - 1$, as the $\mathbf{R}$ affine sub-space putting $\mathbf{R}^n \ni x = (x_1, ..., x_n) \mapsto x_1 i_{j_1} + ... + x_n i_{j_n} + x^0 \in \mathcal{A}_v$, where $j_k \ne j_l$ for each $k \ne l$, x^0 is a marked Cayley-Dickson number, for example, $j_k = k$ for each k, $x^0 = 0$. Moreover, each z_j can be written in the z-representation in accordance with Formulas $2.2.1(1 - 3)$.

We denote by $\mathbf{P} = \mathbf{P}(U)$ the family of all rectifiable paths $\gamma : [a_\gamma, b_\gamma] \to U$ supplied with the metric

$$\rho(\gamma, \omega) := |\gamma(a) - \omega(a_\omega)| + \inf_\phi V_a^b(\gamma(t)) - \omega(\phi(t)) \tag{1}$$

where the infimum is taken by all diffeomorphisms $\phi : [a_\gamma, b_\gamma] \to [a_\omega, b_\omega]$ so that $\phi(a_\gamma) = a_\omega$, $a = a_\gamma < b_\gamma = b$.

Let us introduce a continuous mapping $g : \mathcal{B}(U, \mathcal{A}_v) \times \mathbf{P}(U) \times \mathcal{V}(U, \mathcal{A}_v) \to Y$ such that its values are denoted by $[g; \omega, \gamma; \nu]$, where Y is a module over the Cayley-Dickson algebra $\mathcal{A}_v$, $\omega \in \mathcal{B}(U, \mathcal{A}_v)$, $\gamma \in \mathbf{P}(U)$, $\mathcal{V}(U, \mathcal{A}_v)$ denotes the family of all functions on U with values in the Cayley-Dickson algebra of bounded variation (see §2.3 [33]), $\nu \in \mathcal{V}(U, \mathcal{A}_v)$. For the identity mapping $\nu(z) = id(z) = z$ values of this functional will be denoted shortly by $[g; \omega, \gamma]$. Suppose that this mapping g satisfies the following properties $(G1 - G5)$:

$(G1)$ $[g;\omega,\gamma;\nu]$ is bi- $\mathbf{R}$ homogeneous and $\mathcal{A}_v$ additive by a test function ω and by a function of bounded variation ν;

$(G2)$ $[g;\omega,\gamma;\nu]=[g;\omega,\gamma^1;\nu]+[g;\omega,\gamma^2;\nu]$ for each γ,γ^1 and $\gamma^2\in\mathbf{P}(U)$ such that $\gamma(t)=\gamma^1(t)$ for all $t\in[a_{\gamma^1},b_{\gamma^1}]$ and $\gamma(t)=\gamma^2(t)$ for any $t\in[a_{\gamma^2},b_{\gamma^2}]$ and $a_{\gamma^1}=a_\gamma$ and $a_{\gamma^2}=b_{\gamma^1}$ and $b_\gamma=b_{\gamma^2}$.

$(G3)$ If a rectifiable curve γ does not intersect a support of a test function ω or a function of bounded variation ν, $\gamma([a,b]\cap(supp(\omega)\cap supp(\nu))=\emptyset$, then $[g;\omega,\gamma;\nu]=0$, where $supp(\omega):=cl\{z\in U:\ \omega(z)\neq 0\}$.

Further we put

$(G4)$ $[\partial^{|m|}g(z)/\partial z_0^{m_0}...\partial z_{2^v-1}^{m_{2^v-1}};\omega,\gamma]:=(-1)^{|m|}[g;\partial^{|m|}\omega(z)/\partial z_0^{m_0}...\partial z_{2^v-1}^{m_{2^v-1}},\gamma]$
for each $m=(m_0,...,m_{2^v-1})$, m_j is a non-negative integer $0\le m_j\in\mathbf{Z}$ for each j, $|m|:=m_0+...+m_{2^v-1}$. In the case of a super-differentiable function ω and a generalized function g we also put

$(G5)$ $[(d^kg(z)/dz^k).(h_1,...,h_k);\omega,\gamma]:=(-1)^k[g;(d^k\omega(z)/dz^k).(h_1,...,h_k),\gamma]$
for any natural number $k\in\mathbf{N}$ and Cayley-Dickson numbers $h_1,..,h_k\in\mathcal{A}_v$.

Then g is called the Y valued line generalized function on $\mathcal{B}(U,\mathcal{A}_v)\times\mathbf{P}(U)\times\mathcal{V}(U,\mathcal{A}_v)$. Analogously it can be defined on $\mathcal{D}(U,\mathcal{A}_v)\times\mathbf{P}(U)\times\mathcal{V}(U,\mathcal{A}_v)$. If $Y=\mathcal{A}_v$ we call it simply the line generalized function, while for $Y=L_q(\mathcal{A}_v^k,\mathcal{A}_v^l)$ we call it the line generalized operator valued function, $k,l\ge 1$, omitting "on $\mathcal{B}(U,\mathcal{A}_v)\times\mathbf{P}(U)\times\mathcal{V}(U,\mathcal{A}_v)$" or "line" for short, when it is specified. Their spaces we denote by $L_q(\mathcal{B}(U,\mathcal{A}_v)\times\mathbf{P}(U)\times\mathcal{V}(U,\mathcal{A}_v);Y)$. Thus if g is a generalized function, then Formula $(G5)$ defines the operator valued generalized function $d^kg(z)/dz^k$ with $k\in\mathbf{N}$ and $l=1$.

If g is a continuous function on U, then the formula

$(G6)$ $[g;\omega,\gamma;\nu]=\int_\gamma g(y)\omega(y)d\nu(y)$

defines the generalized function. If $\hat{f}(z)$ is a continuous $L_q(\mathcal{A}_v,\mathcal{A}_v)$ valued function on U, then it defines the generalized operator valued function with $Y=L_q(\mathcal{A}_v,\mathcal{A}_v)$ such that

$(G7)$ $[\hat{f};\omega,\gamma;\nu]=\int_\gamma\{\hat{f}(z).\omega(z)\}d\nu(z)$.

Particularly, for $\nu=id$ we certainly have $d\nu(z)=dz$.

We consider on $L_q(\mathcal{B}(U,\mathcal{A}_v)\times\mathbf{P}(U)\times\mathcal{V}(U,\mathcal{A}_v);Y)$ the strong topology:

$(G8)$ $\lim_l f^l=f$ means that for each marked test function $\omega\in\mathcal{B}(U,\mathcal{A}_v)$ and rectifiable path $\gamma\in\mathbf{P}(U)$ and function of bounded variation $\nu\in\mathcal{V}(U,\mathcal{A}_v)$ the limit relative to the norm in Y exists

$$\lim_l[f^l;\omega,\gamma;\nu]=[f;\omega,\gamma;\nu].$$

8. Poly-functionals. Let $\mathbf{a}_k:\mathcal{B}(U,\mathcal{A}_r)^k\to\mathcal{A}_r$ or $\mathbf{a}_k:\mathcal{D}(U,\mathcal{A}_r)^k\to\mathcal{A}_r$ be a continuous mapping satisfying the following three conditions:

$(P1)$ $[\mathbf{a}_k,\omega^1\otimes...\otimes\omega^k)$ is $\mathbf{R}$ homogeneous

$$[\mathbf{a}_k,\omega^1\otimes...\otimes(\omega^lt)\otimes...\otimes\omega^k)=[\mathbf{a}_k,\omega^1\otimes...\otimes\omega^l\otimes...\otimes\omega^k)t=[\mathbf{a}_kt,\omega^1\otimes...\otimes\omega^k)$$

for each $t\in\mathbf{R}$ and $\mathcal{A}_r$ additive

$[\mathbf{a}_k,\omega^1\otimes...\otimes(\omega^l+\kappa^l)\otimes...\otimes\omega^k)=[\mathbf{a}_k,\omega^1\otimes...\otimes\omega^l\otimes...\otimes\omega^k)+[\mathbf{a}_k,\omega^1\otimes...\otimes\kappa^l\otimes...\otimes\omega^k)$
by any $\mathcal{A}_r$ valued test functions ω^l and κ^l, when other are marked, $l=1,...,k$, i.e. it is k $\mathbf{R}$ linear and k $\mathcal{A}_r$ additive, where $[\mathbf{a}_k,\omega^1\otimes...\otimes\omega^k)$ denotes a value of $\mathbf{a}_k$ on given test $\mathcal{A}_r$ valued functions $\omega^1,...,\omega^k$;

$(P2)$ $[\mathbf{a}_k\alpha, \omega^1 \otimes ... \otimes (\omega^l\beta) \otimes ... \otimes \omega^k) = ([\mathbf{a}_k, \omega^1 \otimes ... \otimes \omega^l \otimes ... \otimes \omega^k)\alpha)\beta = [(\mathbf{a}_k\alpha)\beta, \omega^1 \otimes ... \otimes \omega^l \otimes ... \otimes \omega^k)$ for all real-valued test functions and $\alpha, \beta \in \mathcal{A}_r$;

$(P3)$ $[\mathbf{a}_k, \omega^{\sigma(1)} \otimes ... \otimes \omega^{\sigma(k)}) = [\mathbf{a}_k, \omega^1 \otimes ... \otimes \omega^k)$ for all real-valued test functions and each transposition σ, i.e. bijective surjective mapping $\sigma : \{1, ..., k\} \to \{1, ..., k\}$.

Then $\mathbf{a}_k$ will be called the symmetric k $\mathbf{R}$ linear k $\mathcal{A}_r$ additive continuous functional, $1 \leq k \in \mathbf{Z}$. The family of all such symmetric functionals is denoted by $\mathcal{B}'_{k,s}(U, \mathcal{A}_v)$ or $\mathcal{D}'_{k,s}(U, \mathcal{A}_r)$ correspondingly. A functional satisfying Conditions $(P1, P2)$ is called a continuous k-functional over $\mathcal{A}_r$ and their family is denoted by $\mathcal{B}'_k(U, \mathcal{A}_r)$ or $\mathcal{D}'_k(U, \mathcal{A}_r)$. When a situation is outlined we may omit for short "continuous" or "k $\mathbf{R}$ linear k $\mathcal{A}_v$ additive".

The sum of two k-functionals over the Cayley-Dickson algebra $\mathcal{A}_r$ is prescribed by the equality:

$(P4)$ $[\mathbf{a}_k + \mathbf{b}_k, \omega^1 \otimes ... \otimes \omega^k) = [\mathbf{a}_k, \omega^1 \otimes ... \otimes \omega^k) + [\mathbf{b}_k, \omega^1 \otimes ... \otimes \omega^k)$
for each test functions. Using Formula $(P4)$ each k-functional can be written as

$$[\mathbf{a}_k, \omega^1 \otimes ... \otimes \omega^k) = [\mathbf{a}_{k,0}i_0, \omega^1 \otimes ... \otimes \omega^k) + ... + [\mathbf{a}_{k,2^r-1}i_{2^r-1}, \omega^1 \otimes ... \otimes \omega^k),$$

where $[\mathbf{a}_{k,j}, \omega^1 \otimes ... \otimes \omega^k) \in \mathbf{R}$ is real for all real-valued test functions $\omega^1, ..., \omega^k$ and each j; $i_0,...,i_{2^r-1}$ denote the standard generators of the Cayley-Dickson algebra $\mathcal{A}_r$.

The direct product $\mathbf{a}_k \otimes \mathbf{b}_p$ of two functionals $\mathbf{a}_k$ and $\mathbf{b}_p$ for the same space of test functions is a $k+p$-functional over $\mathcal{A}_r$ given by the following three conditions:

$(P5)$ $[\mathbf{a}_k \otimes \mathbf{b}_p, \omega^1 \otimes ... \otimes \omega^{k+p}) = [\mathbf{a}_k, \omega^1 \otimes ... \otimes \omega^k)[\mathbf{b}_p, \omega^{k+1} \otimes ... \otimes \omega^{k+p})$
for any real-valued test functions $\omega^1, ..., \omega^{k+p}$;

$(P6)$ if $[\mathbf{b}_p, \omega^{k+1} \otimes ... \otimes \omega^{k+p}) \in \mathbf{R}$ is real for any real-valued test functions, then
$[(\mathbf{a}_k N_1) \otimes (\mathbf{b}_p N_2), \omega^1 \otimes ... \otimes \omega^{k+p}) = ([\mathbf{a}_k \otimes \mathbf{b}_p, \omega^1 \otimes ... \otimes \omega^{k+p})N_1)N_2$
for any real-valued test functions $\omega^1, ..., \omega^{k+p}$ and Cayley-Dickson numbers $N_1, N_2 \in \mathcal{A}_r$;

$(P7)$ if $[\mathbf{a}_k, \omega^1 \otimes ... \otimes \omega^k) \in \mathbf{R}$ and $[\mathbf{b}_p, \omega^{k+1} \otimes ... \otimes \omega^{k+p}) \in \mathbf{R}$ are real for any real-valued test functions, then

$$[\mathbf{a}_k \otimes \mathbf{b}_p, \omega^1 \otimes ... \otimes (\omega^l N_1) \otimes ... \otimes \omega^{k+p}) = [\mathbf{a}_k \otimes \mathbf{b}_p, \omega^1 \otimes ... \otimes \omega^{k+p})N_1$$

for any real-valued test functions $\omega^1, ..., \omega^{k+p}$ and each Cayley-Dickson number $N_1 \in \mathcal{A}_r$ for each $l = 1, ..., k+p$.

Therefore, we can now consider a partial differential operator of order u acting on a generalized function $f \in \mathcal{B}'(U, \mathcal{A}_r)$ or $f \in \mathcal{D}'(U, \mathcal{A}_r)$ and with generalized coefficients either $\mathbf{a}_\alpha \in \mathcal{B}'_{|\alpha|}(U, \mathcal{A}_r)$ or all $\mathbf{a}_\alpha \in \mathcal{D}'_{|\alpha|}(U, \mathcal{A}_r)$ correspondingly:

$$Af(x) = \sum_{|\alpha| \leq u} (\partial^\alpha f(x)) \otimes [(\mathbf{a}_\alpha(x)) \otimes 1^{\otimes(u-|\alpha|)}], \tag{1}$$

where $\partial^\alpha f = \partial^{|\alpha|} f(x)/\partial x_0^{\alpha_0}...\partial x_n^{\alpha_n}$, $x = x_0 i_0 + ... x_n i_n$, $x_j \in \mathbf{R}$ for each j, $1 \leq n = 2^r - 1$, $\alpha = (\alpha_0, ..., \alpha_n)$, $|\alpha| = \alpha_0 + ... + \alpha_n$, $0 \leq \alpha_j \in \mathbf{Z}$, $[1, \omega) := \int_U \omega(y)\lambda(dy)$, λ denotes the Lebesgue measure on U, for convenience $1^{\otimes 0}$ means the multiplication on the unit $1 \in \mathbf{R}$. The partial differential equation

$$Af = g \tag{2}$$

in terms of generalized functions has a solution f means by the definition that

$$[Af, \omega^{\otimes(u+1)}) = [g, \omega^{\otimes(u+1)}) \tag{3}$$

for each real-valued test function ω on U, where $\omega^{\otimes k} = \omega \otimes ... \otimes \omega$ denotes the k times direct product of a test functions ω.

9. Theorem. *Let $A = A_u$ be a partial differential operator with generalized over the Cayley-Dickson algebra $\mathcal{A}_r$ coefficients of an even order $u = 2s$ on U such that each $\mathbf{a}_\alpha$ is symmetric for $|\alpha| = u$ and A has the form*

$$Af = (B_{u,1}f)c_{u,1} + ... + (B_{u,k}f)c_{u,k}, \tag{4}$$

where each

$$B_{u,p} = B_{u,p,0} + Q_{u-1,p} \tag{5}$$

is a partial differential operator by variables $x_{m_{u,1}+...+m_{u,p-1}+1},...,x_{m_{u,1}+...+m_{u,p}}$ and of the order u, $m_{u,0} = 0$, $c_{u,k}(x) \in \mathcal{A}_r$ for each k, its principal part

$$B_{u,p,0}f = \sum_{|\alpha|=s} (\partial^{2\alpha} f) \otimes \mathbf{a}_{p,2\alpha}(x) \tag{6}$$

is elliptic, i.e. $\sum_{|\alpha|=s} y^{2\alpha}[\mathbf{a}_{p,2\alpha}, \omega^{\otimes 2s}) \geq 0$ for all $y_{k(1)},...,y_{k(m_{u,p})}$ in $\mathbf{R}$ with $k(1) = m_{u,1} + ... + m_{u,p-1} + 1,...,k(m_{u,p}) = m_{u,1} + ... + m_{u,p}$, $y^\beta := y_{k(1)}^{\beta_{k_1}} ... y_{k(m_{u,p})}^{\beta_{k(m_{u,p})}}$ and $[\mathbf{a}_{p,2\alpha}, \omega^{\otimes 2s}) \geq 0$ for each real test function ω, either $0 \leq r \leq 3$ and f is with values in $\mathcal{A}_r$, or $r \geq 4$ and f is real-valued on real-valued test functions. Then three partial differential operators Υ^s and Υ_1^s and Q of orders s and p with $p \leq u - 1$ with generalized on U coefficients with values in $\mathcal{A}_v$ exist such that

$$[Af, \omega^{\otimes(u+1)}) = [\Upsilon^s(\Upsilon_1^s f) + Qf, \omega^{\otimes(u+1)}) \tag{7}$$

for each real-valued test function ω on U.

10. Corollary. *If $Af = \sum_{j,k}(\partial^2 f(z)/\partial z_k \partial z_j) \otimes a_{j,k}(z) + \sum_j (\partial f(z)/\partial z_j) \otimes b_j(z) \otimes 1 + f(z) \otimes \eta(z) \otimes 1$ is a second order partial differential operator with generalized coefficients in $\mathcal{B}'(U, \mathcal{A}_r)$ or $\mathcal{D}'(U, \mathcal{A}_r)$, where each $a_{j,k}$ is symmetric, f and $\mathcal{A}_r$ are as in §8, then three partial differential operators $\Upsilon + \beta$, $\Upsilon_1 + \beta_1$ and Q of the first order with generalized coefficients with values in $\mathcal{A}_v$ for suitable $v \geq r$ of the same class exist such that*

$$[Af, \omega^{\otimes 3}) = [(\Upsilon + \beta)((\Upsilon_1 + \beta_1)f + Qf), \omega^{\otimes 3}) \tag{1}$$

for each real-valued test function ω on U.

The proofs of Theorem 9 and Corollary 10 are presented in [42].

3.3. Dirac Operators with Generalized Coefficients and PDE

1. Line Integration of Generalized Functions.
Let $C^m_{ab}(V,\mathcal{A}_v)$ denote the $\mathbf{R}$ linear space and right and left $\mathcal{A}_v$ module of all functions $\gamma : V \to \mathcal{A}_v$ such that $\gamma(z)$ and each its derivative $\partial^{|k|}g(z)/\partial z_1^{m_1}...\partial z_n^{m_n}$ for $1 \le |k| \le m$ is absolutely continuous on V (see §§3 [33] and 3.2.7 above). This definition means that $C^{m+1}(V,\mathcal{A}_v) \subset C^m_{ab}(V,\mathcal{A}_v)$, where $C^m(V,\mathcal{A}_v)$ denotes the family of all m times continuously differentiable functions on a domain V either open or canonical closed in $\mathbf{R}^n$, which may be a a real shadow of U as well.

2. Lemma. *Let $\gamma \in C^m_{ab}([a,b],\mathcal{A}_v) \cap \mathbf{P}(U)$ and $\omega \in \mathcal{B}(U,\mathcal{A}_v)$ and $\nu \in C^0_{ab}(U,\mathcal{A}_v)$ for $m = 0$ or $\nu = id$ for $m \ge 1$, where $0 \le m \in \mathbf{Z}$, then a line generalized function $[g;\omega,\gamma|_{[a,x]};\nu]$ is continuous for $m = 0$ or of class C^m by the parameter $x \in [a,b]$ for $m \ge 1$.*

3. Lemma. *If γ is a rectifiable path, then a line generalized function $[g;\omega,\gamma|_{[a,x]}]$ is of bounded variation by the parameter $x \in [a,b]$.*

4. Definition. Let f and η be two line generalized functions on $\mathcal{B}(U,\mathcal{A}_v) \times \mathbf{P}(U) \times \mathcal{V}(U,\mathcal{A}_v)$. We define a line functional with values denoted by

$$[\int_\gamma f d\eta, \omega^1 \otimes \omega) := [\hat{f};\omega^1,\gamma;[\eta;\omega,\kappa]]|_{\kappa=\gamma} = [\hat{f};\omega^1,*;[\eta;\omega,*]](\gamma),$$

where $\gamma \in \mathbf{P}(U)$ is a rectifiable path in U, $\omega,\omega^1 \in \mathcal{B}(U,\mathcal{A}_v)$ are any test functions. The functional $\int_\gamma f d\eta$ is called the non-commutative line integral over the Cayley-Dickson algebra $\mathcal{A}_v$ of line generalized functions f by η. Quite analogously such integral is defined for line generalized functions f and η on $\mathcal{D}(U,\mathcal{A}_v) \times \mathbf{P}(U) \times \mathcal{V}(U,\mathcal{A}_v)$.

5. Theorem. *Let F and Ξ be two generalized functions on U, $F, \Xi \in \mathcal{B}'(U,\mathcal{A}_v)$ or $F, \Xi \in \mathcal{D}'(U,\mathcal{A}_v)$, then the the non-commutative line integral over the Cayley-Dickson algebra $\mathcal{A}_v$ of line generalized functions f by ξ exists, where f is induced by F and ξ by Ξ.*

6. Corollary. *If $F : U \to \mathcal{A}_v$ is a continuous function on U and Ξ is a generalized function on U, then the non-commutative line integral over the Cayley-Dickson algebra $\mathcal{A}_v$ of line generalized functions f by ξ*

$$[\int_\gamma f d\xi, \omega^1 \otimes \omega) \qquad (1)$$

exists, where f is induced by F and ξ by Ξ.

7. Corollary. *If F is a continuous function on U and Ξ is a function of bounded variation on U, then the non-commutative line integral over the Cayley-Dickson algebra $\mathcal{A}_v$ of line generalized functions f by ξ*

$$[\int_\gamma f d\xi, \omega^1 \otimes \omega) \qquad (1)$$

exists, where f is induced by F and ξ by Ξ. Moreover, this integral coincides with the non-commutative line integral from [33] §3 for the unit test functions $\omega(z) = \omega^1(z) = 1$ for each $z \in U$:

$$[\int_\gamma f d\xi, 1 \otimes 1) = \int_\gamma f d\xi. \qquad (2)$$

8. Convolution Formula for Solutions of Partial Differential Equations.
Using convolutions of generalized functions a solution of the equation

$$(\Upsilon^s + \beta) f = g \qquad (C1)$$

in $\mathcal{B}(\mathbf{R}^n, Y)$ or in the space $\mathcal{B}'(\mathbf{R}^n, Y)_l$ is:

$$f = \mathcal{E}_{\Upsilon^s + \beta} * g, \qquad (C2)$$

where $\mathcal{E}_{\Upsilon^s + \beta}$ denotes a fundamental solution of the equation

$$(\Upsilon^s + \beta)\mathcal{E}_{\Upsilon + \beta} = \delta, \qquad (C3)$$

$(\delta, \phi) = \phi(0)$. The fundamental solution of the equation

$$A_0 \mathcal{V} = \delta \qquad (C4)$$

with $A_0 = (\Upsilon^s + \beta)(\Upsilon_1^{s_1} + \beta_1)$ can be written as the convolution

$$\mathcal{V} =: \mathcal{V}_{A_0} = \mathcal{E}_{\Upsilon^s + \beta} * \mathcal{E}_{\Upsilon_1^{s_1} + \beta_1}. \qquad (C5)$$

In view of Formulas [33] $4(7-9)$ each generalized function $\mathcal{E}_{\Upsilon^s + \beta}$ can also be found from the elliptic partial differential equation

$$\Xi_\beta \Psi_{\Upsilon^s + \beta} = \delta \qquad (C6)$$

by the formula:

$$\mathcal{E}_{\Upsilon^s + \beta} = [(\Upsilon^s + \beta)^*] \Psi_{\Upsilon^s + \beta}, \qquad (C7)$$

where

$$\Xi_\beta := (\Upsilon^s + \beta)(\Upsilon^s + \beta)^* \qquad (C8)$$

(see §33 [34]).

9. Anti-derivatives of First Order Partial Differential Operators with Generalized Coefficients.

Theorem. *Let Υ be a first order partial differential operator given by the formula*

$$\Upsilon f = \sum_{j=0}^{n} (\partial f / \partial z_j) \otimes [i_j^* \psi_j(z)] \qquad (1)$$

or

$$\Upsilon f = \sum_{j=0}^{n} (\partial f / \partial z_j) \otimes \phi_j^*(z), \qquad (2)$$

where $supp(\psi_j(z)) = U$ or $supp(\phi_j(z)) = U$ for each j respectively, f and $\psi_j(z)$ or $\psi_j(z)$ are $\mathcal{A}_v$-valued generalized functions in $\mathcal{B}'(U,\mathcal{A}_r)$ or $\mathcal{D}'(U,\mathcal{A}_r)$ on the domain U satisfying Conditions 2.2.1$(D1, D2)$, $alg_{\mathbf{R}}\{[\phi_j,\omega),[\phi_k,\omega),[\phi_l,\omega)\}$ is alternative for all $0 \le j,k,l \le 2^v - 1$ and $alg_{\mathbf{R}}\{[\phi_0,\omega),...,[\phi_{2^v-1},\omega)\} \subset \mathcal{A}_v$ for each real-valued test function ω on U. Then its anti-derivative operator $\mathcal{I}_\Upsilon$ exists such that $\Upsilon\mathcal{I}_\Upsilon f = f$ for each continuous generalized function $f : U \to \mathcal{A}_v$ and it has an expression through line integrals of generalized functions.

The proofs of lemmas 2, 3, theorems 5, 9 and corollaries 6, 7 are written in [43].

3.4. Examples of PDE Integration

1. Example. We consider a consequence of Formulas 3.2.6$(2-6)$. If $q(t)$ is a differentiable function on the real field $\mathbf{R}$ having simple zeros $q(t_j) = 0$ (i.e. zeros of the first order), then

$$\delta(q(t)) = \sum_j \frac{1}{q(t_j)}\delta(t - t_j), \tag{1}$$

where the sum is accomplished by all zeros t_j of the equation $q(t) = 0$ (see Formula II.2.6(IV) [9]). Therefore, if $\gamma(\tau)$ is a C^1 path in U intersecting the surface $\partial U_s \cap \partial U_p$ at the marked point x of index $l = l_{s,p}(x)$, $\gamma(\tau_0) = x$, $0 < \tau_0 < 1$, such that $\gamma(\tau) \in U_s$ for each $\tau < \tau_0$ and $\gamma(\tau) \in U_p$ for each $\tau > \tau_0$ then

$$dg(\gamma(\tau))/d\tau = dg^1(\gamma(\tau))/d\tau + h\sum_{j=1}^{l}\delta(P_j)dP_j(\gamma(\tau))/d\tau, \tag{2}$$

where $\theta(t) = 0$ for $t < 0$ and $\theta(t) = 1$ for $t \ge 0$, $g^1(\gamma(\tau)) = g(\gamma(\tau)) - h\theta(\tau - \tau_0)$,

$$h = \lim_{\tau\downarrow\tau_0} g(\gamma(\tau)) - \lim_{\tau\uparrow\tau_0} g(\gamma(\tau)), \tag{2.1}$$

$$dP_j(\gamma(\tau))/d\tau = \sum_{k=1}^{n}(\partial P_j(z)/\partial z_k)(\partial\gamma_k(\tau)/d\tau)|_{z=\gamma(\tau)} \tag{2.2}$$

(see Example I.2.2.2 [9]). Particularly, if a point x is of index 1, then Formula (2) simplifies:

$$dg(\gamma(\tau))/d\tau = dg^1(\gamma(\tau))/d\tau + h\delta(P)[dP(\gamma(\tau))/d\tau]. \tag{3}$$

Particularly, these formulas can be applied to $d\nu_j$.

Let a partial differential differential operator Q be given by Formula

$$Qf(z) = \sum_{j,k}[(\partial f(z)/\partial z_j)(\partial\phi_j^1(z)^*/\partial z_k)]\phi_k^*(z) + + \sum_j[(\partial f(z)/\partial z_j)\phi_j^1(z)^*]\beta(z) + \sum_k[f(z)(\partial\beta^1(z)/\partial z_k)]\phi_k^*(z)$$

and functions ν_k are found (see Theorems 3.1.2, 3.1.4 and 3.3.9 above). We put in accordance with Formula 3.2.6(6)

$$\nu_k^1(z) = \nu_k(z) - \sum_{s,p} h_{k;s,p}(z)\theta(P_{s,p;1}(z),...,P_{s,p;m_j}(z)), \tag{4}$$

where

$$h_{k;s,p}(z(x)) = h_k(x)|_{x\in\Gamma^{s,p}} := \lim_{y_{1;s,p}\downarrow 0,\dots,y_{n;s,p}\downarrow 0} \nu_k(z(x+y)) - \lim_{y_{1;s,p}\downarrow 0,\dots,y_{n;s,p}\downarrow 0} \nu_k(z(x-y)), \quad (5)$$

where the sum is by s and p with $\partial U_s \cap \partial U_p \neq \emptyset$.
Let a domain W be a canonical closed compact set in the Euclidean space $\mathbf{R}^{n+1}$ embedded into $\mathcal{A}_v$ and contained in a canonical closed compact domain U so that $W = \{z \in U : z_j = 0 \ \forall n < j \leq 2^v - 1\}$. Thus Υ from test and generalized functions on W is extended on test and generalized functions on U. We can put $\nu_j = 0$ for $n < j \leq 2^v - 1$, when $n < 2^v - 1$. Then for the rectifiable path γ (see §2.2.1 above) we get

$$(n+1)^{-1}\sum_{k=0}^{n}[\int_{\gamma|_{[a,t]}} q(y)d\nu_k(y), \omega^1 \times \omega)$$

$$= (n+1)^{-1}\sum_{k=0}^{n}\{[\int_{\gamma} q(y)d\nu_k^1(y), \omega^1 \times \omega)$$

$$+[\hat{q};\omega^1,*;[\sum_{s,p} h_{k;s,p}(z)\sum_{j=1}^{l}\delta(P_j)\sum_{m=1}^{n}(\partial P_j(z)/\partial z_m)(\partial\gamma_m(\tau)/d\tau)|_{z=\gamma(\tau)};\omega,*]](\gamma)\}, \quad (6)$$

where $\gamma \in \{\gamma^\alpha : \alpha \in \Lambda\}$ is taken from the foliation C^1 family of paths (see §2.2.1 and Theorem 3.2.9 above), $z = \gamma(t_z)$, $t_z \in [a_\alpha, b_\alpha]$, $[a,t] = [a_\alpha, t_z]$, $l = l_{s,p}(z)$ denotes an index of a point z in the intersection of boundaries $\Gamma^{s,p} := \partial U_s \cap \partial U_p \neq \emptyset$, ω^1 and ω are real-valued test functions. Since ω^1 is real-valued, we get $\hat{f}(z).\omega^1 = f(z)\omega^1(z)$ and

$$[\hat{q};\omega^1,*;[\sum_{s,p} h_{k;s,p}(z)\sum_{j=1}^{l}\delta(P_j)\sum_{m=1}^{n}(\partial P_j(z)/\partial z_m)(\partial\gamma_m(\tau)/d\tau)|_{z=\gamma(\tau)};\omega,*]](\gamma) =$$

$$= \sum_{s,p}[q(z)\omega^1(z);[\theta(\tau-\tau_{s,p})h_{k;s,p}(z),\omega))|_{z=\gamma(\tau)}, \quad (7)$$

where $\tau_{s,p}$ corresponds to the intersection point $\gamma(\tau_{s,p})$ of γ with $\Gamma^{s,p} \neq \emptyset$. Here the expression $[q,\omega)|_{z=\gamma(\tau)} := \lim_j [q \circ \kappa^j, \omega \circ \kappa^j)$ denotes the restriction of the generalized function from U onto $\gamma([a,b])$, $\kappa^j \in \mathcal{D}(U, \mathcal{A}_v)$ is a sequence of test functions and $\kappa^j(\phi([a,b])) \subset U$ for each $j \in \mathbf{N}$, $\phi \in \mathcal{D}([a,b], \mathcal{A}_v)$, $\bigcap_{j=1}^{\infty} supp(\kappa^j) = \phi([a,b])$, $\lim_j \kappa^j \circ \phi = \gamma$ in $\mathbf{P}(U)$. Therefore, the derivative of the operator $[(n+1)^{-1}\sum_{k=0}^{n}[\int_{\gamma|_{[a,t]}} q(y)d\nu_k(y), \omega^1 \times \omega)$ by the parameter $\tau \in [a,b]$ for the real test functions ω^1 and ω is the following:

$$\partial(n+1)^{-1}\sum_{k=0}^{n}[\int_{\gamma|_{[a,t]}} q(y)d\nu_k(y), \omega^1 \times \omega)/\partial\tau =$$

$$[(n+1)^{-1}]\sum_{k=0}^{n}\{1 \otimes \hat{q}^1(z).(d\nu_k^1(\gamma(\tau))/d\tau)\}+$$

$$+\sum_{s,p}(h^{g'}_{s,p}(z).h_{k;s,p}(z)+\hat{q}^1(z).h_{k;s,p}(z)+h^{g'}_{s,p}(z).(d\nu_k^1(\gamma(\tau))/d\tau))|_{z=\gamma(\tau)}\otimes\delta(\tau-\tau_{s,p}),\omega^1\otimes\omega), \quad (8)$$

where $dg(z)/dz = \hat{q}(z)$ on U in the class of generalized operator $L_q(\mathcal{A}_v, \mathcal{A}_v)$ valued functions, $(dg(z)/dz).1 = q(z)$ on U, $h^{g'}_{s,p}(z) = h(z)$ is given by Formula (2.1) for the derivative operator $dg(z)/dz = g'$ instead of g on each $\Gamma^{s,p} \neq \emptyset$, $\hat{q}^1$ is given by Formula 3.2.6(6) for the function $\hat{q}(z)$ with values in $L_q(\mathcal{A}_v, \mathcal{A}_v)$ instead of $g(z)$. The terms like $\hat{q}^1(z).(d\nu^1_k(\gamma(\tau))/d\tau)$ correspond to the action of the operator valued generalized function $\hat{q}^1(z)$ on the generalized function $(d\nu^1_k(\gamma(\tau))/d\tau)$ which gives a generalized function.

Using Formulas $(6-8)$ for n constant on U and $\psi_j(z)$ or $\phi_j(z)$ respectively non-zero for each $z \in U$ and all $j = 0, ..., n$ we infer that for a continuous or generalized function f

$$\Upsilon \mathcal{I}_\Upsilon f(z) = f(z), \tag{9}$$

where

$$_\Upsilon \int f(z)dz := \{(n+1)^{-1} \sum_{j=0}^{n} \int_{\gamma^\alpha|_{[a_\alpha,t]}} q(z)d\nu^1_j(z), \quad \alpha \in \Lambda,\ a_\alpha \le t \le b_\alpha\}, \tag{10}$$

where $q = (dg/dz).1$ and g is given by the Equation (53) [33], since $f^1 = f$ and $h^{g'}_{s,p} = 0$ in the class of generalized functions f and in the class of continuous functions f, also $h_{k;s,p} = 0$ for $\nu_k = \nu^1_k$ on U.

Formulas $(9, 10)$ show what sort of boundary conditions is sufficient to specify a unique solution for a given domain U with sub-domains U_s. If U is C^1 diffeomorphic to the half-space $H_p := \{z \in \mathcal{A}_v : z_0p_0 + ... + z_{2^v-1}p_{2^v-1} \ge 0\}$, where $p = p_0i_0 + ... + p_{2^v-1}i_{2^v-1}$ is a marked Cayley-Dickson number, $p_0, ..., p_{2^v-1} \in \mathbf{R}$, and sub-domains U_s are not prescribed, then it is sufficient to give the boundary condition $F|_{\partial U} = G$ when a solution is in the class of continuous or generalized functions with the corresponding f and ψ_j or ϕ_j. Indeed, if the functions ν_k along γ^α are defined up to constants μ_k, the differentials are the same $d(\nu_k + \mu_k)(z)|_{z=\gamma^\alpha(\tau)} = d\nu_k|_{z=\gamma^\alpha(\tau)}$ in the anti-derivative operator, when $d\mu_k|_{z=\gamma^\alpha(\tau)} = 0$ for each $\alpha \in \Lambda$ and $\tau \in [a_\alpha, b_\alpha]$.

The operator $\mathcal{I}_\Upsilon$ may be applied also piecewise on each U_s. If a solution F is locally continuous on U and continuous on each sub-domain U_s, then boundary conditions $F|_{\partial U_s} = G^s$ for all $s = 1, ..., m$ may be necessary to specify a solution F. Without boundary conditions the anti-derivative operator applied to f gives the general solution $\mathcal{I}_\Upsilon f$ of the differential equation $\Upsilon F = f$.

If each ν_k is continuously differentiable, which is possible, when each function ψ_k or ϕ_k is continuous, and f is continuous on U, then a solution $F = \{_\Upsilon \int_{\gamma^\alpha|_{[a_\alpha,\tau]}} f(y)dy : \alpha \in \Lambda,\ a_\alpha \le \tau \le b_\alpha\}$ is continuously differentiable by each z_k, $z \in U$, $z = \gamma^\alpha(t_z)$.

One can also mention that the sequence $_m\omega(z_0, ..., z_n) = (2\pi m)^{-(n+1)/2} \exp\{-(z_0^2 + ... + z_n^2)/(2m)\}$ converges to the delta-function on $\mathbf{R}^{n+1}$ embedded into $\mathcal{A}_v$, $m \in \mathbf{N}$. Then the sequence $_m\theta(z_0) = \int_{-m}^{z_0} (2\pi m)^{-1/2} \exp\{-t^2/(2m)\}dt$ converges to the θ function, while each function $_m\theta(z_0)$ is analytic by z_0, since $m \in \mathbf{N}$, $\exp\{-t^2/(2m)\}$ is the analytic function with the infinite radius of convergence of its power series, while $\lim_{m\to+\infty} \int_{-\infty}^{-m} (2\pi m)^{-1/2} \exp\{-t^2/(2m)\}dt = 0$. Then each $_m\omega(z_0, ..., z_n)$ and $\prod_{j=0}^{n} {_m\theta}(z_j)$ can be written in the z-representation over $\mathcal{A}_v$ as the analytic function with the help of Formulas 2.2.1$(1-3)$, where $n \le 2^v - 1$, $z_j \in \mathbf{R}$ for each j, $z = z_0i_0 + ... + z_{2^v-1}i_{2^v-1}$. Thus $\lim_{m\to\infty} {_m\omega}(z_0, ..., z_n) = \delta(z_0, ..., z_n)$ and

$\lim_{m\to\infty}\prod_{j=0}^{n} {}_m\theta(z_j)=\prod_{j=0}^{n}\theta(z_j),\ d\theta(z_0)/dz_0=\delta(z_0)$.

2. Boundary Conditions.
If U is a domain as in §3.2.6, we put $\mathcal{B}(\partial U,Y)=\{f|_{\partial U}:\ f\in\mathcal{B}(U,Y)\}$ and $\mathcal{D}(\partial U,Y)=\{f|_{\partial U}:\ f\in\mathcal{D}(U,Y)\}$ when a boundary ∂U is non-void so that the topologically adjoint linear over $\mathbf{R}$ spaces, left and right $\mathcal{A}_r$ modules, of generalized functions are $\mathcal{B}'(\partial U,Y)$ and $\mathcal{D}'(\partial U,Y)$.

Let us consider a generalized function $f\in\mathcal{B}'(\partial U,Y)$ or $\mathcal{D}'(\partial U,Y)$ and a test function $h\in\mathcal{B}(\partial U,Y)$ or $\mathcal{D}(\partial U,Y)$ respectively. One can take $g\in\mathcal{B}(U,Y)$ or $\mathcal{D}(U,Y)$ and a sequence $q^m\in\mathcal{B}(U,Y)$ or $\mathcal{D}(U,Y)$ with supports non intersecting with the boundary $supp(q^m)\cap\partial U=\emptyset$ such that $(g-q^m)$ tends to zero in $\mathcal{B}(V,Y)$ or $\mathcal{D}(V,Y)$ for each compact subset V in the interior $Int(U)$, when m tends to the infinity, while $\lim_m(g-q^m)=h$ in $\mathcal{B}(\partial U,Y)$ or $\mathcal{D}(\partial U,Y)$ respectively. Here as usually the interior $Int(U)$ is taken in the corresponding topological space $\mathbf{R}^n$ or $\mathcal{A}_r$. Each generalized function is a limit of test functions, consequently, a generalized function $\xi\in\mathcal{B}'(U,Y)$ or $\mathcal{D}'(U,Y)$ exists so that

$$\lim_m[\xi,(g-q^m))=[f,h). \tag{B1}$$

Vise versa if $\xi\in\mathcal{B}'(U,Y)$ or $\mathcal{D}'(U,Y)$ is a generalized function on U, then Formula $(B1)$ defines a generalized function $f\in\mathcal{B}'(\partial U,Y)$ or $\mathcal{D}'(\partial U,Y)$, which we consider as the restriction of ξ on $\mathcal{B}(\partial U,Y)$ or $\mathcal{D}(\partial U,Y)$ correspondingly. In view of the definition of convergence of test and generalized functions Formula $(B1)$ defines the unique restriction f for the given generalized function ξ.

A subsequent use of decomposition of operators into compositions of first order partial differential operators and their anti-derivation operators permits to integrate partial differential equations with locally continuous or generalized coefficients.

The results and definitions of previous sections show that for the differential equation

$$Af=g, \tag{1}$$

where a partial differential operator is written in accordance with Formulas $3.2.3(1,2)$. When ∂U is a C^1-manifold without corner points of index greater than one, the following boundary conditions may be used:

$$f(t)|_{\partial U}=f_0(t'),(\partial^{|q|}f(t)/\partial s_1^{q_1}...\partial s_n^{q_n})|_{\partial U}=f_{(q)}(t') \tag{2}$$

for $|q|\le\alpha-1$, where $s=(s_1,...,s_n)\in\mathbf{R}^n$, $(q)=(q_1,...,q_n)$, $|q|=q_1+...+q_n$, $0\le q_k\in\mathbf{Z}$ for each k, $t\in\partial U$ is denoted by t', f_0, $f_{(q)}$ are given functions. Generally these conditions may be excessive, so one uses some of them or their linear combinations (see (4) below). Frequently, the boundary conditions

$$f(t)|_{\partial U}=f_0(t'),(\partial^l f(t)/\partial\nu^l)|_{\partial U}=f_l(t') \tag{3}$$

for $1\le l\le\alpha-1$ are also used, where ν denotes a real variable along a unit external normal to the boundary ∂U at a point $t'\in\partial U_0$. Using partial differentiation in local coordinates on ∂U and (3) one can calculate in principle all other boundary conditions in (2) almost everywhere on ∂U.

It is possible to describe as an example a particular class of domains and boundary conditions. Suppose that a domain U_1 and its boundary ∂U_1 satisfy Conditions 3.2.6$(D1, i - vii)$ and $g_1 = g\chi_{U_1}$ is a regular or generalized function on $\mathbf{R}^n$ with its support in U_1. Then any function g on $\mathbf{R}^n$ gives the function $g\chi_{U_2} =: g_2$ on $\mathbf{R}^n$, where $U_2 = \mathbf{R}^n \setminus U_1$. Take now new domain U satisfying Conditions 3.2.6$(D1, i - vii)$ and $(D2 - D5)$ below:

$(D2)$ $U \supset U_1$ and $\partial U \subset \partial U_1$;

$(D3)$ if a straight line ξ containing a point w_1 (see 3.2.6(vi)) intersects ∂U_1 at two points y_1 and y_2, then only one point either y_1 or y_2 belongs to ∂U, where $w_1 \in U_1$, $U - w_1$ and $U_1 - w_1$ are convex; if ξ intersects ∂U_1 only at one point, then it intersects ∂U at the same point;

$(D4)$ any straight line ξ through the point w_1 either does not intersect ∂U or intersects the boundary ∂U only at one point.

Take now g with $supp(g) \subset U$, then $supp(g\chi_{U_1}) \subset U_1$. Therefore, any problem (1) on U_1 can be considered as the restriction of the problem (1) defined on U, satisfying $(D1 - D4, i - vii)$. Any solution f of (1) on U with the boundary conditions on ∂U gives the solution as the restriction $f|_{U_1}$ on U_1 with the boundary conditions on ∂U.

Henceforward, we suppose that the domain U satisfies Conditions $(D1, D4, i - vii)$, which are rather mild and natural.

Thus the sufficient boundary conditions are:

$$(\partial^{|\beta|} f(t^{(lj)})/\partial\tau_{\gamma(1)}^{\beta_1} ... \partial\tau_{\gamma(m)}^{\beta_m})|_{\partial U_{(lj)}} = \phi_{\beta,(lj)}(t^{(lj)}) \quad (4)$$

for $|\beta| = |q|$, where $m = |h(lj)|$, $|j| \le \alpha$, $|(lj)| \ge 1$, $\mathbf{a}_j \ne 0$, $q_k = 0$ for $l_k j_k = 0$, $m_k + q_k + h_k = j_k$, $h_k = sign(l_k j_k)$, $0 \le q_k \le j_k - 1$ for $k > n - \kappa$; $\phi_{\beta,(l)}(t^{(l)})$ are known functions on $\partial U_{(l)}$, $t^{(l)} \in \partial U_{(l)}$. In the half-space $t_n \ge 0$ only the partial derivatives by t_n

$$\partial^\beta f(t)/\partial t_n^\beta|_{t_n=0} \quad (5)$$

are necessary for $\beta = |q| < \alpha$ and q as above.

Depending on coefficients of the operator A and the domain U some boundary conditions may be dropped, when the corresponding terms vanish.

Conditions in (5) are given on disjoint for different (l) sub-manifolds $\partial U_{(l)}$ in ∂U and partial derivatives are along orthogonal to them coordinates in $\mathbf{R}^n$, so they are correctly posed.

We recall, that a characteristic surface of a partial differential operator given by Formula 3.2.3(1) is a surface defined as a zero of C^u differentiable function $\phi(x_1, ..., x_n) = 0$ in the Euclidean space or in the Cayley-Dickson algebra such that at each point x of it the condition is satisfied

$$\sum_{|j|=u} \mathbf{a}_j(t(x))(\partial\phi/\partial x_1)^{j_1} ... (\partial\phi/\partial x_n)^{j_n} = 0 \quad (CS)$$

and at least one of the partial derivatives $(\partial\phi/\partial x_k) \ne 0$ is non-zero. Generally a domain U is worthwhile to choose with its interior $Int(U)$ non-intersecting with a characteristic surface $\phi(x_1, ..., x_n) = 0$ (see also [48, 57]).

3. Solutions of Second Order Partial Differential Equations with the Help of the Line Integration over the Cayley-Dickson Algebras.

Mention that the operator $(\Upsilon + \beta)(z_0, ..., z_n)$ acting on a function depending on variables $z_0, ..., z_n$ only can be written as

$$\Psi(z_0, ..., z_{n+1})(f(z)z_{n+1})|_{z_{n+1}=1} = \Upsilon(z_0, ..., z_n)f(z) + f(z)\beta(z) =$$

$$= [\sum_{j=0}^{n+1}(\partial(f(z)z_{n+1})/\partial z_j)\phi_j^*(z)]|_{z_{n+1}=1}, \tag{1}$$

where $\phi_{n+1}^*(z) = \beta(z)$, each function $\phi_j(z)$ and $f(z)$ may depend on $z_0, ..., z_n$ only, omitting for short the direct product $\otimes$ in the case of generalized coefficients in Formula (1) and henceforth. The operator $\Psi(z_0, ..., z_{n+1})(f(z)z_{n+1})$ may be reduced to the form satisfying conditions of Theorems 3.1.2 or 3.3.9 using a suitable change of variables. This procedure gives an anti-derivative operator

$$\mathcal{I}_{\Upsilon+\beta} = \mathcal{I}_\Psi|_{z_{n+1}=1} \tag{1.1}$$

such that

$$(\Upsilon + \beta)\mathcal{I}_{\Upsilon+\beta}f = f \tag{1.2}$$

for a continuous function or a generalized function f. Therefore, we shall consider operators of the form Υ and their compositions and sums.

We take the partial differential equation with piecewise continuous or generalized coefficients

$$A = \Upsilon\Upsilon_1 f(z) + \Upsilon_2 f(z) = g, \tag{2}$$

where

$$\Upsilon_k f(z) = [\sum_{j=0}^{n}(\partial f(z)/\partial z_j)\phi_j^k(z)^*] \tag{3}$$

for $k = 1, 2$ or without this index k, $\phi_j^k(z) = i_j\psi_j^k(z)$ (see §§3.1.1, 3.1.2, 3.1.4 and 3.3.9). For solving it we write the system:

$$\Upsilon_1 f = g_1, \Upsilon g_1 = g - \Upsilon_2 f. \tag{4}$$

In accordance with Equation (2) we have:

$$\Upsilon g_1 + \Upsilon_2(\Upsilon_1^{-1}g_1) = g, \tag{5}$$

where the inverse operator Υ_1^{-1} is the anti-derivation operator $\mathcal{I}_{\Upsilon_1}$ described above in Theorems 3.1.2, 3.1.4 and 3.3.9. If $\Upsilon_2 \neq 0$ we suppose that either (i) g_1 or $[g_1, \omega)$ is real-valued or the algebra $alg_{\mathbf{R}}\{g_1(z), \phi_j^1(z), \phi_k^2(z)\}$ or $alg_{\mathbf{R}}\{[g_1, \omega), [\phi_j^1, \omega), [\phi_k^2, \omega)\}$ for all $j, k = 0, ..., n$ is associative for each $z \in U$ in the case of functions or for every real-valued test function ω in the case of generalized functions correspondingly. Calculating the expression

$$\Upsilon g_1 + \Upsilon_2(\mathcal{I}_{\Upsilon_1}g_1) = (\Upsilon + \beta^3)g_1 = g \tag{6}$$

we get a new operator $(\Upsilon + \beta^3) = \Psi$ in accordance with §§3.1.2, 3.2.3, 3.2.10 and 3.3.9 omitting $\otimes$ and $\otimes\omega$ and $[*, \omega^{\otimes 3})$ for short in the case of generalized coefficients, where

$$\beta^3 = \sum_{j,k=0}^{n} (\partial\nu_j^1/\partial z_k)(\phi_k^2)^*,$$

when a solution $\{\nu_j(z) : j\}$ is chosen real, i.e. each function ν_j is real-valued or a real-valued generalized function on real-valued test functions (see System $(35, 36)$ [33] and §3.3.9).

Generally without supposition (i) we deduce that

$$\Upsilon g_{1,l} + \Upsilon_2(\mathcal{I}_{\Upsilon_1} g_{1,l}) = (\Upsilon + \beta_l^3) g_{1,l} = i_l^* g^l \tag{6.1}$$

for each $l = 0, ..., n$, where $\beta_l^3 = i_l^*[\sum_{j,k=0}^n (i_j(\partial\nu_j^1/\partial z_k))(\phi_k^2)^*]$,

$$\sum_{l=0}^{n} g^l = g, \tag{6.2}$$

$$\sum_{l=0}^{n} g_{1,l} i_l = g_1, \tag{6.3}$$

each $g_{1,l}(z)$ or $[g_{1,l}, \omega)$ is real-valued for each $z \in U$ or every real-valued test function ω respectively. Solving the system $(6.1 - 3)$ with the help of known anti-derivative operators $\mathcal{I}_{\Psi_l}$ one finds g_1, where $\Psi_l = \Upsilon + \beta_l^3$. Thus an anti-derivative operator $\mathcal{J} := \mathcal{J}_{\Upsilon + \Upsilon_2 \mathcal{I}_{\Upsilon_1}}$ exists so that

$$(\Upsilon + \Upsilon_2 \mathcal{I}_{\Upsilon_1}) \mathcal{J}_{\Upsilon + \Upsilon_2 \mathcal{I}_{\Upsilon_1}} g = g. \tag{6.4}$$

In the particular case (i) the equality $\mathcal{J}_{\Upsilon + \Upsilon_2 \mathcal{I}_{\Upsilon_1}} = \mathcal{I}_\Psi$ is satisfied.

Therefore, in the case of either continuous coefficients of operators and g or generalized coefficients and g the general solution is:

$$f = \mathcal{I}_{\Upsilon_1} g_1 = \mathcal{I}_{\Upsilon_1} \mathcal{J} g, \tag{7}$$

where

$$g_1 = \mathcal{J} g. \tag{8}$$

If $\Upsilon_2 = 0$ the formula simplifies to

$$f = \mathcal{I}_{\Upsilon_1} \mathcal{I}_\Upsilon g. \tag{9}$$

Examples of boundary conditions and domains permitting to specify a unique solution are given in §2 above.

4. Example. Let us consider a function and its phrase satisfying Condition 3.2.8$(P3)$. Therefore, we get

$$\Upsilon_1^{-1}(g_1)(z_{a^1}(x)) = -\phi_{f'}(Im(z_{a^1}(x))) + \int_\alpha^x g_1(z)dz + \phi_{g_1}(x') \tag{1}$$

in accordance with Formula 3.2.8(6), where x' and $Im(z_{a^1}(x))$ are written in the x and z-representations respectively using Identities 2.2.1$(1-3)$. In particular, for $\Upsilon^1 = \sigma_1$ we have the coefficients $\psi_l^1(z) = i_l(\mathbf{a}_k(z)w_j i_{2^r k}^*)$ for each $l = 2^r k$ with $k \in \{m_1 + ... + m_{j-1} + 1, ..., m_1 + ... + m_j\}$, while $\psi_l^1(z) = 0$ for all others l for each z. A function a^1 is given by Formula 3.2.8(3) for ψ^1 instead of ψ. Let the first order operator Q be written in its standard form:

$$Qf = \sum_{j=1}^{m} \sum_{k=m_1+...+m_{j_1-1}+1}^{m_1+...+m_{j_1}} \mathrm{s}_k(z)(\partial f/\partial z_{2^r k})(u_j i_{2^r k}), \qquad (2)$$

since $i_{2^r k} w_j = w_j^* i_{2^r k}$ for each $w_j \in \mathcal{A}_r$ and $k \geq 1$, when $r \geq 1$, where $w_j \in \mathcal{A}_r$ and $u_j = u_j(z) \in \mathcal{A}_v$ for each j, $\mathrm{s}_k(z)$ is the real-valued (super-)differentiable function for each k. If $\xi = z_a(y)$, then $(d\xi/dy).[(dy/d\xi).h] = h$ for each Cayley-Dickson number $h \in \mathcal{A}_v$. This implies that these two $\mathbf{R}$-linear $\mathcal{A}_v$-additive operators are related by the equality $(dy/d\xi).h = (d\xi/dy)^{-1}.h$. On the other hand, $(dz_a(y)/dy).1 = a(y) \in \mathcal{A}_v$ and $y \in U \subset \mathcal{A}_v$ in the considered case. The function $z_a(y)$ is defined up to the addendum $z_a(\alpha)$, where $\alpha \in H_{\alpha_0} \cap U$, $H_{\alpha_0} := \{z \in \mathcal{A}_v : Re(z) = Re(\alpha_0)\}$. We can choose $\phi_a(y')$ so that $(dx/dy).(1/a) = 1$ for each y for which $a = a(y) \neq 0$ and inevitably we get $(dy/dx).1 = 1/a(y)$.

In the particular case of σ, σ_1 and Q accomplishing the differentiation with the help of the latter identities we infer that:

$$Q(\sigma_1^{-1} g_1)(x) = -\sum_{j=1}^{m} \sum_{k=m_1+...+m_{j_1-1}+1}^{m_1+...+m_{j_1}} [(d\phi_{f'}(x')/dx).i_{2^r k}$$

$$+\{\hat{g}_1(y) + (d\phi_{g_1}(y')/dy)\}.[(dz_{a^1}(y)/dy)^{-1}.i_{2^r k}](\mathrm{s}_k(y)u_j(y)i_{2^r k}), \qquad (3)$$

where $\hat{g}_1 = d\zeta_1(y)/dy$ for a (super-)differentiable function ζ_1 such that $(d\zeta_1(y)/dy).1 = g_1(y')$, $\psi_l^1(z) = i_l(\mathbf{a}_k(z)w_j i_{2^r k}^*)$ for each $l = 2^r k$ with $k \in \{m_1 + ... + m_{j-1} + 1, ..., m_1 + ... + m_j\}$, while $\psi_l^1(z) = 0$ for all others l for each z. Also $\psi_l(z) = i_l(\mathbf{a}_k(z)w_j^* i_{2^r k}^*)$ for each $l = 2^r k$ with $k \in \{m_1 + ... + m_{j-1} + 1, ..., m_1 + ... + m_j\}$, while $\psi_l(z) = 0$ for all others l and for each z. We introduce the notation:

$$\theta_l(y) = i_l[(dz_a(y)/dy)^{-1}.i_{2^r k}](\mathrm{s}_k(y)u_j(y)i_{2^r k}) \qquad (4)$$

for $l = 2^r k$ and $k \in \{m_1 + ... + m_{j_1-1} + 1, ..., m_1 + ... + m_{j_1}\}$, $\theta_l(y) = 0$ for all others l and for each y;

$$a(x) = -\sum_{j=1}^{m} w_j \sum_{k=m_1+...+m_{j-1}+1}^{m_1+...+m_j} \mathbf{a}_k(x) = \sigma(x) \qquad (5)$$

and

$$\kappa_l(x) = i_l[s_k(x)u_j(x)i_{2^r k}] \qquad (6)$$

for $l = 2^r k$ with $k \in \{m_1 + ... + m_{j_1-1} + 1, ..., m_1 + ... + m_{j_1}\}$, $\kappa_l(x) = 0$ for every other l and for each x.

In the general case

$$\Upsilon_2(\Upsilon_1^{-1}g_1)(x) = -\sum_{j=0}^{2^v-1}[(d\phi_{f'}(x')/dx).i_j$$

$$+\{\hat{g}_1(y) + (d\phi_{g_1}(y')/dy)\}.[(dz_{a^1}(y)/dy)^{-1}.i_j](i_j^*\psi_j(x)), \quad (7)$$

where $\hat{g}_1 = d\zeta_1(y)/dy$ for a (super-)differentiable function ζ_1 such that $(d\zeta_1(y)/dy).1 = g_1(y')$. We shall use the notation:

$$\theta_j(y) = i_j[(dz_a(y)/dy)^{-1}.i_j](i_j^*\psi_j(y)) \quad (8)$$

and for each y and each j;

$$a(x) = \Upsilon(x) \quad (9)$$

and

$$\kappa_j(x) = i_j[i_j^*\psi_j(x)] \quad (10)$$

for each x and j.

Substituting (7) into $3(5)$ we deduce that:

$$g_1(z_a(x)) + g_1(z_\theta(x)) = \zeta(x), \quad (11)$$

where

$$\zeta(x) = -\phi_{g_1}(Im(z_a(x))) - \phi_{g_1}(Im(z_\theta(x))) + \phi_{f'}(Im(z_\kappa(x))) + \int_\alpha^x g(z)dz. \quad (12)$$

For $x = \alpha$ we certainly have $z_a(\alpha) = z_\theta(\alpha)$. Suppose that $a(x) \neq \theta(x)$ identically. The dimension of the Cayley-Dickson algebra $\mathcal{A}_v$ over the real field is not less, than four. Therefore, we can choose a path γ so that γ is orthogonal to θ and κ at each point on γ, that is $\gamma'(t) \perp \theta'(\gamma(t))$ and $\gamma'(t) \perp \kappa'(\gamma(t))$ relative to the real-valued scalar product (RS) for each $t \in [0, 1]$, where $\gamma'(t) := d\gamma(t)/dt$. Then $g_1(z_\theta(x)) = g_1(\alpha)$ and $\phi_{f'}(Im(z_\kappa(x))) = \phi_{f'}(\alpha')$ for each $x = \gamma(t)$. Therefore, along such path γ one has

$$g_1(z_a(x)) + g_1(\alpha) = \zeta(x) = -\phi_{g_1}(Im(z_a(x)) - \phi_{g_1}(\alpha') + \phi_{f'}(\alpha') + \int_\alpha^x g(z)dz \quad (13)$$

for each $x = \gamma(t)$. Expressing $g_1(z)$ from Equation (11), substituting into $26(4)$ and integrating one gets:

$$f(z_{a^1}(x)) = -\phi_f(Im(z_{a^1}(x))) + \int_\alpha^x g_1(z)dz. \quad (14)$$

Particularly, if the operator A is with constant coefficients, then $s_k(x) = 0$ identically for each k, consequently, $\theta = 0$ and $\kappa = 0$ identically and $g_1(z_a(x)) = g_1(\alpha) = \phi_{g_1}(Im(z_a(x)))$ for each x, when f has a right linear derivative by z. Arbitrary integration terms in $(11, 14)$ can be specified from the boundary conditions.

Finally, the restriction from the domain in $\mathcal{A}_v$ onto the initial domain of real variables in the real shadow and the extraction of $\pi_r^v \circ f \in \mathcal{A}_r$ with the help of Formulas $1(1-3)$

gives the reduction of a solution from $\mathcal{A}_v$ to $\mathcal{A}_r$, where $\pi^v_r : \mathcal{A}_v \to \mathcal{A}_r$ is the $\mathbf{R}$-linear projection operator defined as the sum of projection operators $\pi_0 + ... + \pi_{2^r-1}$ given by Formulas 2.2.1$(1-3)$ on $\mathbf{R}i_j$ for $j = 0, ..., 2^r - 1$.

In particular it certainly works for Laplace's operator, the heat kernel, wave operator (d'Alambert operator), Helmholtz' operator, Klein-Gordon's operator (see also [44]).

5. Partial Differential Equations of Order Higher than Two.
The fundamental solution $\Psi_{\Upsilon^m+\beta}$ of the equation

$$\Xi_{2m,\beta}\Psi_{\Upsilon^m+\beta} = \delta, \tag{1}$$

where

$$\Xi_{2m,\beta} = (\Upsilon^m + \beta)(\Upsilon^m + \beta)^* \tag{2}$$

can be obtained using decompositions with the help of operators of the first order $\Upsilon_k + \beta_k$ by induction, if such decomposition exists (see above). Suppose that this decomposition is constructed

$$(\Upsilon^m + \beta)f(z) = (\Upsilon_m + \beta_m)[...[(\Upsilon_2 + \beta_2)[(\Upsilon_m + \beta_m)f(z)]]...], \tag{3}$$

then the fundamental solution can be written as the iterated convolution

$$\Psi_{\Upsilon^m+\beta} = [...[[(\Upsilon_m + \beta_m)^*\mathcal{E}_m] * [(\Upsilon_{m-1} + \beta_{m-1})^*\mathcal{E}_{m-1}]] * ...] * [(\Upsilon_1 + \beta_1)^*\mathcal{E}_1], \tag{4}$$

where $\mathcal{E}_j$ denotes the fundamental solution of the elliptic second order partial differential equation

$$A_j\mathcal{E}_j = \delta, \tag{5}$$

with

$$A_j = (\Upsilon_j + \beta_j)(\Upsilon_j + \beta_j)^*. \tag{6}$$

The fundamental solutions $\mathcal{E}_j$ were written above in §§2 and 28 [44]. Indeed, using Equalities 4$(7-9)$ [44] by induction we have

$$\sum_s (...(a^*_{m,k_m} a^*_{m-1,k_{m-1}})a^*_{m-2,k_{m-2}})...)a^*_{1,k_1})a_{1,l_1})a_{2,l_2})...)a_{m,l_m}$$

$$= Re(a^*_{m,k_m} a_{m,l_m})...Re(a^*_{1,k_1} a_{1,l_1}), \tag{7}$$

where $\sum_s$ denotes the sum by all pairwise transpositions $(k_1, l_1),...,(k_m, l_m)$, $a_{j,k} \in \mathcal{A}_v$. Therefore,

$$\Xi_{2m,\beta}\Psi_{\Upsilon^m+\beta} =$$

$$[...[[(\Upsilon_m+\beta_m)(\Upsilon_m+\beta_m)^*\mathcal{E}_m]*[(\Upsilon_{m-1}+\beta_{m-1})(\Upsilon_{m-1}+\beta_{m-1})^*\mathcal{E}_{m-1}]]*...]*[(\Upsilon_1+\beta_1)(\Upsilon_1+\beta_1)^*\mathcal{E}_1]$$

$$= [...[\delta * \delta] * ...] * \delta = \delta. \tag{8}$$

Vice versa if the fundamental solution $\Psi_{\Upsilon^m+\beta}$ is known, then we get the fundamental solution $\mathcal{E}^m_\beta$ of the equation

$$(\Upsilon^m + \beta)\mathcal{E}^m_\beta = \delta \tag{9}$$

as

$$\mathcal{E}^m_\beta = (\Upsilon^m + \beta)^* \Psi_{\Upsilon^m+\beta} \tag{10}$$

in accordance with $(2, 7)$. Moreover, the equation

$$A_{m+k} f = g \tag{11}$$

with $A_{m+k} = (\Upsilon^m_1 + \beta_1)(\Upsilon^k_2 + \beta_2)$
in $\mathcal{D}(\mathbf{R}^n, \mathcal{A}_v)$ or in the space $\mathcal{D}(\mathbf{R}^n, \mathcal{A}_v)^*_l$, where n is a number of real variables, $2 \leq n \leq 2^v$, has the fundamental solution $\mathcal{V}_{m+k}$:

$$\mathcal{V}_{m+k} = \mathcal{E}_{\Upsilon^m_1+\beta_1} * \mathcal{E}_{\Upsilon^k_2+\beta_2}, \tag{12}$$

where

$$\mathcal{E}_{\Upsilon^m_1+\beta_1} = (\Upsilon^m_1 + \beta_1)^* \Psi_{\Upsilon^m_1+\beta_1} \tag{13}$$

denotes the fundamental solution of the equation

$$(\Upsilon^m_1 + \beta_1)\mathcal{E}_{\Upsilon^m_1+\beta_1} = \delta, \tag{14}$$

consequently,

$$f = \mathcal{V}_{m+k} * g \tag{15}$$

is the solution of Equation (11).

For example, the fourth order partial differential operator

$$A_4 f(z) = \sum_{j=1}^{p} \partial^4 f(z)/\partial z_j^4 - \sum_{j=p+1}^{n} \partial^4 f(z)/\partial z_j^4$$

can be decomposed as the composition of two operators of the second order Υ^2 and Υ^2_1 formally as σ and σ_1 in 3.1.1$(8, 9)$ with the substitution of $\partial f/\partial z_{2^r j}$ on $\partial^2 f/\partial z^2_{2^r j}$ so that in accordance with Theorems 3.2.3 and 3.2.9 this operator A_4 can be presented in the form given by Formulas $(2, 3, 11)$.

On the other hand, fundamental solutions of Δ^k_n and $L^k_{p,q}$ and A^k_2 for certain other second order partial differential operators are known. So combining them with operators of the form $\Upsilon^{m_1}_1 ... \Upsilon^{m_k}_k$ permits to consider fundamental solutions of many partial differential operators of order higher than two as well.

Thus knowing fundamental solutions of the corresponding first or second order operators permits to write fundamental solutions of higher order partial differential operators considered above with the help of the iterated convolutions in a definite order prescribed by the induction process.

6. Non-linear Partial Differential Equations.
We consider the differential equation

$$(\Upsilon^m + \beta + \hat{f}(y)\Upsilon)y = g, \tag{1}$$

where $\Upsilon^m + \beta$ is a partial differential operator as in §3.2.3 of order m, $f(y)$ is a $\mathcal{A}_v$ differentiable function, $y = y(z)$ is an unknown function, $\hat{f}(y)\Upsilon y := \sum_{j=0}^n [\hat{f}(y).(\partial y(z)/\partial z_j)]\phi_j^*(z)$. Suppose that a fundamental solution $\mathcal{E}_{\Upsilon^m+\beta}$ of Equation 5(9) for the operator $(\Upsilon^m + \beta)$ is known. Find at first a fundamental solution $y = \mathcal{V}$ of (1) with $g = \delta$. Then

$$(\Upsilon^m + \beta)\mathcal{V} = \delta - \mu, \tag{2}$$

where $\mu(z) = \hat{f}(y(z))\Upsilon y(z)$. The anti-derivative gives

$$w(y(z)) = {}_\Upsilon\int(\hat{f}(y)\Upsilon y)dz = \int f(y(x))dy(x) = \int f(y)dy = {}_\Upsilon\int \mu(z)dz, \tag{3}$$

then

$$y = w^{-1}({}_\Upsilon\int_{\gamma^\alpha|_{[a,t_z]}} \mu(x)dx), \tag{4}$$

where w^{-1} denotes the inverse function. On the other hand,

$$y = \mathcal{E}_{\Upsilon^m+\beta} * (\delta - \mu) = \mathcal{I}_{\Upsilon^m+\beta}(\delta - \mu), \tag{5}$$

when Υ^m is either of the first order for $m = 1$ or is expressed as a composition of operators of the first order,

$$\Upsilon^m + \beta = (\Upsilon_1 + \beta^1)...(\Upsilon_m + \beta^m) \tag{5.1}$$

so that

$$\mathcal{I}_{\Upsilon^m+\beta} = \mathcal{I}_{\Upsilon_m+\beta^m}...\mathcal{I}_{\Upsilon_1+\beta^1}, \tag{5.2}$$

consequently, $(4, 5)$ imply the equation:

$$\mathcal{E}_{\Upsilon^m+\beta} * (\delta - \mu) = \mathcal{I}_{\Upsilon^m+\beta}(\delta - \mu) = w^{-1}({}_\Upsilon\int_{\gamma^\alpha|_{[a,t_z]}} \mu(x)dx) \tag{6}$$

or

$$w(\mathcal{E}_{\Upsilon^m+\beta} * (\delta - \mu)) = {}_\Upsilon\int_{\gamma^\alpha|_{[a,t_z]}} \mu(x)dx. \tag{7}$$

We have the identity

$(\partial(\mathcal{E} * \Psi)/\partial z_p), \phi) = -((\mathcal{E} * \Psi), \partial\phi/\partial z_p) = ((\mathcal{E} * (\partial\Psi/\partial z_p)), \phi)$

with a generalized function Ψ.

Therefore, differentiating (6) by $z_0,...,z_n$, we infer that:

$$\{\sum_{j=0}^n [\mathcal{E}_{\Upsilon^m+\beta} * (\partial\mu(z)/\partial z_j)]\phi_j^*(z)\}+$$

$$(n+1)^{-1}\sum_{j,k=0}^n [(dw^{-1}(\xi)/d\xi).(\hat{\theta}(z).(\partial\nu_j(z)/\partial z_k))]\phi_k^*(z) = \Upsilon\mathcal{E}_{\Upsilon^m+\beta} \tag{8}$$

or

$$\Upsilon(\mathcal{I}_{\Upsilon^m+\beta}\mu)+$$

$$(n+1)^{-1}\sum_{j,k=0}^{n}[(dw^{-1}(\xi)/d\xi).(\hat{\theta}(z).(\partial\nu_j(z)/\partial z_k))]\phi_k^*(z) = \Upsilon(\mathcal{I}_{\Upsilon^m+\beta}\delta), \quad (9)$$

where

$$[((dg(z)/dz).\phi_j(z)) \otimes \phi_j^*(z), \omega^{\otimes 2}) = [\mu(z) \otimes 1, \omega^{\otimes 2}) \quad (9.1)$$

for each real-valued test function ω and each j, $\hat{\theta}(z) = dg(z)/dz$, $\xi = {}_{\Upsilon}\int_{\gamma^\alpha|_{[a,t_z]}} \mu(x)dx$ (see also §§3.1.2, 3.1.4, 3.3.9 and 3.4.3). If Υ and β are independent of z_j, i.e. $\phi_j(z) = 0$ is zero identically on U, then $\partial(\mathcal{I}_{\Upsilon^m+\beta}\mu)/\partial z_j = \mathcal{I}_{\Upsilon^m+\beta}(\partial\mu/\partial z_j)$ (see also Note 23.1 [44]). Otherwise the derivative $\partial(\mathcal{I}_{\Upsilon^m+\beta}\mu)/\partial z_j$ is given by Formulas 4(6) and 23(8) [44] and 3.4.3(1). The function w is known from (3) after the line integration by the variable y, so Equation (8) is linear by $(\partial\mu(z)/\partial z_j)$, $j = 0, ..., n$. It can be solved as in [28]. Calculating μ from (8) or (9) we get the fundamental solution:

$$\mathcal{V} = \mathcal{E}_{\Upsilon^m+\beta} * (\delta - \mu) \quad (10)$$

and the (particular) solution of (1) is:

$$y = \mathcal{V} * g. \quad (11)$$

When $[\mathcal{E}_{\Upsilon^m+\beta}, \omega)$ is real for each real-valued test function ω or $\Upsilon f = (\partial f/\partial z_0)\phi_0(z)$ with a real-valued function $\phi_0(z)$ and the inverse relative to the convolution generalized function is known $\mathcal{E}^{-1}_{\Upsilon^m+\beta}$ such that

$$\mathcal{E}^{-1}_{\Upsilon^m+\beta} * \mathcal{E}_{\Upsilon^m+\beta} = \delta, \quad (12)$$

then Equation (8) simplifies:

$$\Upsilon\mu(z) + (n+1)^{-1}\sum_{j,k=0}^{n}\mathcal{E}^{-1}_{\Upsilon^m+\beta} * [[(dw^{-1}(\xi)/d\xi).(\hat{\theta}(z).(\partial\nu_j(z)/\partial z_k))]\phi_k^*(z)]$$

$$= \sum_{j=0}^{n}[\mathcal{E}^{-1}_{\Upsilon^m+\beta} * (\partial\mathcal{E}_{\Upsilon^m+\beta}/\partial z_j)]\phi_j^*(z), \text{ consequently,} \quad (13)$$

$$\Upsilon\mu(z) + \nu(\mu) = b(z), \quad (14)$$

where $\nu(\mu) := (n+1)^{-1}\sum_{j,k=0}^{n}\mathcal{E}^{-1}_{\Upsilon^m+\beta} * [[(dw^{-1}(\xi)/d\xi).(\hat{\theta}(z).(\partial\nu_j(z)/\partial z_k))]\phi_k^*(z)]$ and $b(z) = \mathcal{E}^{-1}_{\Upsilon^m+\beta} * (\Upsilon\mathcal{E}_{\Upsilon^m+\beta})$.

If equation (1) is solved, then it provides a solution of more general equation:

$$(\Upsilon^m + \beta + \hat{f}((\Upsilon)^{k-1}\xi)(\Upsilon)^k)\xi = g \quad (15)$$

finding ξ from the equation $(\Upsilon)^{k-1}\xi = y$, where $(\Upsilon)^k$ denotes the k-th power of the operator Υ.

If $\phi_j(z) = i_j\psi_j(z)$ for each j, then functions $\{\nu_j(z) : j\}$ can be chosen real-valued or real-valued generalized functions on real valued test functions (see System $(35, 36)$ [33] and §23 [44]). In such case the equality

$$\sum_{j,k=0}^{n}[(dw^{-1}(\xi)/d\xi).(\hat{\theta}(z).(\partial\nu_j(z)/\partial z_k))]\phi_k^*(z)$$

$$=\sum_{j,k=0}^{n}[(dw^{-1}(\xi)/d\xi).(\mu(z)(\partial\nu_j(z)/\partial z_k)i_j)]i_k^*\psi_k(z)]$$

is satisfied. For $\Upsilon f = (\partial f/\partial z_0)\phi_0(z)$ with a real-valued function $\phi_0(z)$ these equations simplify, since $\hat{\theta}.h = \mu(z)h$ for each $h \in \mathbf{R}$ and $z \in U$ and

$$(n+1)^{-1}\sum_{j,k=0}^{n}\mathcal{E}^{-1}_{\Upsilon^m+\beta} * [[(dw^{-1}(\xi)/d\xi).(\hat{\theta}(z).(\partial\nu_j(z)/\partial z_k))]\phi_k^*(z)] =$$

$$= \mathcal{E}^{-1}_{\Upsilon^m+\beta} * [(dw^{-1}(\xi)/d\xi).\mu(z)].$$

Thus the results of this section over the Cayley-Dickson algebras enrich the technique of integration of partial differential equations in comparison with the complex field.

4. Multidimensional Laplace Transforms over Cayley-Dickson Algebras and PDE

1. Definitions. Transforms in $\mathcal{A}_r$ Cartesian coordinates.

Denote by $\mathcal{A}_r$ the Cayley-Dickson algebra, $0 \le r$, which may be, in particular, $\mathbf{H} = \mathcal{A}_2$ the quaternion skew field or $\mathbf{O} = \mathcal{A}_3$ the octonion algebra. For unification of the notation we put $\mathcal{A}_0 = \mathbf{R}$, $\mathcal{A}_1 = \mathbf{C}$. A function $f : \mathbf{R}^n \to \mathcal{A}_r$ we call a function-original, where $2 \le r, n \in \mathbf{N}$, if it fulfills the following conditions $(1-5)$.

(1). The function $f(t)$ is almost everywhere continuous on $\mathbf{R}^n$ relative to the Lebesgue measure λ_n on $\mathbf{R}^n$.

(2). On each finite interval in $\mathbf{R}$ each function $g_j(t_j) = f(t_1,...,t_n)$ by t_j with marked all other variables may have only a finite number of points of discontinuity of the first kind, where $t = (t_1,...,t_n) \in \mathbf{R}^n$, $t_j \in \mathbf{R}$, $j = 1,...,n$. Recall that a point $u_0 \in \mathbf{R}$ is called a point of discontinuity of the first type, if there exist finite left and right limits $\lim_{u\to u_0,u<u_0} g(u) =: g(u_0-0) \in \mathcal{A}_r$ and $\lim_{u\to u_0,u>u_0} g(u) =: g(u_0+0) \in \mathcal{A}_r$.

(3). Every partial function $g_j(t_j) = f(t_1,...,t_n)$ satisfies the Hölder condition: $|g_j(t_j+h_j) - g_j(t_j)| \le A_j|h_j|^{\alpha_j}$ for each $|h_j| < \delta$, where $0 < \alpha_j \le 1$, $A_j = const > 0$, $\delta_j > 0$ are constants for a given $t = (t_1,...,t_n) \in \mathbf{R}^n$, $j = 1,...,n$, everywhere on $\mathbf{R}^n$ may be besides points of discontinuity of the first type.

(4). The function $f(t)$ increases not faster, than the exponential function, that is there exist constants $C_v = const > 0$, $v = (v_1,...,v_n)$, $a_{-1}, a_1 \in \mathbf{R}$, where $v_j \in \{-1,1\}$ for every $j = 1,...,n$, such that

$|f(t)| < C_v \exp((q_v,t))$ for each $t \in \mathbf{R}^n$ with $t_jv_j \ge 0$ for each $j = 1,...,n$, $q_v = (v_1a_{v_1},...,v_na_{v_n})$; where

(5). $(x,y) := \sum_{j=1}^{n} x_jy_j$ denotes the standard scalar product in $\mathbf{R}^n$.

Certainly for a bounded original f it is possible to take $a_{-1} = a_1 = 0$.

Each Cayley-Dickson number $p \in \mathcal{A}_r$ we write in the form

(6). $p = \sum_{j=0}^{2^r-1} p_j i_j$, where $\{i_0, i_1, ..., i_{2^r-1}\}$ is the standard basis of generators of $\mathcal{A}_r$ so that $i_0 = 1$, $i_j^2 = -1$ and $i_0 i_j = i_j = i_j i_0$ for each $j > 0$, $i_j i_k = -i_k i_j$ for each $j > 0$ and $k > 0$ with $k \ne j$, $p_j \in \mathbf{R}$ for each j.

If there exists an integral

(7). $F^n(p) := F^n(p;\zeta) := \int_{\mathbf{R}^n} f(t) e^{-<p,t)-\zeta} dt$,

then $F^n(p)$ is called the noncommutative multiparameter (Laplace) transform at a point $p \in \mathcal{A}_r$ of the function-original $f(t)$, where $\zeta - \zeta_0 = \zeta_1 i_1 + ... + \zeta_{2^r-1} i_{2^r-1} \in \mathcal{A}_r$ is the parameter of an initial phase, $\zeta_j \in \mathbf{R}$ for each $j = 0, 1, ..., 2^r - 1$, $\zeta \in \mathcal{A}_r$, $n = 2^r - 1$, $dt = \lambda_n(dt)$,

(8). $< p, t) = p_0(t_1 + ... + t_{2^r-1}) + \sum_{j=1}^{2^r-1} p_j t_j i_j$, we also put

(8.1). $u(p,t;\zeta) =< p,t) + \zeta$.

For vectors $v, w \in \mathbf{R}^n$ we shall consider a partial ordering

(9). $v \prec w$ if and only if $v_j \le w_j$ for each $j = 1, ..., n$ and a k exists so that $v_k < w_k$, $1 \le k \le n$.

2. Transforms in $\mathcal{A}_r$ Spherical Coordinates.

Now we consider also the non-linear function $u = u(p,t;\zeta)$ taking into account non commutativity of the Cayley-Dickson algebra $\mathcal{A}_r$. Put

$$u(p,t) := u(p,t;\zeta) := p_0 s_1 + M(p,t) + \zeta_0, \qquad (1)$$

where

$$M(p,t) = M(p,t;\zeta) = (p_1 s_1 + \zeta_1)[i_1 \cos(p_2 s_2 + \zeta_2) + i_2 \sin(p_2 s_2 + \zeta_2)$$

$$\cos(p_3 s_3 + \zeta_3) + ... + i_{2^r-2} \sin(p_2 s_2 + \zeta_2) ... \sin(p_{2^r-2} s_{2^r-2} + \zeta_{2^r-2}) \cos(p_{2^r-1} s_{2^r-1}$$

$$+\zeta_{2^r-1}) + i_{2^r-1} \sin(p_2 s_2 + \zeta_2) ... \sin(p_{2^r-2} s_{2^r-2} + \zeta_{2^r-2}) \sin(p_{2^r-1} s_{2^r-1} + \zeta_{2^r-1})] \quad (2)$$

for the general Cayley-Dickson algebra with $2 \le r < \infty$,

$$s_j := s_j(n;t) := t_j + ... + t_n \qquad (2.1)$$

for each $j = 1, ..., n$, $n = 2^r - 1$, so that $s_1 = t_1 + ... + t_n$, $s_n = t_n$. More generally, let

$$u(p,t) = u(p,t;\zeta) = p_0 s_1 + w(p,t) + \zeta_0, \qquad (3)$$

where $w(p,t)$ is a locally analytic function, $Re(w(p,t)) = 0$ for each $p \in \mathcal{A}_r$ and $t \in \mathbf{R}^{2^r-1}$, $Re(z) := (z + \tilde{z})/2$, $\tilde{z} = z^*$ denotes the conjugated number for $z \in \mathcal{A}_r$. Then the more general non-commutative multiparameter transform over $\mathcal{A}_r$ is defined by the formula:

$$F_u^n(p;\zeta) := \int_{\mathbf{R}^n} f(t) \exp(-u(p,t;\zeta)) dt \qquad (4)$$

for each Cayley-Dickson numbers $p \in \mathcal{A}_r$ whenever this integral exists as the principal value of either Riemann or Lebesgue integral, $n = 2^r - 1$. This non-commutative multiparameter transform is in $\mathcal{A}_r$ spherical coordinates, when $u(p,t;\zeta)$ is given by Formulas $(1,2)$.

At the same time the components p_j of the number p and ζ_j for ζ in $u(p,t;\zeta)$ we write in the p- and ζ-representations respectively such that

$$h_j = (-hi_j + i_j(2^r-2)^{-1}\{-h + \sum_{k=1}^{2^r-1} i_k(hi_k^*)\})/2 \tag{5}$$

for each $j = 1, 2, ..., 2^r - 1$,

$$h_0 = (h + (2^r-2)^{-1}\{-h + \sum_{k=1}^{2^r-1} i_k(hi_k^*)\})/2, \tag{6}$$

where $2 \le r \in \mathbf{N}$, $h = h_0 i_0 + ... + h_{2^r-1} i_{2^r-1} \in \mathcal{A}_r$, $h_j \in \mathbf{R}$ for each j, $i_k^* = \tilde{i}_k = -i_k$ for each $k > 0$, $i_0 = 1$, $h \in \mathcal{A}_r$. Denote $F_u^n(p;\zeta)$ in more details by $\mathcal{F}^n(f,u;p;\zeta)$.

Henceforth, the functions $u(p,t;\zeta)$ given by $1(8, 8.1)$ or $(1, 2, 2.1)$ are used, if another form (3) is not specified. If for $u(p,t;\zeta)$ concrete formulas are not mentioned, it will be undermined, that the function $u(p,t;\zeta)$ is given in $\mathcal{A}_r$ spherical coordinates by Expressions $(1, 2, 2.1)$. If in Formulas $1(7)$ or (4) the integral is not by all, but only by $t_{j(1)}, ..., t_{j(k)}$ variables, where $1 \le k < n$, $1 \le j(1) < ... < j(k) \le n$, then we denote a noncommutative transform by $F_u^{k;t_{j(1)},...,t_{j(k)}}(p;\zeta)$ or $\mathcal{F}^{k;t_{j(1)},...,t_{j(k)}}(f,u;p;\zeta)$. If $j(1) = 1$,...,$j(k) = k$, then we denote it shortly by $F_u^k(p;\zeta)$ or $\mathcal{F}^k(f,u;p;\zeta)$. Henceforth, we take $\zeta_m = 0$ and $t_m = 0$ and $p_m = 0$ for each $1 \le m \notin \{j(1), ..., j(k)\}$ if something other is not specified.

Properties of such transforms for originals and generalized originals over Cayley-Dickson algebras and their applications to solutions of PDE and with boundary conditions as well were described in [34, 35].

3. Examples. PDE. Let

$$Af(t) = \sum_{j=1}^{n} (\partial^2 f(t)/\partial t_j^2) c_j \tag{1}$$

be the operator with constant coefficients $c_j \in \mathcal{A}_r$, $|c_j| = 1$, by the variables $t_1, ..., t_n$, $n \ge 2$. We suppose that c_j are such that the minimal subalgebra $alg_{\mathbf{R}}(c_j, c_k)$ containing c_j and c_k is alternative for each j and k and $|(...(c_1^{1/2}c_2^{1/2})...)c_n^{1/2}| = 1$. Since

$$\partial f(t)/\partial t_j = \sum_{k=1}^{n} (\partial f(t(s))/\partial s_k)(\partial s_k/\partial t_j) = \sum_{k=1}^{j} \partial f(t(s))/\partial s_k, \tag{2}$$

the operator A takes the form

$$Af = \sum_{j=1}^{n} (\sum_{1 \le k, b \le j} (\partial^2 f(t(s))/\partial s_k \partial s_b)) c_j, \tag{3}$$

where $s_j = t_j + ... + t_n$ for each j. Therefore, by Theorem 12 and Formulas $25(SO)$ and $28(6)$ [35] we get:

$$\mathcal{F}^n(Af;u;p;\zeta) = \sum_{j=1}^{n}\{[\mathsf{R}_{e_j}(p)]^2 F_u^n(p;\zeta)\}c_j \tag{4}$$

for $u(p,t;\zeta)$ either in $\mathcal{A}_r$ spherical or $\mathcal{A}_r$ Cartesian coordinates with the corresponding operators $\mathsf{R}_{e_j}(p)$ (see also Formulas $25(1.1, 1.2)$ [35]).
On the other hand,

$$\mathcal{F}^n(\delta;u;p;\zeta) = e^{-u(p,0;\zeta)} = e^{-u(0,0;\zeta)} \tag{5}$$

in accordance with Formula 20(2) [35]. The delta function $\delta(t)$ is invariant relative to any invertible linear operator $C : \mathbf{R}^n \to \mathbf{R}^n$ with the determinant $|\det(C)| = 1$, since

$$\int_{\mathbf{R}^n} \delta(Cx)\phi(x)dx = \int_{\mathbf{R}^n} \delta(y)\phi(C^{-1}y)|\det(C)|dy = \phi(C^{-1}0) = \phi(0).$$

Thus

$$\mathcal{F}^n(C(Af);u;p;\zeta) = \mathcal{F}^n(Af;u;p;\zeta) \tag{5}$$

for any Fundamental solution f, where $Cg(t) := g(Ct)$, $Af = \delta$. If $C : \mathbf{R}^n \to \mathbf{R}^n$ is an invertible linear operator and $\xi = Ct$, $q = Cp$, $\zeta' = C\zeta$, then $t = C^{-1}\xi$, $p = C^{-1}q$ and $\zeta = C^{-1}\zeta'$. In the multiparameter noncommutative transform $\mathcal{F}^n$ there are the corresponding variables (t_j, p_j, ζ_j). This is accomplished in particular for the operator $C(t_1, ..., t_n) = (s_1, ..., s_n)$. The operator C^{-1} transforms the right side of Formula (4), when it is written in the $\mathcal{A}_r$ spherical coordinates, into $\sum_{j=1}^n\{(p_0 + q_j\mathsf{S}_{e_j})^2 F_u^n(q;\zeta)\}c_j$. The Cayley-Dickson number $q = q_0 + q_1 i_1 + ... + q_n i_n$ can be written as $q = q_0 + q_M M$, where $|M| = 1$, M is a purely imaginary Cayley-Dickson number, $q_M \in \mathbf{R}$, $q_M M = q_1 i_1 + ... + q_n i_n$, since $q_0 = Re(q)$. After a suitable automorphism $\theta : \mathcal{A}_r \to \mathcal{A}_r$ we can take $\theta(q) = q_0 + q_M i_1$, so that $\theta(x) = x$ for any real number. The functions $[\sum_{j=1}^n q_j^2 \mathsf{S}_{e_j}^2 c_j]$ and $[\sum_{j=1}^n p_j^2 \mathsf{S}_{e_j}^2 c_j]$ are even by each variable q_j and p_j respectively. Therefore, we deduce in accordance with (5) and $2(3, 4)$ and Corollary 6.1 [34] with parameters $p_0 = 0$ and $\zeta = 0$ and $c_j \in \{-1, 1\}$ for each j that

$$(\mathcal{F}^n)^{-1}(1/[\sum_{j=1}^{n}\{\sum_{1\le k,b\le j} p_k\mathsf{S}_{e_k}p_b\mathsf{S}_{e_b}\}c_j];u;y;\zeta) = -[g, e^{N([y],[q])}) \tag{6}$$

in the $\mathcal{A}_r$ spherical coordinates, where $g = 1/[\sum_{j=1}^n q_j^2 c_j]$, or

$$(\mathcal{F}^n)^{-1}(1/[\sum_{j=1}^{n}\{p_j^2\mathsf{S}_{e_j}^2\}c_j];u;y;\zeta) = -[g, e^{N([y],[p])}) \tag{6.1}$$

in the $\mathcal{A}_r$ Cartesian coordinates, where $g = 1/[\sum_{j=1}^n p_j^2 c_j]$, $N = y/|y|$ for $y \ne 0$, $N = i_1$ for $y = 0$, $y = y_1 i_1 + ... + y_n i_n \in \mathcal{A}_r$, $[y] = (y_1, ..., y_n) \in \mathbf{R}^n$, $([y], [q]) = \sum_{j=1}^n y_j q_j$, since $\mathsf{S}_{e_k}^2 \cos(\phi + \zeta_k) = \cos(\phi + \zeta_k + \pi) = -\cos(\phi + \zeta_k)$ and $\mathsf{S}_{e_k}^2 \sin(\phi + \zeta_k) = \sin(\phi + \zeta_k + \pi) = -\sin(\phi + \zeta_k)$ for each k.

Particularly, we take $c_j = 1$ for each $j = 1, ..., k_+$ and $c_j = -1$ for any $j = k_+ + 1, ..., n$, where $1 \le k_+ \le n$. Thus the inverse Laplace transform for $q_0 = 0$ and $\zeta = 0$ in accordance with Formulas $2(1-4)$ reduces to

$$(\mathcal{F}^n)^{-1}(1/[\sum_{j=1}^{n}\{\sum_{1\le k,b\le j} p_k \mathsf{S}_{e_k} p_b \mathsf{S}_{e_b}\}c_j]; u; y; \zeta) =$$

$$(2\pi)^{-n}\int_{\mathbf{R}^n} \exp(\mathbf{i}(q_1y_1+...+q_ny_n))(1/[\sum_{j=1}^{k_+} q_j^2 - \sum_{j=k_++1}^{n} q_j^2])dq_1...dq_n \qquad (7)$$

in the $\mathcal{A}_r$ spherical coordinates and

$$(\mathcal{F}^n)^{-1}(1/[\sum_{j=1}^{n} p_j^2 \mathsf{S}_{e_j}^2 c_j]; u; y; \zeta) =$$

$$(2\pi)^{-n}\int_{\mathbf{R}^n} \exp(\mathbf{i}(p_1y_1+...+p_ny_n))(1/[\sum_{j=1}^{k_+} p_j^2 - \sum_{j=k_++1}^{n} p_j^2])dp_1...dp_n \qquad (7.1)$$

in the $\mathcal{A}_r$ Cartesian coordinates,
since for any even function its cosine Fourier transform coincides with the Fourier transform.

The inverse Fourier transform $(F^{-1}g)(x) = (2\pi)^{-n}(Fg)(-x) =: \Psi_n$ of the functions $g = 1/(\sum_{j=1}^{n} z_j^2)$ for $n \ge 3$ and $\mathcal{P}(1/(\sum_{j=1}^{2} z_j^2))$ for $n = 2$ in the class of the generalized functions is known (see [9] and §§9.7 and 11.8 [57]) and gives

$$\Psi_n(z_1,...,z_n) = C_n(\sum_{j=1}^{n} z_j^2)^{1-n/2} \qquad (8)$$

for $3 \le n$, where $C_n = -1/[(n-2)\sigma_n]$, $\sigma_n = 4\pi^{n/2}/\Gamma((n/2)-1)$ denotes the surface of the unit sphere in $\mathbf{R}^n$, $\Gamma(x)$ denotes Euler's gamma-function, while

$$\Psi_2(z_1,z_2) = C_2 \ln(\sum_{j=1}^{2} z_j^2) \qquad (9)$$

for $n = 2$, where $C_2 = 1/(4\pi)$.

For the function

$$P(x) = \sum_{j=1}^{k_+} x_j^2 - \sum_{j=k_++1}^{n} x_j^2 \qquad (10)$$

with $1 \le k_+ < n$ the generalized functions $(P(x)+\mathbf{i}0)^\lambda$ and $(P(x)-\mathbf{i}0)^\lambda$ are defined for any $\lambda \in \mathbf{C} = \mathbf{R}\oplus\mathbf{i}\mathbf{R}$ (see Chapter 3 in [9]). The function P^λ has the cone surface $P(z_1,...,z_n) = 0$ of zeros, so that for the correct definition of generalized functions corresponding to P^λ the generalized functions

$$(P(x)+c\mathbf{i}0)^\lambda = \lim_{0<c\epsilon,\epsilon\to 0}(P(x)^2+\epsilon^2)^{\lambda/2}\exp(\mathbf{i}\lambda arg(P(x)+\mathbf{i}c\epsilon)) \qquad (11)$$

with either $c = -1$ or $c = 1$ were introduced. Therefore, the identity

$$F(\Psi_{k_+,n-k_+})(x) = -(\sum_{j=1}^{k_+} x_j^2 - \sum_{j=k_++1}^{n} x_j^2)[F(\Psi_{k_+,n-k_+})(x)]^2 \qquad (12)$$

or

$$F(\Psi) = -1/(P(x) + c\mathbf{i}0) \tag{13}$$

follows, where $c = -1$ or $c = 1$.

The inverse Fourier transform in the class of the generalized functions is:

$$F^{-1}((P(x) + c\mathbf{i}0)^{\lambda})(z_1, ..., z_n) =$$

$$= \exp(-\pi c(n-k_+)\mathbf{i}/2)2^{2\lambda+n}\pi^{n/2}\Gamma(\lambda+n/2)(Q(z_1, ..., z_n)-c\mathbf{i}0)^{-\lambda-n/2)}/[\Gamma(-\lambda)|D|^{1/2}] \tag{14}$$

for each $\lambda \in \mathbf{C}$ and $n \geq 3$ (see §IV.2.6 [9]), where $D = \det(g_{j,k})$ denotes a discriminant of the quadratic form $P(x) = \sum_{j,k=1}^{n} g_{j,k}x_jx_k$, while $Q(y) = \sum_{j,k=1}^{n} g^{j,k}x_jx_k$ is the dual quadratic form so that $\sum_{k=1}^{n} g^{j,k}g_{k,l} = \delta_l^j$ for all j, l; $\delta_l^j = 1$ for $j = l$ and $\delta_l^j = 0$ for $j \neq l$. In the particular case of $n = 2$ the inverse Fourier transform is given by the formula:

$$F^{-1}((P(x)+c\mathbf{i}0)^{-1})(z_1, z_2) = -4^{-1}|D|^{-1/2}\exp(-\pi c(n-k_+)\mathbf{i}/2)\ln(Q(z_1, ..., z_n)-c\mathbf{i}0). \tag{15}$$

Making the inverse Fourier transform F^{-1} of the function $-1/(P(x)+\mathbf{i}0)$ in this particular case of $\lambda = -1$ we get two complex conjugated fundamental solutions

$$\Psi_{k_+,n-k_+}(z_1, ..., z_n) = -\exp(\pi c(n-k_+)\mathbf{i}/2)\Gamma((n/2)-1)(Q(z_1, ..., z_n)+c\mathbf{i}0)^{1-(n/2)}/(4\pi^{n/2}) \tag{16}$$

for $3 \leq n$ and $1 \leq k_+ < n$, while

$$\Psi_{1,1}(z_1, z_2) = 4^{-1}\exp(\pi c(n-k_+)\mathbf{i}/2)\ln(Q(z_1, z_2) + c\mathbf{i}0) \tag{17}$$

for $n = 2$, where either $c = 1$ or $c = -1$.

Generally for the operator A given by Formula (1) we get $P(x) = P_0(x) + P_i(x)$, where $P_0(x) = \sum_{j=1}^{n} x_j^2 Re(c_j)$ and $P_i(x) = \sum_{j=1}^{n} x_j^2 Im(c_j)$ are the real and imaginary parts of P, $Im(z) = z - Re(z)$ for any Cayley-Dickson number z. Take $\mathbf{l} = i_{2^r}$ and consider the form $P(x) + \epsilon c\mathbf{l}$ with $\epsilon \neq 0$ and either $c = 1$ or $c = -1$, then $P(x) + \epsilon c\mathbf{l} \neq 0$ for each $x \in \mathbf{R}^n$. We put

$$(P(x) + c\mathbf{l}0)^{\lambda} = \lim_{0<c\epsilon,\epsilon\to 0}(P(x)^2 + \epsilon^2)^{\lambda/2}\exp(\mathbf{i}\lambda Arg(P(x) + \mathbf{l}c\epsilon)). \tag{18}$$

Consider $\lambda \in \mathbf{R}$, the generalized function $(P(x)^2 + \epsilon^2)^{\lambda/2}\exp(\mathbf{i}\lambda Arg(P(x) + \mathbf{l}c\epsilon))$ is non-degenerate and for it the Fourier transform is defined. The limit $\lim_{0<c\epsilon,\epsilon\to 0}$ gives by our definition the Fourier transform of $(P(x) + c\mathbf{l}0)^{\lambda}$. Since

$$c_j(\beta_j + \sum_{1\leq k\leq n, k\neq j} c_j^{-1}c_k\beta_k) = \sum_{j=1}^{n} c_j\beta_j \tag{19}$$

for all $\beta_j \in \mathbf{R}$ and any $1 \leq j \leq n$ in accordance with the conditions imposed on c_j at the beginning of this section and $\mathbf{i}N_j = N_j\mathbf{i}$ for each j, the Fourier transform with the generator $\mathbf{i}$ can be accomplished subsequently by each variable using Identity (19). The transform $x_j \mapsto (c_j)^{1/2}x_j$ is diagonal and $|(...((c_1^{1/2}c_2^{1/2})...)c_n^{1/2}| = 1$, so we can put $|D| = 1$.

Each Cayley-Dickson number can be presented in the polar form $z = |z|e^{\phi M}$, $\phi \in \mathbf{R}$, $|\phi| \leq \pi$, M is a purely imaginary Cayley-Dickson number $|M| = 1$, $Arg(z) = (\phi +$

$2\pi k)M$ has the countable number of values, $k \in \mathbf{Z}$ (see §3 in [26, 25]). Therefore, we choose the branch $z^{1/2} = |z|^{1/2}\exp((Argz)/2)$, $|z|^{1/2} > 0$ for $z \neq 0$, with $|Arg(z)| \leq \pi$, $Arg(M) = M\pi/2$ for each purely imaginary M with $|M| = 1$.

Taking initially $c_j \in \mathbf{R}$ and considering the complex analytic extension of formulas given above in each complex plane $\mathbf{R} \oplus N_j\mathbf{R}$ by c_j for each j by induction from 1 to n, when c_j is not real in the operator A, $Im(c_j) \in \mathbf{R}N_j$, we get the fundamental solutions for A with the form $(P(x)+c\mathbf{1}0)^\lambda$ instead of $(P(x)+c\mathbf{i}0)^\lambda$ with multipliers $(...(c_1^{c/2}c_2^{c/2})...)c_n^{c/2}$ instead of $\exp(\pi c(n-k_+)\mathbf{i}/2)$ as above and putting $|D| = 1$. Thus

$$\Psi(z_1,...,z_n) = -\Gamma((n/2)-1)(P^*(z_1,...,z_n)-c\mathbf{1}0)^{1-(n/2)}[(...(c_1^{c/2}c_2^{c/2})...)c_n^{c/2}]^*/(4\pi^{n/2}) \quad (20)$$

for $3 \leq n$, while

$$\Psi(z_1,z_2) = 4^{-1}[c_1^{c/2}c_2^{c/2}]^* Ln(P^*(z_1,z_2) - c\mathbf{1}0) \quad (21)$$

for $n = 2$, since $c_j^* = c_j^{-1}$ for $|c_j| = 1$,

$$y_jq_j = y_j(c_j^{c/2})^* q_j c_j^{1/2},$$

while

$$(...(dc_1^{c/2}q_1 dc_2^{c/2}q_2)...)dc_n^{c/2}q_n] = dq_1...dq_n[(...(c_1^{c/2}c_2^{c/2})...)c_n^{c/2}]$$

and

$$|(...(c_1^{c/2}c_2^{c/2})...)c_n^{c/2}| = 1.$$

4. PDE with Polynomial Real Coefficients. Let

$$A = \sum_{|\alpha|\leq m} a_\alpha(q)\partial_q^\alpha, \quad (1)$$

$$a_\alpha(q) = \sum_\beta a_{\alpha,\beta}q^\beta,$$

$$q^\beta := q_1^{\beta_1}...q_n^{\beta_n},$$

$a_{\alpha,\beta}$ and f have values as in §28 [35], and Af be an original. Using the transform in the $\mathcal{A}_r$ Cartesian coordinates we take $q_j = t_j$ for each j, while using the transform in $\mathcal{A}_r$ spherical coordinates we choose $q_j = s_j(t)$ for each j. Then

$$\mathcal{F}^n(Af;u;p;\zeta) = \sum_\beta (-1)^{|\beta|} \mathsf{S}_\beta(p)\partial_p^\beta$$

$$[\sum_\beta a_{\alpha,\beta}([p_0 + p_1\mathsf{S}_{e_1}]^{\alpha_1} p_2^{\alpha_2}\mathsf{S}_{e_2}^{\alpha_2}...p_n^{\alpha_n}\mathsf{S}_{e_n}^{\alpha_n})]F^n(p;\zeta) = G^n(p;\zeta) \quad (2)$$

in the $\mathcal{A}_r$ spherical coordinates and

$$\mathcal{F}^n(Af;u;p;\zeta) = \sum_\beta (-1)^{|\beta|} \mathsf{S}_\beta(p)\partial_p^\beta$$

$$(\sum_{\beta} a_{\alpha,\beta}[p_0 + p_1 S_{e_1}]^{\alpha_1}[p_0 + p_2 S_{e_2}]^{\alpha_2}...[p_0 + p_n S_{e_n}]^{\alpha_n})F^n(p;\zeta) = G^n(p;\zeta) \quad (2.1)$$

in the $\mathcal{A}_r$ Cartesian coordinates (see Theorems 12 and 13 [35]). It may happen that the second differential equation is simpler than the initial one:

$$Af = g. \quad (3)$$

For example, when coefficients depend only on one variable t_n, then the second differential equation is ordinary and linear.

5. Noncommutative Transforms of Products and Convolutions of Functions in the $\mathcal{A}_r$ Spherical Coordinates.

For any Cayley-Dickson number $z = z_0 i_0 + ... + z_{2^r-1} i_{2^r-1}$ we consider projections

$$\theta_j(z) = z_j, z_j \in \mathbf{R} or \mathbf{C}_{\mathbf{i}} or \mathbf{H}_{J,K,L}, j = 0, ..., 2^r - 1, \theta_j(z) = \pi_j(z) i_j^*, \quad (1)$$

given by Formulas $2(5,6)$ and $33(17)$ [35]. We define the following operators

$$\mathcal{R}_{\alpha,j}(F^n(p;\zeta)) := F^n(p_0, (-1)^{\alpha_1} p_1, ..., (-1)^{\alpha_{j+1-\delta_{j,n}}} p_{j+1-\delta_{j,n}}, p_{j+2-\delta_{j,n}},$$

$$..., p_n; \zeta_0, (-1)^{\alpha_1}\zeta_1 + \pi\alpha_1/2, ..., (-1)^{\alpha_{j+1-\delta_{j,n}}}\zeta_{j+1-\delta_{j,n}} + \pi\alpha_{j+1-\delta_{j,n}}/2, \zeta_{j+2-\delta_{j,n}}, ..., \zeta_n) \quad (2)$$

on images F^n, $2^{r-1} \le n \le 2^r - 1$, $j = 0, ..., n$. For α_j and $\beta_j \in \{0,1\}$ their sum $\alpha_j + \beta_j$ is considered by $(mod\ 2)$, i.e. in the ring $\mathbf{Z}_2 = \mathbf{Z}/(2\mathbf{Z})$, for two vectors α and $\beta \in \{0,1\}^{2^r-1}$ their sum is considered componentwise in $\mathbf{Z}_2$. Let

$$\mathcal{F}^n(f;u;p;\zeta) = \sum_{j=0}^{n}\sum_{k=0}^{2^r-1} \theta_j(\mathcal{F}^n(\theta_k(f);u;p;\zeta)) i_k i_j, \quad (3)$$

also $F_j^n(p;\zeta) := \sum_{k=0}^{2^r-1} \theta_j(\mathcal{F}^n(\theta_k(f);u;p;\zeta)) i_k$ for an original f, where $u(p,t;\zeta)$ is given by Formulas $2(1,2,2.1)$ [35]. If f is real or $\mathbf{C}_{\mathbf{i}}$ or $\mathbf{H}_{J,K,L}$ -valued, then $F_j^n = \theta_j(F^n)$.

Theorem. *If f and g are two originals, then*

$$\mathcal{F}^n(fg;u;p;\zeta) =$$

$$= \sum_{j=0}^{n} \sum_{\alpha,\beta\in\{0,1\}^n} (-1)^{\alpha_{j+1}(1-\delta_{j+1,n})} (\mathcal{R}_{\alpha,j}(F_j^n(p - q_0;\zeta - \eta)) * (\mathcal{R}_{\beta,j}(G_j^n(p + q_0 - p_0;\eta)) i_j, \quad (4)$$

$$\mathcal{F}^n(f * g;u;p;\zeta) =$$

$$= \sum_{j=0}^{n} \sum_{\alpha,\beta\in\{0,1\}^n} (-1)^{\alpha_{j+1}(1-\delta_{j+1,n})} (\mathcal{R}_{\alpha,j}(F_j^n(p;\zeta - \eta)) (\mathcal{R}_{\beta,j}(G_j^n(p;\eta)) i_j, \quad (4.1)$$

whenever $\mathcal{F}^n(fg)$, $\mathcal{F}^n(f)$, $\mathcal{F}^n(g)$ exist, where $1 \le n \le 2^r - 1$, $2 \le r$; $\alpha_k + \beta_k = 1\ (mod\ 2)$ for $k \le j$ or $k = j + 1 = n$, $\alpha_k + \beta_k = 0\ (mod\ 2)$ for $k = j + 1 < n$ and $\alpha_k = \beta_k = 0$ for $k > j + 1$ in the j-th addendum on the right of Formulas $(4, 4.1)$; the convolution is by

$(p_1, ..., p_n)$ *in* (4), *at the same time* $q_0 \in \mathbf{R}$ *and* $\eta \in \mathcal{A}_r$ *are fixed.*

6. Moving Boundary Problem.

Let us consider a boundary problem

$$Af = g \quad (1)$$

in the half-space $t_n \geq \phi(t_n)$, where $\phi(0) = 0$ and $\phi(t_n) < t_n$ for each $0 \leq t_n \in \mathbf{R}$. Suppose that the function $t_n - \phi(t_n) =: \psi(t_n)$ is differentiable and bijective. For example, if $0 < v < 1$ and $\phi(t_n) = vt_n$, then the boundary is moving with the speed v. Make the change of variables $y_n = \psi(t_n)$, $y_1 = t_1$,...,$y_{n-1} = t_{n-1}$, then $t_n = \psi^{-1}(y_n)$ and $dt_n = ds_n = (dt_n/dy_n)dy_n$ and due to Theorem 25 [35] we infer that

$$\mathcal{F}^n(\sum_{|\alpha|\leq m} \mathbf{b}_\alpha \partial_s^\alpha \chi_{y_n\geq 0} f(t); p; \zeta) = \sum_{|\alpha|\leq m, 0\leq q_n\leq \alpha_n - 1} \mathbf{b}_\alpha(\delta_{0,\alpha_n} - 1)$$

$$(p_0 + \mathsf{S}_{e_1}p_1)^{\alpha_1} p_2^{\alpha_2} ... p_{n-1}^{\alpha_{n-1}} p_n^{\alpha_n - q_n - 1} \mathsf{S}_{\alpha - \alpha_1 e_1 - (q_n+1)e_n} \mathcal{F}^{n-1,y^n}(\partial_{t_n}^{q_n} w(y), u(p, (y^n); \zeta); p; \zeta)$$

$$+ \sum_{|\alpha|\leq m} \mathbf{b}_\alpha (p_0 + \mathsf{S}_{e_1}p_1)^{\alpha_1} p_2^{\alpha_2} ... p_n^{\alpha_n} \mathsf{S}_{\alpha - \alpha_1 e_1} \mathcal{F}^n(\chi_{y_n\geq 0}(y) w(y); p; \zeta) = G^n(p; \zeta) \quad (2)$$

in the $\mathcal{A}_r$ spherical coordinates and

$$\mathcal{F}^n(\sum_{|\alpha|\leq m} \mathbf{a}_\alpha \partial_t^\alpha \chi_{y_n\geq 0} f(t); p; \zeta) = \sum_{|\alpha|\leq m, 0\leq q_n\leq \alpha_n - 1} \mathbf{a}_\alpha(\delta_{0,\alpha_n} - 1)$$

$$(p_0 + \mathsf{S}_{e_1}p_1)^{\alpha_1}(p_0 + p_2\mathsf{S}_{e_2})^{\alpha_2} ... (p_0 + p_{n-1}\mathsf{S}_{e_{n-1}})^{\alpha_{n-1}}(p_0 + p_n\mathsf{S}_{e_n})^{\alpha_n - q_n - 1}$$

$$\mathcal{F}^{n-1,y^n}(\partial_{t_n}^{q_n} w(y), u(p, (y^n); \zeta); p; \zeta)$$

$$+ \sum_{|\alpha|\leq m} \mathbf{a}_\alpha (p_0 + \mathsf{S}_{e_1}p_1)^{\alpha_1}(p_0 + p_2\mathsf{S}_{e_2})^{\alpha_2} ... (p_0 + p_n\mathsf{S}_{e_n})^{\alpha_n} \mathcal{F}^n(\chi_{y_n\geq 0}(y) w(y); p; \zeta) = G^n(p; \zeta) \quad (2.1)$$

in the $\mathcal{A}_r$ Cartesian coordinates, where $w(y) := f(t(y))(dt_n/dy_n)$.

Expressing $\mathcal{F}^n(\chi_{y_n\geq 0}(y)w(y); p; \zeta)$ through $G^n(p; \zeta)$ and the boundary terms $\mathcal{F}^{n-1,y^n}(\partial_{t_n}^{q_n} w(y), u(p, (y^n); \zeta); p; \zeta)$ as in §28.3 and making the inverse transform 8(4) or 8.1(1) [35], or using the integral kernel ξ as in §28.5 [35], one gets a solution $w(y)$ or $f(t) = w(y(t))(dy_n(t_n)/dt_n)$.

7. PDE with Discontinuous Coefficients.

Consider a domain U and its subdomains $U \supset U_1 \supset ... \supset U_k$ satisfying Conditions 28$(D1, D4, i - vii)$ [35] so that coefficients of an operator A (see 28(2) [35]) are constant on $Int(U_k)$ and on $V_1 = U \setminus Int(U_1)$, $V_2 = U_1 \setminus Int(U_2)$,...,$V_k = U_{k-1} \setminus Int(U_k)$ and are allowed to be discontinuous at the common borders $\partial V_j \cap \partial U_j$ for each $j = 1, ..., k$. Each function $f\chi_{U_j}$ is an original on U or a generalized function with the support $supp(f\chi_{U_j}) \subset U_j$ if f is an original or a generalized function on U. Choose operators A^j with constant coefficients on U^j and $A^j|_{Int(V_j)} = 0$, where $U^0 = U$, so that $A|_{U_k} = A^k$,..., $A|_{U_j} =$

$A^j + ... + A^k$,..., $A|_U = A^0 + ... + A^k$. Therefore, in the class of originals or generalized functions on U the problem (see $28(1,2)$ [35]) can be written as

$$Af = g, \tag{1}$$

or

$$A^0 f\chi_{V_1} = g\chi_{V_1}, ..., A^{k-1} f\chi_{V_k} = g\chi_{V_k}, A^k f\chi_{U_k} = g\chi_{U_k}, \tag{2}$$

since $\chi_{V_1} + ... + \chi_{V_k} + \chi_{U_k} = \chi_U$. Thus the equivalent problem is:

$$A^0 f^0 = g^0, A^1 f^1 = g^1, ..., A^k f^k = g^k \tag{3}$$

with $f^k = f\chi_{U_k}$, $g^k = g\chi_{U_k}$, also $f^j = f\chi_{V_{j+1}}$, $g^j = g\chi_{V_{j+1}}$ for each $j = 0, ..., k-1$. On ∂U take the boundary condition in accordance with $28(5.1)$ [35]. With any boundary conditions in the class of originals or generalized functions on additional borders $\partial U_j \setminus \partial U$ given in accordance with $28(5.1)$ [35] a solution f^j on U^j exists, when the corresponding condition $8(3)$ is satisfied (see Theorems 8 and 28.1 [35]).

Each problem $A^j f^j = g^j$ can be considered on U_j, since $supp(g^j) \subset U_j$. Extend f^j by zero on $U \setminus V_j$ for each $0 \le j \le k-1$. When the right side of $28(6)$ [35] is non-trivial, then f^j is non-trivial. If f^{j-1} is calculated, then the boundary conditions on $\partial U^j \setminus \partial U$ can be chosen in accordance with values of f^{j-1} and its corresponding derivatives $(\partial^\beta f^{j-1}/\partial \nu^\beta)|_{(\partial U^j \setminus \partial U)}$ for some $\beta < ord(A^j)$ in accordance with the operator A^j and the boundary conditions $28(5.1)$ [35] on the boundary $\partial U^j \setminus \partial U$. Having found f^j for each $j = 0, ..., k$ one gets the solution $f = f^0 + ... + f^k$ on U of Problem (1) with the boundary conditions $28(5.1)$ [35] on ∂U.

5. Integration of Vector Sobolev Type PDE Over Octonions

1. Definitions. Let $C^m(U, \mathcal{A}_v)$ denote the family of all m times continuously differentiable functions $f : U \to \mathcal{A}_v$ by real variables $x_0, ..., x_{p-1}$ for a domain U in $\mathcal{A}_r$ such that $Int_{\mathbf{R}^p}(G) \subset G \subset cl_{\mathbf{R}^p}[Int(G)]$, where G is the real shadow of either the domain U or its section by a plane containing zero of real dimension p, $Int_{\mathbf{R}^p}(G)$ denotes the interior of G in the Euclidean space $\mathbf{R}^p$, $cl_{\mathbf{R}^p} G$ denotes the closure of G in $\mathbf{R}^p$. There is considered the embedding $\theta : \mathbf{R}^p \hookrightarrow \mathcal{A}_r$ corresponding to variables in G such that $\theta(x_0, ..., x_{p-1}) = x_0 i_{j(0)} + ... + x_{p-1} i_{j(p-1)}$ with $0 \le j(0) < ... < j(p-1) \le 2^r - 1$, where $x_0, ..., x_{p-1} \in \mathbf{R}$.

Let also $C_0^m(H_n, \mathcal{A}_v)$ denote the subspace of functions m times continuously differentiable by $t_0, ..., t_{n-1}$ such that $\sigma_t^k f|_{t_0=0} = 0$ for each $k = 0, ..., m-1$ when $m \ge 1$, where $H_n := \{z \in \mathcal{A}_r : z_0 \ge 0, z = z_0 i_0 + ... + z_{n-1} i_{n-1}\}$, $z_0, ..., z_{n-1} \in \mathbf{R}$, $\{i_0, ..., i_{2^r-1}\}$ is the set of standard generators of the Cayley-Dickson algebra $\mathcal{A}_r$, $t = t_0 i_0 + ... + t_{n-1} i_{n-1} \in H_n$, $1 \le n \le 2^r$, $t = \theta_0(t_0, ..., t_{n-1})$. The space of all functions $u(x,t)$ on $U \times H_n$ with values in $\mathcal{A}_v$ such that $u(x,t)$ by t belongs to $C_0^m(H_n, \mathcal{A}_v)$ and by x to $C^l(U, \mathcal{A}_v)$ so that $\partial^{m_0 + ... + m_{n-1} + l_0 + ... + l_{p-1}} u(x,t) / \partial_{t_0}^{m_0} ... \partial_{t_{n-1}}^{m_{n-1}} \partial_{x_0}^{l_0} ... \partial_{x_{p-1}}^{l_{p-1}}$ is continuous in (x,t) for each $m_0 + ... + m_{n-1} \le m$ and $l_0 + ... + l_{p-1} \le l$ is denoted by $C_0^m(H_n, C^l(U, \mathcal{A}_v))$, where $0 \le m_j$ and $0 \le l_k$ for each j and k.

2. Note. We take for definiteness a canonical closed domain V in $\hat{\mathcal{A}}_r$ with $V \cap \mathcal{A}_r$ satisfying Conditions 2.2.1$(D1, D2)$ so that $\infty \in V$, where $\hat{\mathcal{A}}_r = \mathcal{A}_r \cup \{\infty\}$ denotes the one-point compactification of $\mathcal{A}_r$, $2 \le r < \infty$.

A path $\gamma :< a, b > \to U$ is called locally rectifiable, if it is rectifiable on each compact segment $[c, e] \subset < a, b >$, where $< a, b >= [a, b] := \{t \in \mathbf{R} : a \le t \le b\}$ or $< a, b >= [a, b) := \{t \in \mathbf{R} : a \le t < b\}$ or $< a, b >= (a, b] := \{t \in \mathbf{R} : a < t \le b\}$ or $< a, b >= (a, b) := \{t \in \mathbf{R} : a < t < b\}$.

A domain V is called foliated by locally rectifiable paths $\{\gamma^\alpha : \alpha \in \Lambda\}$ if $\gamma :< a_\alpha, b_\alpha > \to V$ for each α and it satisfies the following three conditions:

$(F1)$ $\bigcup_{\alpha \in \Lambda} \gamma^\alpha(< a_\alpha, b_\alpha >) = V$ and
$(F2)$ $\gamma^\alpha(< a_\alpha, b_\alpha >) \cap \gamma^\beta(< a_\beta, b_\beta >) = \emptyset$ for each $\alpha \ne \beta \in \Lambda$.
Moreover, if the boundary $\partial V = cl(V) \setminus Int(V)$ of the domain V is non-void then
$(F3)$ $\partial V = (\bigcup_{\alpha \in \Lambda_1} \gamma^\alpha(a_\alpha)) \cup (\bigcup_{\beta \in \Lambda_2} \gamma^\beta(b_\beta))$,
where $\Lambda_1 = \{\alpha \in \Lambda :< a_\alpha, b_\beta >= [a_\alpha, b_\beta >\}$, $\Lambda_2 = \{\alpha \in \Lambda :< a_\alpha, b_\beta >=< a_\alpha, b_\beta]\}$.

For the canonical closed subset V one has $cl(V) = V = cl(Int(V))$, where $cl(V)$ denotes the closure of V in $\hat{\mathcal{A}}_r$ and $Int(V)$ denotes the interior of V in $\mathcal{A}_r$. For convenience one can choose C^1 foliation, i.e. each γ^α is of class C^1. When V is with non-void boundary we choose a foliation family such that $\bigcup_{\alpha \in \Lambda} \gamma(a_\alpha) = \partial V_1$, where a set ∂V_1 is open in the boundary ∂V and so that $w|_{\partial V_1}$ would be a sufficient initial condition to characterize a unique branch of an anti-derivative

$$w(x) = \mathcal{I}_\sigma f(x) = {}_\sigma\int_{{}_0x}^{x} f(z)dz,$$

where ${}_0x \in \partial V_1$, $x \in V$, $\gamma^\alpha(t_0) = {}_0x$, $\gamma^\alpha(t) = x$ for some $\alpha \in \Lambda$, t_0 and $t \in < a_\alpha, b_\alpha >$,

$${}_\sigma\int_{{}_0x}^{x} f(z)dz = {}_\sigma\int_{\gamma^\alpha|_{[t_0,t]}} f(z)dz.$$

Particularly, it is possible to take V such that $H_n \subset V$ and supply the domain H_n (see §1) with the natural foliation by straight lines perpendicular to the plane ∂H_n in H_n relative to standard Euclidean scalar product induced from the real shadow $\mathbf{R^n}$.

We consider three first order partial differential operators with Cayley-Dickson coefficients:

$$\sigma_{s,t} f(t) := \sum_{j=0}^{2^r-1} (\partial f(t)/\partial t_j)[i_j^* \psi_{s,j}(t)], \tag{1}$$

$r \le 2$, $\psi_{s,j}(t)$ are $\mathcal{A}_r$-valued sufficiently smooth functions on the domain U in for each j and s, also u and f are sufficiently smooth $\mathcal{A}_{r+1}$ valued functions. Henceforth, the operator $\sigma_{0,t}$ is taken with constant coefficients $\psi_{0,j} = 1$ for each $j = 0, ..., n-1$ and $\psi_{0,j} = 0$ for any $j \ge n$, $1 \le n \le 2^r$.

3. Definition. For two continuous functions f and w on an open neighborhood W of H_n in $\mathcal{A}_r$ with values in the Cayley-Dickson algebra $\mathcal{A}_r$ their convolution $f(x, t) * w(x, t)$ is defined by the formula:

$$f(x,t) * w(x,t) := {}_{\sigma_0}\int_{t'}^{t} f(x, t-\tau)w(x,\tau)d\tau, \tag{1}$$

where $t' = t_1 i_1 + ... + i_{n-1}t_{n-1}$, $t = t_0 i_0 + t' \in H_n$, while $u, w \in C^0(H_n, \mathcal{A}_v)$ with $t \in H_n$, the operator σ_0 is given by Formula $1(1)$, while ${}_{\sigma_0}\int_{t'}^{t}$ is the anti-derivative operator as in §§2 and 3.1.2, 3.1.4.

Then we put $f_0 = 1$, $f_1 = f$, $f_2 = f * f$ and $f_n = f * f_{n-1}$ for each natural number $n \geq 3$. Let also $T = \{t \in H_n : t_j \in [0, T_j]\ \forall j = 0, ..., n-1\}$ denote a domain, where $0 < T_j$ for each j.

4. Lemma. *Let a convergence radius of a series $\sum_{k=1}^{\infty} z^k b_k$ be $0 < R \leq \infty$, $b_k \in \mathcal{A}_r$ for each natural number k, where $r \leq 3$. If $f \in C(W, \mathcal{A}_r)$ and $\prod_{j=1}^{n} T_j < \infty$ (see §3), then the series $F(t) := \sum_{k=1}^{\infty} f_k(t)b_k$ converges for each $t \in T$ and belongs to $C(T, \mathcal{A}_r) := C^0(T, \mathcal{A}_r)$.*

Proof. For a bounded domain V the norm on $C(V, \mathcal{A}_r) := C^0(V, \mathcal{A}_r)$ is standardly defined by the formula:

$$\|f\|_{C(V,\mathcal{A}_r)} := \sup_{z \in V} |f(z)|. \tag{1}$$

The Cayley-Dickson algebra $\mathcal{A}_r$ is 2^r-dimensional as the linear space over the real field $\mathbf{R}$ when r is a natural number, where $\{i_0, i_1, ..., i_{2^r-1}\}$ is its standard basis of generators such that $i_0 = 1$, while $i_j^2 = -1$ and $i_j i_k = -i_k i_j$ for each $j \neq k \geq 1$. Therefore the topological space T is compact relative to the norm topology inherited from $\mathcal{A}_r$, since T is closed and bounded in $\mathcal{A}_r$. Thus each continuous function $f : T \to \mathcal{A}_r$ is bounded on T.

In view of Formulas $2.4(30, 31, 35, 36)$ [33], (2.46) in §2.7 [25], $5(1)$ and (1) the estimate is valid:

$$\|f * w\|_{C(T,\mathcal{A}_r)} \leq \|f\|_{C(T,\mathcal{A}_r)} \|w\|_{C(T,\mathcal{A}_r)} \frac{\sum_{j=0}^{n-1} T_j}{n} C_1 \exp(C_2 R^{2^r+2}) < \infty, \tag{2}$$

where $R = \sqrt{T_0^2 + ... T_{n-1}^2}$, while $C_1 > 0$ and $C_2 > 0$ are constants independent of f, w, T_0,...,T_{n-1}, since $\psi_{0,j} = 1$ for each $j = 0, ..., n-1$ and $\psi_{0,j} = 0$ for $j \geq n$. On the other hand, from $d^n z^n / dz^n = n! I$ for each $z \in \mathcal{A}_r$, Inequality (2) and §2.7 [25], we deduce by induction that

$$\|f_k\|_{C(T,\mathcal{A}_r)} \leq \|f\|_{C(T,\mathcal{A}_r)}^k \frac{\beta_T^{k-1}}{(k-1)!} < \infty \tag{3}$$

for each natural number $k \geq 2$, where

$$\beta_T := \frac{\sum_{j=0}^{n-1} T_j}{n} C_1 \exp(C_2 R^{2^r+2}), \tag{4}$$

while I denotes the unit operator.

From the conditions of this lemma it follows that

$$R = \frac{1}{\overline{\lim}_{k \to \infty} |b_k|^{1/k}} \tag{5}$$

according to Cauchy's formula for a radius of convergence of a power series and hence there exists a constant $C_3 > 0$ such that $|b_k| \le C_3 R^{-k}$ for each $k \ge 1$. From the convergence of the series

$$1 + \sum_{k=1}^{\infty} \frac{\alpha^k}{k!} = e^{\alpha} \tag{6}$$

for each $\alpha \in \mathcal{A}_r$ the statement of this lemma follows and

$$\|F\|_{C(T,\mathcal{A}_r)} \le \frac{\|f\|_{C(T,\mathcal{A}_r)} e^{\delta} C_3}{R}, \tag{7}$$

where $\delta = \|f\|_{C(T,\mathcal{A}_r)} \beta_T / R$.

5. Lemma. *If either $b_k, c_k \in \mathcal{A}_r$ with $r \le 2$ or $b_k, c_k \in \mathbf{R}$ with $r = 3$ for each $k \in \mathbf{N}$, functions $f, w \in C(W, \mathcal{A}_r)$ are such that $[g_m(z), b_k] = 0$ for each $z \in W$ and each natural numbers k, l, whilst $\prod_{j=1}^{n} T_j < \infty$ (see §3), the power series $\sum_{k=1}^{\infty} b_k z^k$ and $\sum_{k=1}^{\infty} z^k c_k$ have positive radii R_1 and R_2 of convergence on $\mathcal{A}_r$, then*

$$(\sum_{k=1}^{\infty} f_k(t) b_k) * (\sum_{l=1}^{\infty} g_l(t) c_l) = \sum_{k,l=1}^{\infty} (f_k(t) * g_l(t))(b_k c_l) \tag{1}$$

is in $C(T, \mathcal{A}_r)$.

Proof. For each $z \in W$ functions f and g have values in the algebra $\mathcal{A}_r$ with $r \le 3$, while $\mathcal{A}_r$ is associative for $r \le 2$ and alternative for $r = 3$. The double power series $\sum_{k,l=1}^{\infty} (z^k b_k)(x^l c_l)$ has positive radius of convergence in $\mathcal{A}_r$. According to Lemma 3 the double series $\sum_{k,l=1}^{\infty} f_k(t) * g_l(t)(b_k c_l)$ converges in norm of $C(T, \mathcal{A}_r)$. The algebra $C(W, \mathcal{A}_r)$ is associative when $r \le 2$, it is alternative for $r = 3$, while $b_k, c_k \in \mathbf{R}$ for each k when $r = 3$ by the conditions of this lemma. On the other hand, for each natural number n the equality is satisfied:

$$(\sum_{k=1}^{n} f_k(t) b_k) * (\sum_{l=1}^{n} g_l(t) c_l) = \sum_{k,l=1}^{n} ((f_k(t) * g_l(t)) b_k) c_l = \sum_{k,l=1}^{\infty} f_k(t) * g_l(t)(b_k c_l)$$

according to the condition $[g_l(z), b_k] = g_l(z) b_k - b_k g_l(z) = 0$ for every $k, l \in \mathbf{N}$ and $z \in W$ and due to and Formulas $4(3-7)$. Taking the limit when n tends to the infinity one gets Formula (1).

6. Definitions. A Banach space X over the quaternion skew filed $\mathcal{A}_2 = \mathbf{H}$ or the octonion algebra $\mathcal{A}_3 = \mathbf{O}$ is considered as the direct sum $X_0 i_0 \oplus ... \oplus X_{2^r-1} i_{2^r-1}$ with $r = 2$ or $r = 3$ respectively, where $X_0, ..., X_{2^r-1}$ are pairwise isomorphic real Banach spaces such that $b x_j = x_j b$ for each $x_j \in X_j$ and $b \in \mathcal{A}_r$.

An operator $A : X \to Y$ is called $\mathbf{R}$-homogeneous $\mathcal{A}_r$-additive, if

$$A(\alpha x + \beta y) = \alpha A x + \beta A y$$

for each $\alpha, \beta \in \mathbf{R}$ and $x, y \in X$,
where X and Y are Banach spaces over $\mathcal{A}_r$. The family of all $\mathbf{R}$-homogeneous $\mathcal{A}_r$-additive

bounded operators from X into Y is denoted by $L_q(X,Y)$. If $X = Y$ it will be written shortly also $L_q(X)$ instead of $L_q(X,X)$.

An operator $T : X \to Y$ is called left $\mathcal{A}_r$-linear if $T(zx_0) = zTx_0$ for each $x_0 \in X_0$ and $z \in \mathcal{A}_r$.

For a Banach space X over $\mathcal{A}_0 = \mathbf{R}$ or $\mathcal{A}_1 = \mathbf{C}$ or the quaternion skew field $\mathcal{A}_2 = \mathbf{H}$ or the octonion algebra $\mathcal{A}_3 = \mathbf{O}$ we suppose that an $\mathbf{R}$-linear $\mathcal{A}_r$-additive operator ($r = 0, 1, 2, 3$ respectively) $A(t)$ depends continuously on the parameter $t \in W$, where W is an open neighborhood of H_n in $\mathcal{A}_r$ with $n \le 8$ such that

$$\|A\|_{C_b(T)} := \sup_{t\in T} |A(t)| < \infty, \tag{1}$$

where $T = \{t \in H_n : t_j \in [0, T_j]\ \forall j = 0, ..., n-1\}, 0 \le T_j \le \infty$. The family of all such operators we denote by $C_b(T, L_q(X))$ or for short $C_b(T)$. Particularly $A(t)$ may be the operator of the point-wise (left or right) multiplication on a continuous $\mathcal{A}_r$-valued function.

Henceforth we also suppose that over the octonion algebra $\mathbf{O}$ the minimal subalgebra $(AC)\ alg_{\mathbf{O}} A$ containing an operator A is alternative.

Then $C(W, L_q(X))$ will denote the family of all operators $A : W \to L_q(X)$ continuous in the variable $z \in W$.

We denote by $u(x,t) * w(x,t)$ the convolution:

$$u(x,t) * w(x,t) := {}_{\sigma_0}\int_{t'}^{t} u(x, t-\tau) w(x,\tau) d\tau, \tag{2}$$

where $t' = t_1 i_1 + ... + i_{n-1} t_{n-1}$, $t = t_0 i_0 + t' \in H_n$, when $w \in C^0(W, X)$ and $u \in C(W, L_q(X))$ with $t \in T$. Then we put

$$A(t) * u(t) = {}_{\sigma_0}\int_{t'}^{t} A(t-\tau)(u(\tau)) d\tau, \tag{3}$$

when $u \in C^0(W, \mathcal{A}_v)$, $X = C^0(H_n, \mathcal{A}_v)$, $0 \le r \le \min(3, v)$, $A(t) \in C(W, L_q(X))$, where the integral ${}_{\sigma_0}\int$ is considered in an open domain $W \supset \{z \in \mathcal{A}_r : Re(z) \ge 0\}$, whilst $t', t \in T$. We also define

$$A(t) * B(t) := {}_{\sigma_0}\int_{t'}^{t} A(t-\tau) B(\tau) d\tau \tag{4}$$

for every $A, B \in C(W, L_q(X))$ and by induction

$$A_k(t) := A(t) * A_{k-1}(t), \tag{5}$$

where $A_2(t) = A(t) * A(t)$, $A_1(t) = A(t)$, $A_0(t) = I$, I denotes the unit operator.

7. Lemma. *Let a convergence radius of a series* $\sum_{k=1}^{\infty} z^k b_k$ *be* $0 < R \le \infty$, $b_k \in \mathcal{A}_r$ *for each natural number* k *with* $r \le 3$. *If* $A \in C(W, L_q(X)) \cap C_b(T)$ *and* $\prod_{j=1}^{n} T_j < \infty$, *then the series* $F(t) := \sum_{k=1}^{\infty} A_k(t) b_k$ *converges for each* $t \in W_1$ *for a sufficiently small open neighborhood* W_1 *of* T *in* $\mathcal{A}_r$ *and belongs to* $C(W_1, L_q(X)) \cap C_b(T, L_q(X))$. *The algebra* $C_b(T, L_q(X))$ *is complete relative to its norm. Moreover, the algebra* $C_b(T, L_q(X))$ *over*

$\mathbf{H}$ *supplied with the convolution multiplication of operators is associative.*

Proof. The quaternion skew field is associative, consequently,

$$(A(t) * B(t)) * u(t) = A(t) * (B(t) * u(t)) \tag{1}$$

for every $A, B \in C(W, L_q(X)) \cap C_b(T)$ and $u \in C(W, \mathcal{A}_r) \cap X$, $X = C^0(H_n, \mathcal{A}_r)$ when $r \leq 2$ and $B(t) * u(t) \in C(W, \mathcal{A}_r) \cap C^0(H_n, \mathcal{A}_r)$ (see Theorems 2.1 and 2.2 in [33]). For $r = 3$ we consider $u \in C^0(H_n, \mathbf{R})$. That is Formulas $6(3)$ and $6(4)$ are correct in this sense (see also Lemmas 4 and 5 above). Certainly each $\mathbf{R}$-homogeneous $\mathcal{A}_r$-additive operator is also $\mathbf{R}$-linear. The set $C(W, L_q(X)) \cap C_b(T, L_q(X))$ is supplied with the addition of operators

$$(A + B)(t)x = A(t)x + B(t)x$$

for each $A, B \in C(W, L_q(X)) \cap C_b(T, L_q(X))$, $t \in T$ and $x \in X$, multiplication of them on constants $b \in \mathcal{A}_r$ such that

$$(bA)(t)x = (bI) \circ (A(t)x) \tag{LM}$$

and

$$(A(t)b)x = (A(t)x)b \tag{RM}$$

for each $x \in X$, where the Banach space X is over $\mathcal{A}_r$, $r \leq 3$, I denotes the unit operator. Therefore, $C_b(T, L_q(X))$ is the left and right module over $\mathcal{A}_r$. Moreover, one gets $(bcI)A = bI(cIA)$ and $A(bcI) = (AbI)cI$ for each $A \in C(W, L_q(X)) \cap C_b(T, L_q(X))$ and $b, c \in \mathcal{A}_2$, when $r = 2$ (see also [54] and Condition $6(AC)$ above).

As the multiplication of operators in the algebra $C_b(T, L_q(X))$ is considered their convolution. At the same time from Formulas $4(1-7)$ it follows that

$$\|A * B\|_{C_b(T)} \leq \|A\|_{C_b(T)} \|B\|_{C_b(T)} \frac{\sum_{j=0}^{n-1} T_j}{n} C_1 \exp(C_2 R^{2^r+2}) < \infty, \tag{2}$$

where $R = \sqrt{T_0^2 + ...T_{n-1}^2}$, while $C_1 > 0$ and $C_2 > 0$ are constants independent of A, B, T_0,...,T_{n-1}. Then from Formulas $6(3,4)$ and (1) we infer that $A * (B * C) = (A * B) * C$ for each $A, B, C \in C(W_1, L_q(X)) \cap C_b(T, L_q(X))$, when $r = 2$, consequently, the algebra $C_b(T, L_q(X))$ over $\mathcal{A}_2 = \mathbf{H}$ is associative, where W_1 is a sufficiently small open neighborhood of T in $\mathcal{A}_r$.

Formula (2) above imply that

$$\|A_k\|_{C_b(T)} \leq \|A\|^k_{C_b(T)} [\frac{\sum_{j=0}^{n-1} T_j}{n} C_1 \exp(C_2 R^{2^r+2})]^{k-1} / (k-1)! \tag{3}$$

for each $k \geq 2$, hence $F \in C_b(T, L_q(X))$.

The space $L_q(X, X)$ of all bounded $\mathbf{R}$-homogeneous $\mathcal{A}_r$-additive operators S from X into X is complete relative to the operator norm

$$|S| := \sup_{0 \neq y \in X} \|Sy\| / \|y\| < \infty, \tag{4}$$

since X is the Banach space. Therefore, the algebra $C_b(T, L_q(X))$ is complete relative to its norm $6(1)$.

8. Lemma. *If either $b_k, c_k \in \mathcal{A}_r$ with $r \le 2$ or $b_k, c_k \in \mathbf{R}$ with $r = 3$ for each $k \in \mathbf{N}$ and a Banach space X is over $\mathcal{A}_r$, operators $A, B \in C(W, L_q(X)) \cap C_b(T, L_q(X))$ are such that $[B_l, b_k I] = 0$ for each natural numbers k, l, the power series $\sum_{k=1}^{\infty} b_k z^k$ and $\sum_{k=1}^{\infty} z^k c_k$ have positive radii of convergence on $\mathcal{A}_r$, the subalgebra $alg_{\mathbf{O}}(A, B)$ containing A and B is alternative for $r = 3$, then*

$$(\sum_{k=1}^{\infty} A_k(t) b_k) * (\sum_{l=1}^{\infty} B_l(t) c_l) = \sum_{k,l=1}^{\infty} (A_k(t) * B_l(t))(b_k c_l) \qquad (1)$$

is in $C(W_1, L_q(X)) \cap C_b(T, L_q(X))$ for sufficiently small open neighborhood W_1 of T in $\mathcal{A}_r$.

Proof. The double power series $\sum_{k,l=1}^{\infty} (z^k b_k)(x^l c_l)$ has positive radius of convergence in $\mathcal{A}_r$. In view of Theorems 3.1.2 and 3.1.4 $\sum_{k,l=1}^{\infty} A_k(t) * B_l(t)(b_k c_l)$ converges in norm of $C_b(T, L_q(X))$. For each natural number n we have

$$(\sum_{k=1}^{n} A_k(t) b_k) * (\sum_{l=1}^{n} B_l(t) c_l) = \sum_{k,l=1}^{n} ((A_k(t) * B_l(t)) b_k) c_l = \sum_{k,l=1}^{\infty} A_k(t) * B_l(t)(b_k c_l)$$

due to the condition $[B_l, b_k I]x = B_l b_k x - b_k B_l x = 0$ for every natural numbers k, l and each vector $x \in X$ and Formulas $7(LM), (RM)$, since the algebra $C_b(T, L_q(X))$ is associative when $r \le 2$, while for $r - 3$ the subalgebra $alg_{\mathbf{O}}(A, B)$ is alternative and $b_k, c_k \in \mathbf{R}$ for each k. Taking the limit when n tends to the infinity and using Lemma 5 and Definitions 6 we get Formula (1).

9. Definition. An operator $B*$ is of class $\mathcal{E} = \mathcal{E}_r$ if it has the form

$$B* = I + A(t)*,$$

where $A \in C(W, L_q(X)) \cap C_b(T, L_q(X))$, X is a Banach space over $\mathcal{A}_r$, I denotes the unit operator, W is an open neighborhood of T in $\mathcal{A}_r$.

Henceforward $C_b(T, L_q(X))$ is supplied with the algebraic structure over $\mathcal{A}_r$ as in §7.

10. Lemma. *If X is a Banach space over $\mathcal{A}_r$, $0 \le r \le 3$, an operator A is in $C(W, L_q(X)) \cap C_b(T, L_q(X))$ and $\prod_{j=1}^{n} T_j < \infty$, then the inverse of the operator $I + A(t)*$ is given by the formulas:*

$$(I + A(t)*)^{-1} = I + \hat{A}(t), \text{ where } \hat{A}(t) = \sum_{k=1}^{\infty} (-1)^k A_k(t), \qquad (1)$$

is in $C_b(T, L_q(X))$.

Proof. The series for $\hat{A}$ converges in norm of $C_b(T, L_q(X))$ due to Lemma 8. At the same time the algebra $C_b(T, L_q(X))$ is associative for $r \leq 2$ or the subalgebra $alg_{\mathbf{O}}(A)$ is alternative for $r = 3$ due to Condition 8(AC), consequently, $(I + A(t)*)\sum_{k=1}^{n}(-1)^k A_k(t) = I - A_{k+1}(t)$ and hence Formula (1) is valid, since $\lim_k \|A_k\|_{C_b(T)} = 0$ as follows from Formula 7(3).

11. Definition. For a real number α the power of the operator $B* = I + A(t)* \in \mathcal{E}$ is defined by the formulas:

$$(B*)^\alpha = I + U_\alpha(t)*, \text{ where } U_\alpha(t) := \sum_{k=1}^{\infty} \binom{\alpha}{k} A_k(t), \tag{1}$$

$\binom{\alpha}{k} = \alpha ...(\alpha - k + 1)/k!$ for $k \geq 2$, $\binom{\alpha}{1} = \alpha$ and $\binom{\alpha}{0} = 1$ so that $(B * -I)^0 := I$.

12. Lemma. *If $B* \in \mathcal{E}$ and $D* = I + C(t)* \in \mathcal{E}$ for $\mathcal{E} = \mathcal{E}_r$ over $\mathcal{A}_r$ with $r \leq 3$, then the following identities are valid:*

$$(B*)^a = \sum_{k=0}^{\infty} \binom{a}{k} (B * -I)^k, \tag{1}$$

$$(B*)^a (B*)^b = (B*)^{a+b}, \tag{2}$$

$$((B*)^b)^{1/b} = B * \text{ for } b \neq 0, \tag{3}$$

$$(B*)^{-1} = \sum_{k=0}^{\infty} \binom{-0.5}{k} [(B*)^2 - I]^k \text{ for each } a, b \in \mathbf{R} \tag{4}$$

and

$$(B*)^{-1}(D*)^{-1} = \frac{1}{\pi}\sum_{k=0}^{\infty}(-1)^k \sum_{j=0}^{k} \frac{\Gamma(k-j+0.5)\Gamma(j+0.5)}{j!(k-j)!} [(B*)^2 - I]^j [(D*)^2 - I]^{k-j}. \tag{5}$$

Proof. From Definitions 6 and 11 and Lemma 7 we deduce by induction that $(B*)^k \in \mathcal{E}$ and

$$A_k(t) * u(t) = (B * -I)^k u(t)$$

for an operator $B* = I + A(t)* \in \mathcal{E}$, every natural number k and a function $u \in C^0(T, \mathcal{A}_r)$, since the quaternion skew field $\mathbf{H}$ is associative, while for $r = 3$ the subalgebra $alg_{\mathbf{O}}(A)$ is alternative according to Condition 8(AC).

If $x \in \mathcal{A}_r$ is a marked Cayley-Dickson number, then a minimal subalgebra $alg_{\mathbf{R}} x$ containing x and over the real field $\mathbf{R}$ is embeddable into the complex field $\mathbf{C}$, that is $alg_{\mathbf{R}} x \hookrightarrow \mathbf{C}$. Therefore, the generating functions of binomial coefficients $\binom{a}{k}$ are prescribed by the formula:

$$(1 + x)^a = \sum_{k=0}^{\infty} \binom{a}{k} x^k$$

for each $x \in \mathcal{A}_r$ with $|x| < 1$, when the parameter a is real, since the real field $\mathbf{R}$ is the center of the Cayley-Dickson algebra $\mathcal{A}_r$ for each $r \geq 2$. Thus $\binom{a}{k} \in \mathbf{R}$ for each real a and each nonnegative integer k. Particularly, there is the well-known identity related with the standard Γ-function:

$$\binom{-0.5}{k} = \frac{(-1)^k \Gamma(k+0.5)}{\sqrt{\pi} k!}.$$

Therefore, Formulas $(1-5)$ follow from Definitions 6, 9 and Lemmas 7, 8, 10.

13. Definition. Let $A*$ and $B*$ be two operators belonging to $\mathcal{E}$ over $\mathcal{A}_r$ with $r \leq 2$.

For each number $z = z_1 + z_2 i_{2^r}$ in $\mathcal{A}_{r+1} = \mathcal{A}_r \oplus \mathcal{A}_r i_{2^r}$ with $z_1, z_2 \in \mathcal{A}_r$ let a new operator $[(A,B)z]$ be defined by:

$$[(A,B)z]f := (A*f)z_1 + i_{2^r}((B*f)z_2) \tag{1}$$

on $X \oplus X i_{2^r}$, where X is a Banach space over $\mathcal{A}_r$, $f \in X \oplus X i_{2^r}$. It is supposed that the minimal subalgebra $alg_{\mathbf{O}}\{(A,B)z : z \in \mathcal{A}_{r+1}\}$ containing operators $(A,B)z$ is alternative when $r = 2$, where $\mathbf{O} = \mathcal{A}_3$.

14. Lemma. *Suppose that $r \leq 2$ and $z \in \mathcal{A}_{r+1}$ is a non zero element (number). Then the operator $(A,B)z$ from Definition 13 is invertible.*

Proof. For $z \neq 0$ the inverse element (number) z^{-1} exists. The the subalgebra $alg_{\mathcal{A}_{r+1}}\{(A,B)z : z \in \mathcal{A}_{r+1}\}$ is associative for $r \leq 1$ and alternative for $r = 2$ according to the suppositions of §13. Let $A* = I + A(t)*$ and $B* = I + B(t)*$, then $[(A,B)z]f = fz + (A(t)*f)z_1 + i_{2^r}((B(t)*f)z_2)$. Therefore, the operator $(A,B)z$ can be written in the form:

$$[(A,B)z]f = \{If + [(A(t)*f)z_1 + i_{2^r}((B(t)*f)z_2)]z^{-1}\}z = \{(I + F(z,t)*)f\}z, \tag{2}$$

where $I + F(z,t)* \in \mathcal{E}_{r+1}$ for each $z \in \mathcal{A}_{r+1} \setminus \{0\}$, $\mathcal{E}_{r+1}$ denotes the class of operators $\mathcal{E}$ over $\mathcal{A}_{r+1}$. Then one gets $F(z,t)* \in C_b(T, L_q(X \oplus X i_{2^r}))$ for each non zero number $z \in \mathcal{A}_{r+1}$. Moreover, this operator is bounded

$$\|F(z,t)\|_{C_b(T)} \leq \max(\|A\|_{C_b(T)}, \|B\|_{C_b(T)}) =: K(A,B,T), \tag{3}$$

since $|z_1 z^{-1}| \leq 1$ and $|z_2 z^{-1}| \leq 1$. Therefore according to Lemma 10 and Formulas $7(LM), (RM)$ the inverse operator exists

$$(I + F(z,I)*)^{-1} = I + \Phi(z,t)*, \tag{4}$$

where a function $\Phi(z,t)$ is:

$$\Phi(z,t) = \sum_{k=1}^{\infty} (-1)^k F_k(z,t). \tag{5}$$

Thus we get

$$((A,B)z)^{-1} f = [(I + \Phi(z,t)*)f]z^{-1} \tag{6}$$

for each $z \neq 0$, since the octonion algebra $\mathcal{A}_3 = \mathbf{O}$ is alternative, while $r \leq 2$.

15. Corollary. *The function* Φ *given by series 14*(5) *is continuous on* $(\mathcal{A}_{r+1} \setminus \{0\}) \times T$ *and*

$$\|\Phi(z,t)\|_{C_b(T)} \leq K \exp(K[\frac{\sum_{j=0}^{n-1} T_j}{n} C_1 \exp(C_2 R^{2^r+2})]) \qquad (1)$$

with $K = K(A,B,T)$ *and* $C_b(T) = C_b(T, L_q(X \oplus X i_{2^r}))$, *where* $z \in (\mathcal{A}_{r+1} \setminus \{0\})$ *and* $t \in T$.

This follows from Lemma 14 and Estimates 7(3) and 14(3).

16. Corollary. *If* $A*, B* \in \mathcal{E}$, $f(y,t) : \mathcal{A}_{r+1} \times T \to \mathcal{A}_r$, $r \leq 2$, f *as a function in* t *is in* $C(W, \mathcal{A}_r)$, *where* W *is an open neighborhood of* T *in* $\mathcal{A}_r$, $y, z \in \mathcal{A}_{r+1} \setminus \{0\}$, $n \in \mathbf{Z}$, *then the limits* $\lim_{y \to z} f(y,t)$, $\lim_{y \to z} A * f(y,t)$ *and* $\lim_{y \to z}[((A,B)z)*]^n f(y,t)$ *simultaneously either exist or not in* $C(T, \mathcal{A}_{r+1})$.

This is implied by Lemmas 12 and 14.

17. PDE. We consider the partial differential equation

$$[P_1(\sigma_{0,t})\sigma_{1,x_1}^2 + P_2(\sigma_{0,t})\sigma_{2,x_2}^2]u(x,t) = 0, \text{ where} \qquad (1)$$

$$P_s(\sigma_t) = \sigma_t^l + \sum_{k=0}^{l-1} \sigma_t^k a_{k,s} \qquad (2)$$

with coefficients $a_{k,s} \in \mathcal{A}_r$, $l \in \mathbf{N}$, σ_t denotes the Dirac type operator by the variable $t \in H_n$, $x_s \in \mathcal{A}_r$, $s \in \{1,2\}$, $\sigma_t^0 = 1$, $u(x,t) \in \mathcal{A}_{r+1}$,

$$\sigma_{s,t} f(t) = \sum_{j=0}^{2^r-1} (\partial f(t)/\partial t_j)[i_j^* \psi_{s,j}(t)], \qquad (3)$$

$r \leq 2$, $(\sigma_{s,t}^k z) f(t) := (\sigma_{s,t}^k f) z$ for each $z \in \mathcal{A}_{r+1}$, u and f are sufficiently smooth $\mathcal{A}_{r+1}$-valued functions on the domain U, where $\psi_{s,j} \in C^1(U, \mathbf{R})$ are real-valued continuously differentiable functions for each j and s, particularly they may be $\mathcal{A}_r$ (super-)differentiable functions, $\psi_{s,j}(z) \neq 0$ for each z either in H_n (for $s = 0$) or U (for $s = 1,2$) and each $j = 0, ..., n_s - 1$, $s \in \{0,1,2\}$, $\psi_{0,j} = 1$ for each $j = 0, ..., n-1$ with $n = n_0$, $\psi_{0,j} = 0$ for any $j \geq n$, $1 \leq n_s \leq 2^r$ for each s. For $s = 1$ and $s = 2$ more general Dirac type operators also are considered henceforth as well

$$\sigma_{s,z} f(z) = \sum_{j=0}^{n} (\partial f/\partial z_j)\phi_{s,j}^*(z), \qquad (3')$$

where $\phi_{s,j}(z) \neq \{0\}$ for each $z \in U$ and $\phi_{s,j}(z) \in C^0(U, \mathcal{A}_r)$ for each $j = 0, ..., n$ such that $Re(\phi_{s,j}(z)\phi_{s,k}^*(z)) = 0$ for each $z \in U$ and each $0 \leq j \neq k \leq n$, where a domain U satisfies Conditions 2.2.1$(D1, D2)$, $1 \leq n = n_s \leq 2^r$, $r \leq 2$. Suppose also that a system $\{\phi_{s,0}(z), ..., \phi_{s,n}(z)\}$ is either for $n = 2^r - 1$, or can be completed by Cayley-Dickson numbers $\phi_{s,n+1}(z), ..., \phi_{s,2^r-1}(z)$, such that

$$alg_{\mathbf{R}}\{\phi_{s,j}(z), \phi_{s,k}(z), \phi_{s,l}(z)\} \qquad (\alpha)$$

is alternative for all $0 \le j, k, l \le 2^r - 1$ and

$$alg_{\mathbf{R}}\{\phi_{s,0}(z), ..., \phi_{s,2^r-1}(z)\} = \mathcal{A}_r \qquad (\beta)$$

for each $z \in U$. Let G be a real shadow of U. Suppose that initial conditions are:

$$\sigma^k_{0,t} u(x,t)|_{t_0=0} = 0 \qquad (4)$$

for each $k = 0, ..., l-1$.

18. Theorem. *Suppose that* $u(x,t)$ *and* $v(x,t) \in C^l(W, C^2(G, \mathcal{A}_r)) \cap C^l_0(H_n, C^2(G, \mathcal{A}_r))$ *satisfy the system of equations*

$$(A*)^{-1}\sigma^2_{1,x_1} u(x,t) = (B*)^{-1}\sigma_{1,x_1}\sigma_{2,x_2} v(x,t) \qquad (1)$$

and

$$(B*)^{-2}\sigma^2_{2,x_2} u(x,t) = -(A*)^{-1}(B*)^{-1}\sigma_{1,x_1}\sigma_{2,x_2} v(x,t) \qquad (2)$$

for each $t \in H_n$ *and* $x_1, x_2 \in U$, *where* $x = x_1 + x_2 i_{2^r} \in U + U i_{2^r}$,

$$A* = (\hat{A}*)^{-1/2}, B* = (\hat{B}*)^{-1/2}, \qquad (3)$$

$$\hat{A}* = I + \mathcal{A}(t)*, \hat{B}* = I + \mathcal{B}(t)* \qquad (4)$$

,

$$\mathcal{A}(t) * u(x,t) := \sum_{k=0}^{l-1} [u(x,t) * \frac{(t_0 i_0)^{l-k-1} + ... + (t_{n-1} i_{n-1})^{l-k-1}}{(l-k-1)! n}] a_{k,1}, \qquad (5)$$

$$\mathcal{B}(t) * u(x,t) := \sum_{k=0}^{l-1} [u(x,t) * \frac{(t_0 i_0)^{l-k-1} + ... + (t_{n-1} i_{n-1})^{l-k-1}}{(l-k-1)! n}] a_{k,2}. \qquad (6)$$

Then they satisfy Equations 17$(1,4)$. *Moreover, operators* $A*$ *and* $B*$ *are left* $\mathcal{A}_{r+1}$ *linear.*

Proof. We have $\psi_{0,j} = 1$ for each $j = 0, ..., n-1$ (see Formula (3)). In the class of z analytic functions $f(z)$ the integral ${}_{\sigma_0}\int_\gamma f(z)dz$ along a rectifiable path γ in a domain satisfying Conditions 2.2.1$(D1, D2)$ in $\mathcal{A}_r$ depends only on initial $\gamma(0)$ and final points $\gamma(1)$ according to Formulas 2.4$(30, 35, 36)$ [33] and such property of the non commutative line integral over $\mathcal{A}_r$ in conformity with Theorem 2.15 [25]. Particularly, this is accomplished in the domain $W_0 := \{z \in \mathcal{A}_r : Re(z) \ge 0\}$. We consider by induction functions of the form:

$$g_{m+1}(t) := {}_{\sigma_0}\int_0^t g_m(\tau) d\tau$$

for $m = 0, 1, 2...$, where $g_0(z) = 1$ for each z. Then we deduce that particular solutions are:

$$g_m(z) = \frac{(z_0 i_0)^m + ... + (z_{n-1} i_{n-1})^m}{n}, \qquad (7)$$

since $z_0, ..., z_{2^r-1} \in \mathbf{R}$, the real field $\mathbf{R}$ is the center of the Cayley-Dickson algebra $\mathcal{A}_r$ for each $r \ge 2$, $i^m_k i^*_k = i^{m-1}_k$ for each $k = 0, ..., 2^r - 1$ and $1 \le m$. Functions g_m are given by such polynomials up to additives $\phi_m(z)$ such that $\phi_m(z) \in \ker(\sigma_0)$, i.e. $\sigma_0\phi_m(z) = 0$ for each z. Functions g_m satisfy the conditions $(g_m(t) - g_m(t'))|_{t_0=0} = 0$ (see the notation of §6) and $\sigma_0 g_m(z) = g_{m-1}(z)$, consequently,

$$\sigma_0^k(g_m(z) - g_m(z'))|_{z_0=0} = 0 \qquad (8)$$

for every $z \in W_0$ with $z_0 = 0$, $k = 0, ..., l-1$ and $1 \le l$. Applying Formulas $(7,8)$ and $7(3)$, integrating ${}_{\sigma_0}\int$ of Equation (1) l times by $t \in W_0$ and taking into account the initial conditions $17(4)$ we infer that

$$\hat{A} * [\sigma_{1,x_1}^2 u(x,t)] + \hat{B} * [\sigma_{2,x_2}^2 u(x,t)] = 0 \qquad (9)$$

so that $\hat{A}*$ and $\hat{B}*$ of class $\mathcal{E}$ are given by Formulas $(4-6)$. Then Lemma 14 provides operators $A*$ and $B*$ with which Equation (9) takes the form:

$$(A*)^{-2}[\sigma_{1,x_1}^2 u(x,t)] + (B*)^{-2}[\sigma_{2,x_2}^2 u(x,t)] = 0. \qquad (10)$$

Since $r \le 2$, the algebra $alg_{\mathcal{A}_r}(A*, B*)$ is alternative. Therefore equation (10) implies that functions u and v satisfying the system of PDE $(1,2)$ provide a solution u of PDE $17(1)$.

From Formula (5) it follows that $((\hat{A}*)z)f = (\hat{A}*f)z$ for each real valued continuous function f on H_n and $z \in \mathcal{A}_{r+1}$, since the real field $\mathbf{R}$ is the center of the algebra $\mathcal{A}_q$ for each $q \ge 2$. Applying Formulas (3) and $12(1)$ one gets that $((A*)z)f = z(A*f)$. This means that $A*$ and analogously $B*$ are the left $\mathcal{A}_{r+1}$ linear operators.

19. Corollary. *If the conditions of Theorem 18 are fulfilled and:*
Operators $A*$, $B*$, σ_{1,x_1} and σ_{2,x_2} all pairwise commute:

$$[A*, B*] = 0, [A*, \sigma_{1,x_1}] = 0, [A*, \sigma_{2,x_2}] = 0,$$

$$[B*, \sigma_{1,x_1}] = 0, [B*, \sigma_{2,x_2}] = 0, [\sigma_{1,x_1}, \sigma_{2,x_2}] = 0, \qquad (1)$$

then the equalities

$$(A*)^{-1}\sigma_{1,x_1}u(x,t) = (B*)^{-1}\sigma_{2,x_2}v(x,t) \qquad (2)$$

and

$$(B*)^{-1}\sigma_{2,x_2}u(x,t) = -(A*)^{-1}\sigma_{1,x_1}v(x,t) \qquad (3)$$

for each $t \in H_n$ and $x_1, x_2 \in U$ imply that a function $u(x,t)$ satisfies Equations $17(1,4)$.

Proof. This follows immediately from Conditions (1) and $18(1,2)$ applying integrations ${}_{\sigma_1}\int$ and ${}_{\sigma_2}\int$ of both sides of Equalities $18(1)$ and $18(2)$ respectively, since the equality

$$(B*)^{-2}\sigma_{2,x_2}^2 u(x,t) = -(B*)^{-1}(A*)^{-1}\sigma_{1,x_1}\sigma_{2,x_2}v(x,t)$$

is equivalent to

$$(B*)^{-1}\sigma_{2,x_2}^2 u(x,t) = -(A*)^{-1}\sigma_{1,x_1}\sigma_{2,x_2}v(x,t)$$

due to Lemmas 10, 14 and Corollary 16.

20. Remark. From definitions of operators above it can be lightly seen that in the following examples Conditions (1) of Corollary 19 are fulfilled:

(1). coefficients $a_{k,1}$ and $a_{k,2}$ are real for each k, whilst $\sigma_1 = \sigma_2$ is arbitrary, while either (α) $\psi_{0,0} = 0$ with zero $a_{k,1}$ and $a_{k,2}$ for each odd k or (β) $\psi_{0,j} = 0$ for each $j \geq 1$;
(2). either $\sigma_{1,x_1} = \partial/\partial x_{1,0}$ or $\sigma_{2,x_2} = \partial/\partial x_{2,0}$, also coefficients $a_{k,1}$ and $a_{k,2}$ are real for each k while either (α) $\psi_{0,0} = 0$ with zero $a_{k,1}$ and $a_{k,2}$ for each odd k or (β) $\psi_{0,j} = 0$ for each $j \geq 1$;
(3). $\sigma_{1,x_1} = \partial/\partial x_{1,0}$ and $\sigma_{2,x_2} = \partial/\partial x_{2,0}$, coefficients $a_{k,1}$ and $a_{k,2}$ are real for each k, but σ_0 is arbitrary;
(4). $P_1(\sigma_0) = P_2(\sigma_0)$ is arbitrary whilst $\sigma_{1,x_1} = \partial/\partial x_{1,0}$ and $\sigma_{2,x_2} = \partial/\partial x_{2,0}$;
(5). the operators P_1, P_2, σ_0, σ_1 and σ_2 are over $\mathbf{C} = \mathcal{A}_1$.

The advantage of the system of PDE $19(2,3)$ is that it is of lower order, than the initial PDE $17(1)$.

21. Definitions. Put

$$\sigma_{a,b} f(x,t) = a\sigma_{1,x_1} f(x,t) + b i_{2^r}^* (\sigma_{2,x_2} f(x,t)) \tag{1}$$

for $r \leq 2$ and $x_1, x_2 \in U$ with $x = x_1 + x_2 i_{2^r}$, where $f \in C^0(H_n, C^1(U + U i_{2^r}, \mathcal{A}_{r+1}))$, a, b are real parameters. If a function $f(x,t)$ on $(U + U i_{2^r}) \times H_n$ with values in $\mathcal{A}_{r+1}$ is such that a function on the right exists:

$$[\sigma_{(A,B)*,w} f(x,t)] = a(A*)^{-1} \sigma_{1,x_1} f(x,t) + b i_{2^r}^* [(B*)^{-1} (\sigma_{2,x_2} f(x,t))], \tag{2}$$

then a function $f(x,t)$ will be called differentiable by operator $\sigma_{(A,B)*,w}$ along a vector $w = w(x)$ at a point $x = x_1 + x_2 i_{2^r}$.

For vector spaces X and Y over $\mathcal{A}_{r+1}$ we call an operator $T : X \to Y$ left linear (at $x \in X$) by $a + b i_{2^r}^* \in \mathbf{R} \oplus \mathbf{R} i_{2^r}^*$ if $T(a x_0 + b i_{2^r}^* x_0) = (a + b i_{2^r}^*)(T x_0)$ for every $x_0 \in X_0$ (at x for each $x_j \in X_j$ in place of x_0 correspondingly, where $j = 0, ..., 2^{r+1} - 1$) and $a, b \in \mathbf{R}$, where $1 \leq r \leq 2$.

An operator T is left alternative by $a + b i_{2^r}^* \in \mathbf{R} \oplus \mathbf{R} i_{2^r}^*$ at $x \in X$ if $(a + b i_{2^r}^*) T((a + b i_{2^r}^*) x) = T((a + b i_{2^r}^*)^2 x)$ for each $a + b i_{2^r}^* \in \mathbf{R} \oplus \mathbf{R} i_{2^r}^*$. If the latter is accomplished for each $x \in X$, then T will be called left alternative by $a + b i_{2^r}^* \in \mathbf{R} \oplus \mathbf{R} i_{2^r}^*$.

22. Notation. We put $w = a w_1 + b w_2 i_{2^r}$ with $w_s = \sum_{j=0}^{2^r - 1} w_{s,j}$ for $s = 1, 2$ and $r \leq 2$, where for every j and s either $w_{s,j}(x_s) = i_j^* \psi_{s,j}(x_s)$ or $w_{s,j}(x_s) = \phi_{s,j}(x_s)$ all (see §1 and Formulas $17(3,3')$), $a, b \in \mathbf{R}$, where $1 \leq r \leq 2$, $w \in C^1(U + U i_{2^r}, \mathcal{A}_{r+1})$.

23. Theorem. *Suppose that $\sigma_1 = \sigma_2$ and operators $A*$, $B*$, σ_1 all pairwise commute on a neighborhood of x and Conditions 2.2.1$(D1, D2)$, 17$(1 - 3')$ are fulfilled. Let a function f be in $C^0(W, C^1(U + U i_{2^r}, \mathcal{A}_{r+1}))$, where W is an open neighborhood of H_n in $\mathcal{A}_r$. A function $f(x,t) = u(x,t) + i_{2^r} v(x,t)$ is differentiable by operator $\sigma_{(A,B)*,w}$ along a vector $w = w(x) = a w_1 + b w_2 i_{2^r}$ for each $a, b \in \mathbf{R}$ and $t \in H_n$ at a point $x = x_1 + x_2 i_{2^r}$ with $x_1, x_2 \in U$ and $\sigma_{(A,B)*,w}$ is left linear and alternative by $a + b i_{2^r}^* \in \mathbf{R} \oplus \mathbf{R} i_{2^r}$ at f if* *and only if functions $u(x,t)$ and $v(x,t)$ satisfy the system of PDE 19$(2,3)$.*

Proof. Since $f \in C^0(W, C^1(U + Ui_{2^r}, \mathcal{A}_{r+1}))$ and $\phi_{s,j} \in C^1(U + Ui_{2^r}, \mathcal{A}_{r+1})$ for each s, then $u, v \in C^0(W, C^1(U + Ui_{2^r}, \mathcal{A}_r))$, whilst $\sigma_x f(x,t)$ is a continuous function on $U + Ui_{2^r}$. Operators $A*$ and $B*$ defined above are independent of x. It is well-known that the algebraic identity $i_{2^r} p = p^* i_{2^r}$ is satisfied for each $p \in \mathcal{A}_r$ when $r \geq 1$.

From the definitions of the operators $(A,B)z$, the first order partial differential operator $\sigma_{a,b}$ and the antiderivative operator ${}_\sigma \int$ it follows, that

$$\sigma_{a,b} \int_\gamma (A,B) * f(\xi,t) d\xi = a_{\sigma_1} \int_{\gamma_1} (A * f(\xi,t)) d\xi_1 + bi_{2^r \sigma_2} \int_{\gamma_2} (B * f(\xi,t)) d\xi_2 \quad (1)$$

for each rectifiable path $\gamma_s : [0,1] \to U$, $\gamma = \gamma_1 + \gamma_2 i_{2^r} \in U + Ui_{2^r}$, where $\xi = \xi_1 + \xi_2 i_{2^r} \in \gamma([0,1])$, $\xi_1, \xi_2 \in U$ (see §2.2.1 and Theorems 3.1.2, 3.1.4 above). On the other hand, the formula

$$\sigma_y \{ {}_\sigma \int_x^y g(\xi) d\xi \} = g(y) \quad (2)$$

is satisfied for each $x, y \in (U + Ui_{2^r})$ and a continuous function g on $U + Ui_{2^r}$ with values in $\mathcal{A}_{r+1}$ according to Theorems 2.4.1 and 2.5.2 [33]. From the latter formula we deduce that

$${}_\sigma \int_x^y [\sigma g(\xi)] d\xi = g(y) - g(x) + \phi(y) - \phi(x) \quad (3)$$

for each continuously differentiable function $g \in C^1(U + Ui_{2^r}, \mathcal{A}_{r+1})$, where $\phi \in ker(\sigma)$. Thus Condition 21(2) and Formulas $(1-3)$ imply that

$$\lim_{h \to 0} \sigma_{(A,B)*,w} \{ \sigma_{a,b} \int_x^{x+hw} ((A,B) * [f(x+hw,t) - f(x,t)]) d\xi \} = f(x,t), \quad (4)$$

where $h \in \mathbf{R} \setminus \{0\}$ and $w \in \mathcal{A}_{r+1} \setminus \{0\}$ and $x_1, x_2 \in U$. The operator $\sigma_{(A,B)*,w}$ is the composition of two $\mathbf{R}$-linear $\mathcal{A}_{r+1}$-additive operators $\sigma_{(A,B)*,w} = (C*)^{-1} \circ \sigma_{a,b}$ with an operator C prescribed by operators A and B and Formula 21(2) with the help of Lemma 12. Therefore the anti-derivation operator is provided by the formula:

$$\sigma_{(A,B)*,w} \int_\gamma f(\xi,t) d\xi = {}_{\sigma_{a,b}} \int_\gamma (C*) f(\xi,t) d\xi = {}_{\sigma_{a,b}} \int_\gamma (A,B) * f(\xi,t) d\xi. \quad (5)$$

By Theorem 18 the operators $A*$ and $B*$ are left $\mathcal{A}_{r+1}$ linear. On the other hand, Formulas 17$(3,3')$ imply that σ_1 and σ_2 are left $\mathcal{A}_r$ linear and left $\mathcal{A}_{r+1}$ alternative, since $r \leq 2$.

Suppose that the limit in Formula (4) exists at x for each $t \in H_n$ and the operator $\sigma_{(A,B)*,w}$ is left linear and alternative by $a + bi_{2^r}^* \in \mathbf{R} \oplus \mathbf{R} i_{2^r}$ at f. If $a^2 + b^2 > 0$, then $a + bi_{2^r}^* \neq 0$ and $(a + bi_{2^r}^*)^{-1} = (a + bi_{2^r})/(a^2 + b^2)$, since $i_{2^r}^* = -i_{2^r}$ for $r \geq 1$. From Lemma 14, Corollaries 15 and 16, Formulas 21(1) and $(1,2)$ we deduce that $\sigma_{(A,B)*,w} f(x,t) = (A*)^{-1} \sigma_{1,x_1}(u(x,t) + i_{2^r} v(x,t))$ for $a = 1$ and $b = 0$, $\sigma_{(A,B)*,w} f(x,t) = i_{2^r}^* [(B*)^{-1} \sigma_{2,x_2}(u(x,t) + i_{2^r} v(x,t))]$ for $a = 0$ and $b = 1$. Taking into consideration the conditions $\sigma_1 = \sigma_2$, the alternativity of the octonions algebra, also that $A*$, $B*$, σ_1 all pairwise commute on a neighborhood of x, we infer that functions $u(x,t)$ and $v(x,t)$ satisfy the system of PDE 19$(2,3)$.

Suppose now that functions $u(x,t)$ and $v(x,t)$ satisfy the system of PDE 19(2, 3) and $f = u + i_{2^r} v$. The operators $A*$ and $B*$ are invertible as Lemma 12 asserts, consequently, there exist $\sigma_{1,x_1} u(x,t)$, $\sigma_{2,x_2} u(x,t)$, $\sigma_{1,x_1} v(x,t)$ and $\sigma_{2,x_2} u(x,t)$, hence $\sigma_{a,b} f(x,t)$ exists for each real numbers a, b as well. Since operators $A*$ and $B*$ are left $\mathcal{A}_{r+1}$ linear, then $(A*)^{-1}$ and $(B*)^{-1}$ are left $\mathcal{A}_{r+1}$ linear as well. Formulas 19(2, 3) imply that $\sigma_{a,b} f$ is left linear and alternative by $a + bi_{2^r}^* \in \mathbf{R} \oplus \mathbf{R} i_{2^r}$ at f. Then using Formulas $(1, 2, 4, 5)$ we infer that the operator given by Formula $21(2)$ exists at f.

24. Note. Certainly for $r = 1$ the left alternativity of an operator $A : X \to Y$ by $a + bi_{2^r}^* \in \mathbf{R} \oplus \mathbf{R} i_{2^r}$ is implied by the left linearity of A by $a + bi_{2^r}^* \in \mathbf{R} \oplus \mathbf{R} i_{2^r}$ and associativity of the quaternion skew field (see §21). But for $r = 2$ this condition is indispensable.

Analogously to §21 are defined powers of operators $\sigma_{a,b}^k f$ and $\sigma_{(A,B)*,w}^k f$ for a natural number k.

When elements w_1 and w_2 are specified and the operator $\sigma_{a,b}$ or $\sigma_{(A,B)*,w}$ is left linear and alternative (at f) by $a + bi_{2^r}^* \in \mathbf{R} \oplus \mathbf{R} i_{2^r}$ we write it shortly as σ or $\sigma_{(A,B)*}$ (at f) respectively.

25. Proposition. *If $C* \in \mathcal{E}$, $w \in C^{k-1}(U + U i_{2^r}, \mathcal{A}_{r+1})$, then $\sigma_{(A,B)*,w}^k f$ and $\sigma_{(A,B)*,w}^k (C * f)$ exist or not simultaneously. If $C*$ and $\sigma_{(A,B)*,w}$ are left linear (and alternative) at f by $a + bi_{2^r}^* \in \mathbf{R} \oplus \mathbf{R} i_{2^r}$, then the operator $\sigma_{(A,B)*,w} \circ C*$ is left linear (and alternative) at f by $a + bi_{2^r}^* \in \mathbf{R} \oplus \mathbf{R} i_{2^r}$.*

Proof. The first statement follows by induction from Corollary 16, Definition 21, Note 24 and Formula $21(2)$. The second statement is implied by the fact that the composition of left linear (and left alternative) by $a + bi_{2^r}^* \in \mathbf{R} \oplus \mathbf{R} i_{2^r}$ operators is also such.

26. Definition. A function $f(x,t)$ on a domain $(U + U i_{2^r}) \times H_n$ we call (σ, A, B) pseudo-analytic if there exists $\sigma_{(A,B)*,w} f(x,t) \in C^0((U + U i_{2^r}) \times H_n, \mathcal{A}_{r+1})$ and the operator $\sigma_{(A,B)*,w}$ is left linear and alternative by $a + bi_{2^r}^* \in \mathbf{R} \oplus \mathbf{R} i_{2^r}$ at f for each $x \in U + U i_{2^r}$ and $t \in H_n$, while $f(x,t)$ is supposed to have a continuous extension on $(U + U i_{2^r}) \times W$, where W is an open neighborhood of H_n in $\mathcal{A}_r$.

27. Proposition. *If functions f and g are (σ, A, B) pseudo-analytic on a domain $(U + U i_{2^r}) \times H_n$, $a, b \in \mathcal{A}_{r+1}$, $1 \le r \le 2$, then*

(1). $fa + gb$ is (σ, A, B) pseudo-analytic on $(U + U i_{2^r}) \times H_n$;

(2). if $g(x,t) \in \mathbf{R} \oplus \mathbf{R} i_{2^r}$ for each $x \in U + U i_{2^r}$ and $t \in H_n$, then the convolution $g * f$ is (σ, A, B) pseudo-analytic on $(U + U i_{2^r}) \times H_n$;

(3). if an operator $C* \in \mathcal{E}_{r+1}$ is left linear and alternative by $a + bi_{2^r}^* \in \mathbf{R} \oplus \mathbf{R} i_{2^r}$, then their convolution $C * f$ is also (σ, A, B) pseudo-analytic on $(U + U i_{2^r}) \times H_n$.

Proof. For two open neighborhoods W_f and W_g of H_n in $\mathcal{A}_r$ their intersection $W = W_f \cap W_g$ is an open neighborhood of H_n as well.

(1). The first statement follows from definitions and alternativity of the octonion algebra.

(2). If the operator $\sigma_{(A,B)*,w}$ is left linear and alternative by $a + bi^*_{2^r} \in \mathbf{R} \oplus \mathbf{R}i_{2^r}$ at f for each $x \in U + Ui_{2^r}$ and $t \in H_n$, then the anti-derivation operator ${}_{\sigma_{(A,B)*,w}} \int_\gamma f(\xi, t) d\xi$ is also such due to Formulas $21(1)$ and $23(5)$. On the other hand, the real algebra $\mathbf{R} \oplus \mathbf{R}i_{2^r}$ is isomorphic with the complex field $\mathbf{C}$ supplied with the purely imaginary generator i_{2^r}. From Definition 17 of the first order partial differential operators σ_s with $s = 0, 1, 2$ it follows that these operators are left $\mathcal{A}_{r+1}$ linear and alternative by $a + bi^*_{2^r} \in \mathbf{R} \oplus \mathbf{R}i_{2^r}$. Then we infer that the operators $A*$ and $B*$ are also left linear and alternative by $a + bi^*_{2^r} \in \mathbf{R} \oplus \mathbf{R}i_{2^r}$ (see also §21 above). Therefore, the convolution

$$g(x,t) * f(x,t) := {}_{\sigma_0} \int_{t'}^{t} g(x, t-\tau) f(x,\tau) d\tau$$

is left linear and alternative by $a + bi^*_{2^r} \in \mathbf{R} \oplus \mathbf{R}i_{2^r}$ in g and f as well. Thus $\sigma_{(A,B)*,w}$ is also such at $g * f$ as the composition of two left linear and alternative by $a + bi^*_{2^r} \in \mathbf{R} \oplus \mathbf{R}i_{2^r}$ operators at g and f and $g * f$ respectively for each $x \in U + Ui_{2^r}$ and $t \in H_n$.

(3). Since $C* \in \mathcal{E}_{r+1}$ is left linear and alternative by $a + bi^*_{2^r} \in \mathbf{R} \oplus \mathbf{R}i_{2^r}$, then $C * f$ in $C*$ and f and hence $\sigma_{(A,B)*,w}(C * f)$ at $C * f$ are left linear and alternative by $a + bi^*_{2^r} \in \mathbf{R} \oplus \mathbf{R}i_{2^r}$.

28. Proposition. *If conditions of Corollary 19 are satisfied with $\sigma_1 = \sigma_2$ and a function $f(x,t) = u(x,t) + i_{2^r}v(x,t)$ is (σ, A, B) pseudo-analytic on a domain $(U + Ui_{2^r}) \times H_n$, where functions u and v are with values in $\mathcal{A}_r$, $1 \le r \le 2$, then functions u and v satisfy the equation*

$$(A*)^{-2}\sigma^2_{1,x_1} w + (B*)^{-2}\sigma^2_{2,x_2} w = 0 \qquad (1)$$

for each $x \in U + Ui_{2^r}$ and $t \in H_n$.

Proof. Equation (1) follows from Formulas $19(2,3)$ and Theorem 23, since the operators $A*$ and $B*$ are invertible due to Lemmas 10 and 12, whilst $(A*)^{-1}$, $(B*)^{-1}$ and σ_1 commute by the conditions imposed in this proposition.

29. Definition. A function u on $(U + Ui_{2^r}) \times H_n$ with values in $\mathcal{A}_r$ satisfying Equation $28(1)$ we call (σ, A, B) pseudo-harmonic.

30. Theorem. *Let a function $u(x,t)$ be (σ, A, B) pseudo-analytic on a domain $(U + Ui_{2^r}) \times H_n$, $1 \le r \le 2$, and let the section V of $U + Ui_{2^r}$ by the plane $\mathbf{R} \oplus \mathbf{R}i_{2^r}$ be simply connected. Let also $\sigma_1 = \sigma_2$ and $A*$, $B*$, σ_1 pairwise commute on $(U + Ui_{2^r}) \times H_n$. Then there exists a (σ, A, B) pseudo-harmonic function $v(x,t)$ with values in $\mathcal{A}_r$ such that a function $f(x,t) + i_{2^r}v(x,t)$ is (σ, A, B) pseudo-analytic on a domain $(U + Ui_{2^r}) \times H_n$.*

Proof. In view of Theorem 23 and Definition 26 a function $u(x,t)$ satisfies the system of equations:

$$(A*)^{-1}\sigma_{1,x_1} u(x,t) = 0 \qquad (1)$$

and

$$(B*)^{-1}\sigma_{2,x_2} u(x,t) = 0 \qquad (2)$$

on $(U+Ui_{2^r})\times H_n$. Let a function v satisfy the system of equations on $(U+Ui_{2^r})\times H_n$:

$$(B*)^{-1}\sigma_{2,x_2}v(x,t)=0 \tag{3}$$

and

$$(A*)^{-1}\sigma_{1,x_1}v(x,t)=0. \tag{4}$$

The latter two equations imply that v is (σ, A, B) pseudo-harmonic, since $\sigma_1=\sigma_2$, the operators $A*$, $B*$ and σ_1 pairwise commute on $(U+Ui_{2^r})\times H_n$, whilst operators $A*$ and $B*$ are invertible, the quaternion skew field $\mathcal{A}_2=\mathbf{H}$ is associative. Then Formulas $(1-4)$ lead to the conclusion that the function $f=u+i_{2^r}v$ satisfies the system $19(2,3)$ and hence it is (σ, A, B) pseudo-analytic on the domain $(U+Ui_{2^r})\times H_n$.

31. Remark. Let $\mathcal{A}_r$ be a Cayley-Dickson algebra over the real field $\mathbf{R}$ and let i_{2^r} be its doubling generator so that $\mathcal{A}_{r+1}=\mathcal{A}_r\oplus\mathcal{A}_r i_{2^r}$. The sum is direct, that is $\mathcal{A}_r\cap\mathcal{A}_r i_{2^r}=\{0\}$. Then there exist linear over the real field $\mathbf{R}$ projections $\pi:\mathcal{A}_{r+1}\to\mathcal{A}_r$ and $\eta:\mathcal{A}_{r+1}\to\mathcal{A}_r i_{2^r}$ such that $\pi(z)=z_1$ and $\pi(z)=i_{2^r}z_2$, where $z=z_1+i_{2^r}z_2$ with $z_1,z_2\in\mathcal{A}_r$, while $\mathcal{A}_{r+1}$ and $\mathcal{A}_r$ are considered as linear spaces over the real field $\mathbf{R}$. Therefore, $i_{2^r}^*(\eta(z))=z_2$ is the $\mathbf{R}$ linear projection operator.

32. Proposition. *If conditions of Theorem 18 and Corollary 19 are satisfied with $\sigma_1=\sigma_2$, then each $\mathcal{A}_r$ valued solution $u(x,t)$ of PDE 17(1) on $(U+Ui_{2^r})\times H_n$ with the initial conditions 17(4) can be considered as a part of a (σ, A, B) pseudo-analytic function f with values in $\mathcal{A}_{r+1}$.*

Proof. In view of Theorem 18, Corollary 19 and Theorem 30 if a function $u(x,t)$ with values in $\mathcal{A}_r$ satisfies Equations $17(1,4)$, then there exists a (σ, A, B) pseudo-analytic function $f=u+i_{2^r}v$ on the domain $(U+Ui_{2^r})\times H_n$. The operators $A*$, $B*$ and σ are left linear and alternative by $a+bi_{2^r}^*\in\mathbf{R}\oplus\mathbf{R}i_{2^r}$, consequently, $(A*)^{-1}$, $(B*)^{-1}$ and ${}_\sigma\int$ also posses these properties. Therefore, the function $g=v+i_{2^r}u$ also is (σ, A, B) pseudo-analytic. Thus one gets either $u=\pi(f)$ or $u=i_{2^r}^*(\eta(g))$ in the notation of §31.

6. Integration of Vector Hydrodynamical PDE Over Octonions

1. First Order Partial Differential Operators. We consider an arbitrary first order partial differential operator σ given by the formula

$$\sigma f=\sum_{j=0}^{2^r-1} i_j^*(\partial f/\partial z_{\xi(j)})\psi_j, \tag{1}$$

where f is a differentiable $\mathcal{A}_r$-valued function on the domain U satisfying

For an ordered product $\{{}_1f\ldots{}_kf\}_{q(k)}$ of differentiable functions ${}_sf$ we put

$${}^s\sigma\{{}_1f\ldots{}_kf\}_{q(k)}=\sum_{j=0}^{n} i_j^*\{{}_1f\ldots(\partial_s f/\partial z_{\xi(j)})\ldots{}_kf\}_{q(k)}\psi_j, \tag{2}$$

where a vector $q(k)$ indicates on an order of the multiplication in the curled brackets (see also §2 [26, 25]), so that

$$\sigma\{{}_1f...{}_kf\}_{q(k)} = \sum_{s=1}^{k} {}^s\sigma\{{}_1f...{}_kf\}_{q(k)}. \tag{3}$$

Symmetrically other operators

$$\hat{\sigma} f = \sum_{j=0}^{2^r-1} (\partial f/\partial z_{\xi(j)}) i_j \psi_j, \tag{4}$$

are defined. Therefore, these operators are related:

$$(\sigma f)^* = \hat{\sigma}(f^*). \tag{5}$$

2. Integral Operators. We consider integral operators of the form:

$$\mathsf{K}(x,y) = \mathsf{F}(x,y) + {}_\sigma\int_x^\infty \mathsf{F}(z,y)\mathsf{N}(x,z,y)dz, \tag{1}$$

where σ is an $\mathbf{R}$-linear partial differential operator as in §1 and ${}_\sigma\int$ is the non-commutative line integral (anti-derivative operator) over the Cayley-Dickson algebra $\mathcal{A}_r$ from §§4.2.5 [27], where F and K are continuous functions with values in the Cayley-Dickson algebra $\mathcal{A}_r$ or more generally in the real algebra $Mat_s(\mathcal{A}_r)$ of $s \times s$ matrices with entries in $\mathcal{A}_r$. For definiteness we take ${}_{\partial/\partial x_0}\int g(z)dz = \int g(z)dz$ and ${}_\sigma\int g(z)dz$ calculated with the help of the left algorithm 2.6 [26, 25] or 1.2.6(LI) [27].

3. Proposition. *Let*

$$\lim_{z\to\infty} {}^1\sigma_z^k\, {}^2\sigma_x^s\, {}^2\sigma_z^n \mathsf{F}(z,y)\mathsf{K}(x,z) = 0 \tag{1}$$

for each x, y in a domain U satisfying Conditions 2.2.1($D1, D2$) with $\infty \in U$ and every non-negative integers $0 \le k, s, n \in \mathbf{Z}$ such that $k + s + n \le m$. Suppose also that ${}_\sigma\int_x^\infty \partial_x^\alpha \partial_y^\beta \partial_z^\omega [\mathsf{F}(z,y)\mathsf{K}(x,z)]dz$ converges uniformly by parameters x, y on each compact subset $W \subset U \subset \mathcal{A}_r^2$ for each $|\alpha| + |\beta| + |\omega| \le m$, where $\alpha = (\alpha_0, ..., \alpha_{2^r-1})$, $|\alpha| = \alpha_0 + ... + \alpha_{2^r-1}$, $\partial_x^\alpha = \partial^{|\alpha|}/\partial x_0^{\alpha_0}...\partial x_{2^r-1}^{\alpha_{2^r-1}}$. Then the non-commutative line integral ${}_\sigma\int_x^\infty \mathsf{F}(z,y)\mathsf{K}(x,z)dz$ satisfies the identities:

$$\sigma_x^m\, {}_\sigma\int_x^\infty \mathsf{F}(z,y)\mathsf{K}(x,z)dz = {}^2\sigma_x^m\, {}_\sigma\int_x^\infty \mathsf{F}(z,y)\mathsf{K}(x,z)dz + A_m(\mathsf{F},\mathsf{K})(x,y), \tag{2}$$

$${}^1\sigma_z^m\, {}_\sigma\int_x^\infty \mathsf{F}(z,y)\mathsf{K}(x,z)dz = (-1)^m\, {}^2\sigma_z^m\, {}_\sigma\int_x^\infty \mathsf{F}(z,y)\mathsf{K}(x,z)dz + B_m(\mathsf{F},\mathsf{K})(x,y), \tag{3}$$

where

$$A_m(\mathsf{F},\mathsf{K})(x,y) = -\, {}^2\sigma_x^{m-1}[\mathsf{F}(x,y)\mathsf{K}(x,z)]|_{z=x} + \sigma_x\, A_{m-1}(\mathsf{F},\mathsf{K})(x,y) \tag{4}$$

for $m \ge 2$,

$$B_m(\mathsf{F},\mathsf{K})(x,y) = (-1)^m\, {}^2\sigma_z^{m-1}\mathsf{F}(x,y)\mathsf{K}(x,z)|_{z=x} + [{}^1\sigma_z\, B_{m-1}(\mathsf{F}(z,y),\mathsf{K}(x,z))]|_{z=x} \tag{5}$$

for $m \geq 2$,

$$A_1(\mathsf{F},\mathsf{K})(x,y) = B_1(\mathsf{F},\mathsf{K})(x,y) = -\mathsf{F}(x,y)\mathsf{K}(x,x), \tag{6}$$

σ_x *is an operator* σ *acting by the variable* $x \in U \subset \mathcal{A}_r$.

4. Method of non-Commutative Integration of Vector PDE.

Let an equation over the Cayley-Dickson algebra $\mathcal{A}_r$ be given in the non-commutative line integral form:

$$\mathsf{K}(x,y) = \mathsf{F}(x,y) + \mathsf{p}\ {}_\sigma\int_x^\infty \mathsf{F}(z,y)N(x,z,y)dz, \tag{1}$$

where K, F and N are continuous integrable functions as in §2,
$\mathsf{p} \in \mathbf{R} \setminus \{0\}$ is a non-zero real constant. These functions may depend on additional parameters $t, \tau, \ldots$.

The first step consists in a concretization of a function N and its expression throughout F and K. Suppose that an operator

$$(I - \mathsf{A}_x)\mathsf{F}(x,y) = \mathsf{K}(x,y) \tag{2}$$

is invertible, so that $(I - \mathsf{A}_x)^{-1}$ is continuous, where I denotes the unit operator, while
$\mathsf{A}_x = -\mathsf{p}\ {}_\sigma\int_x^\infty \mathsf{F}(z,y)\mathsf{N}(x,z,y)dz$
is an operator acting by variables x.

On the second step two $\mathbf{R}$-linear differential or partial differential operators L_s over the Cayley-Dickson algebra $\mathcal{A}_r$ are given:

$$L_s f = \sum_j i_j^*(L_{s,j} f), \tag{3}$$

where f is a differentiable function in the domain of L_s, $L_{s,j}g$ is real-valued function for each $ord(L_s)$ times differentiable real-valued function g in the domain of $L_{s,j}$ for every j, where $ord(L_s)$ denotes the order of the partial differential operator L_s. On a function F the conditions either:

$$L_s\mathsf{F} = 0 \tag{4}$$

for each $s = 1, 2$, or

$$\sum_{j\in\Psi_k} i_j^*[\,{}_jc(L_{s,0}\mathsf{F}) + L_{s,j}\mathsf{F}] = 0 \tag{5}$$

for each $s = 1, 2$ and $1 \leq k \leq m$ are imposed, where ${}_jc$ are constants ${}_jc \in \mathcal{A}_r$, $\Psi_k \subset \{0, 1, ..., 2^r - 1\}$ for each k, $\bigcup_k \Psi_k = \{0, 1, ..., 2^r - 1\}$, $\Psi_k \cap \Psi_l = \emptyset$ for each $k \neq l$, $1 \leq m \leq 2^r$. Coefficients ${}_jc$ or operators $L_{s,j}$ may be zero for some j.

On the third step a function K is calculated from Equation (2).

Acting by the operator L_s from the left on (2) and using Conditions either (4) or (5) one gets either

$$L_s[(I - A_x)\mathsf{K}] = 0 \tag{6}$$

or the equalities:

$$\sum_{j\in\Psi_k} i_j^*\{\,{}_jcL_{s,0}[(I - \mathsf{A}_x)\mathsf{K}] + L_{s,j}[(I - \mathsf{A}_x)\mathsf{K}]\} = 0 \tag{7}$$

for each $k = 1, ..., m$.
Therefore, using Conditions either (4) or (5) we infer that

$$(I - \mathsf{A}_x)(L_s \mathsf{K}) = R_s(\mathsf{K}) for s = 1, 2, \tag{8}$$

where an operator

$$R_s(f) = (I - \mathsf{A}_x)(L_s f) - L_s[(I - \mathsf{A}_x) f] \tag{9}$$

is formed with the help of commutators $[A, B] = AB - BA$ and anti-commutators $\{A, B\} = AB + BA$ of operators $(I - (\mathsf{A}_x)_0)$, $(\mathsf{A}_x)_k$, $L_{s,j}$, $k, j = 0, ..., 2^r - 1$. A function N and operators L_1 and L_2 are chosen such that

$$R_s(\mathsf{K}) = (I - \mathsf{A}_x) M_s(\mathsf{K}) for s = 1, 2, \tag{10}$$

where $M_s(\mathsf{K})$ are functionals of K which generally may be non-$\mathbf{R}$-linear. In view of Condition (2) the function K must satisfy partial differential equations

$$L_s \mathsf{K} - M_s(\mathsf{K}) = 0 for s = 1, 2, \tag{11}$$

which generally may be non-$\mathbf{R}$-linear. Thus each solution K of the $\mathbf{R}$-linear integral equation (1) is also the solution of partial differential equations (11). Frequently particular cases are considered, when Equations (4) correspond to an eigenvalue problem for $s = 1$ and to an evolution in time problem for $s = 2$. Generally operators $L_{s,0}, ..., L_{s,2^r-1}$ can be chosen $\mathcal{A}_r$ vector independent and ${}_jc \in \mathbf{R} i_{k(j)}$ for each j and $s = 1, 2$ with $k = k(j) \in \{0, 1, .., 2^r - 1\}$.

The Euclidean space $\mathbf{R}^{2^r}$ is the real shadow of the Cayley-Dickson algebra $\mathcal{A}_r$, that is, by the definition $\mathcal{A}_r$ considered as the $\mathbf{R}$ linear space is isomorphic with $\mathbf{R}^{2^r}$. The Lebesgue (non-negative) measure μ on the Borel σ-algebra $\mathcal{B}(\mathbf{R}^{2^r})$ of the Euclidean space $\mathbf{R}^{2^r}$ induces the Lebesgue measure on $\mathcal{B}(\mathcal{A}_r)$. Therefore, the Hilbert space $X = L^2(\mathcal{A}_r, \mu, \mathcal{A}_r)$ of all Lebesgue measurable functions $f : \mathcal{A}_r \to \mathcal{A}_r$ with integrable square module $|f|^2$, i.e.

$$\|f\|^2 := \int_{\mathcal{A}_r} |f(z)|^2 \mu(dz) < \infty, \tag{HN}$$

and with the $\mathcal{A}_r$ valued scalar product

$$(f, g) := \int_{\mathcal{A}_r} f^*(z) g(z) \mu(dz) \tag{SP}$$

exists. Analogously the Hilbert space $X = L^2(\mathcal{A}_r, \mu, Mat_s(\mathcal{A}_r))$ is defined with $\sum_{j,k=0}^{s} f^*_{j,k}(z) g_{j,k}(z)$ instead of $f^*(z)g(z)$ in the integral in Formula (SP), where $f_{j,k} \in \mathcal{A}_r$ denotes a matrix element at the intersection of row j with column k of a matrix $f \in Mat_s(\mathcal{A}_r)$.

Let A be an $\mathbf{R}$ linear $\mathcal{A}_r$ additive operator $A : \mathsf{D}(X) \to Y$, where $\mathsf{D}(A)$ is a domain of A dense in X, $\mathsf{D}(A) \subset X$, X and Y are Hilbert spaces over $\mathcal{A}_r$. Then its adjoint operator A^* is defined on a domain consisting of all those vectors $y \in Y$ such that for some vector $z \in X$ the equality $(x, z) = (Ax, y)$ is valid for all $x \in \mathsf{D}(A)$. For such $y \in Y$ we put $A^* y = z$. If $A^* = A$ we say that A is self-adjoint. Thus $(Ax, y) = (x, A^* y)$ for all $x \in \mathsf{D}(A)$ and $y \in \mathsf{D}(A^*)$.

If an operator A is self-adjoint $A = A^*$, then $(Ax, y) = (x, Ay)$ for all $x, y \in \mathsf{D}(A)$.

When $\mathsf{D}(A)$ is dense in X and $(Ax, y) = (x, Ay)$ for all $x, y \in \mathsf{D}(A)$, we say that A is symmetric. A self-adjoint operator is maximal symmetric.

We define a self-adjoint operator A to be positive, when $(Ax, x) \geq 0$ for each $x \in \mathsf{D}(A)$.

An $\mathbf{R}$ linear $\mathcal{A}_r$ additive operator A is called invertible if it is densely defined and one-to-one and has dense range $\mathcal{R}(A)$.

The operator A^*A is self-adjoint and positive. If A is invertible, then $(A^*A)^{-1}A^* \subseteq A^{-1}$ (see also [4, 31]).

If an expression of the form

$$\sum_k [(I - A_x)\ {}_kf(x, y)]\ {}_kg(y) = u(x, y) \tag{12}$$

will appear on a domain U, which need to be inverted we consider the case when

$$(I - A_x) \tag{RS}$$

is either right strongly $\mathcal{A}_r$ linear, or right $\mathcal{A}_r$ linear and ${}_kf \in X_0$ for each k, or $\mathbf{R}$ linear and ${}_kg(y) \in \mathbf{R}$ for each k and every $y \in U$, at each point $x \in U$.

When an operator $(I - A_x)$ is invertible and Condition (RS) is satisfied, Equation (12) can be resolved:

$$\sum_k {}_kf(x, y)\ {}_kg(y) = (I - A_x)^{-1}u(x, y). \tag{13}$$

If condition (RS) is not fulfilled, the corresponding system of equations in real components $(A_x)_{j,s}$, ${}_kf_s$ and ${}_kg_s$ can be considered.

If $A : X \to X$ is a bounded $\mathbf{R}$ linear operator on a Banach space X with the norm $\|A\| < 1$, then

$(I - A)^{-1} = \sum_{n=0}^{\infty} A^n$.

The anti-derivative operator $g \mapsto {}_\sigma\int_{0x}^{x} g(z)dz$ is compact from $C^0(V, \mathcal{A}_r)$ into $C^0(V, \mathcal{A}_r)$ for a compact domain V in $\mathcal{A}_r$, where $C^0(V, \mathcal{A}_r)$ is the Banach space over $\mathcal{A}_r$ of all continuous functions $g : V \to \mathcal{A}_r$ supplied with the supremum norm $\|g\| := \sup_{x\in V} |g(x)|$, ${}_0x$ is a marked point in V, $x \in V$. Therefore, an operator A_x will be invertible with a suitable choice of a function F satisfying the system of $\mathbf{R}$ linear partial differential equations (4) or (5) and a real non-zero parameter $\mathsf{p} \neq 0$.

5. Theorem. *A solution of PDE*

$$({}_3\sigma_t + {}_1\sigma_x^3 + 3\ {}_2\sigma_y\ {}_1\sigma_x^2 + 3\ {}_2\sigma_y^2\ {}_1\sigma_x + {}_2\sigma_y^3)\mathsf{K}(x, y)$$

$$+6({}_1^1\sigma_x + {}_2^1\sigma_y)[\mathsf{K}(x, y)(\ {}_1\sigma_x\mathsf{K}(x, x))] - \mathsf{K}(x, y)\{[{}_1\sigma_z,\ {}_1\sigma_x]\mathsf{K}(x, z)]|_{z=x}\}$$

$$-[{}_1^1\sigma_x,\ {}_1^2\sigma_x][\mathsf{K}(x, y)\mathsf{K}(x, x)] = 0 \tag{1}$$

with ${}_1\psi_0 = {}_2\psi_0 = 0$ *over the Cayley-Dickson algebra* $\mathcal{A}_r$ *with* $2 \leq r \leq 3$ *is given by Formulas of §4 and*

$$L_s\mathsf{K}(x, y) = L_s\mathsf{p}\ {}_\sigma\int_x^\infty \mathsf{F}(z, y)\mathsf{K}(x, z)dz, \tag{2}$$

$$L_1 = {}_1\sigma_x^2 - {}_2\sigma_y^2 \quad (3)$$

$${}_1\sigma_z^2 F(z,y) = {}_2\sigma_y^2 F(z,y), \quad (4)$$

$$L_{2,j} f = [{}_3\psi_{\xi_3(j)}\partial_{t_j} + ({}_1\psi_j\, \partial_{x_{\xi_1(j)}} + 3\, {}_2\psi_j \partial_{y_{\xi_2(j)}})\, {}_1\sigma_x^2 + ({}_2\psi_j\, \partial_{y_{\xi_2(j)}} + 3\, {}_1\psi_j \partial_{x_{\xi_1(j)}})\, {}_2\sigma_y^2] f, \quad (5)$$

where

$$L_{2,j} \mathsf{F}(x,y) = 0 foreachj, \quad (6)$$

where the operator ${}_3\sigma_t$ *is with real constant coefficients* ${}_3\psi_j$ *and a transposition* $\xi_3 \in S_{2^r}$ *with* $\mathsf{p} = 1$ *whenever the appearing integrals uniformly converge by parameters on compact sub-domains and the operator* $(I - A_x)$ *is invertible and* $\mathsf{F} \in Mat_s(\mathbf{R})$ *and* $\mathsf{K} \in Mat_s(\mathcal{A}_r)$*,* $s \in \mathbf{N}$ *when* $r = 2$*,* $s = 1$ *when* $r = 3$*.*

6. Theorem. *PDE*

$$L_2 \mathsf{K}(x,y) = 3\mathsf{p}[(\sigma_x + \sigma_y)\mathsf{K}(x,y)]\mathsf{K}^2(x,x) + 3\mathsf{p}\mathsf{K}(x,y)[{}^2\sigma_x(\mathsf{K}(x,x)\mathsf{K}(x,x))] \quad (1)$$

with $\psi_0 = {}_1\psi_0 = 0$ *over the Cayley-Dickson algebra* $\mathcal{A}_r$*,* $2 \le r \le 3$*, has a solution given by Formulas of* §*4 and*

$$\mathsf{K}(x,y) = \mathsf{F}(x,y) + \frac{\mathsf{p}}{4}\, {}_\sigma\int_x^\infty (\, {}_\sigma\int_x^\infty \mathsf{F}(u,y)[\mathsf{F}(z,u)\mathsf{K}(x,z)]dz)du, \quad (2)$$

where p *is a non-zero real constant;*

$$L_1 = \sigma_x - {}_1\sigma_y with \psi_0 = {}_1\psi_0 = 0; \quad (3)$$

$$L_1 \mathsf{F}(x,y) = 0, \quad (4)$$

where $\mathsf{F}(x,y) = \mathsf{G}(\frac{(a,x>+({}_1a,y>}{2})$ and

$$(a,x> := \sum_{j=0}^{2^r-1} a_j x_j i_j \quad (5)$$

for $a, x \in \mathcal{A}_r$*,* $x = \sum_{j=0}^{2^r-1} x_j i_j$*,* $x_j \in \mathbf{R}$ *for each* j*,* $a_j = \psi_j$ *for* $\psi_j \ne 0$ *and* $a_j = 1$ *for* $\psi_j = 0$*;*

$$L_{2,j}\mathsf{F} = 0 foreachj = 0, ..., 2^r - 1, \quad (6)$$

where

$$L_{2,j} = {}_2\psi_j \partial_{t_j} + \psi_j \partial_{x_j} \sigma_x^2 + 3\, {}_1\psi_j \partial_{y_j} \sigma_x^2 + 3\psi_j\, {}_1\sigma_y^2 \partial_{x_j} + {}_1\psi_j \partial_{y_j}\, {}_1\sigma_y^2, \quad (7)$$

when the appearing integrals uniformly converge by parameters and the operator $(I - A_x)$ *is invertible and* $\mathsf{F} \in Mat_s(\mathbf{R})$ *and* $\mathsf{K} \in Mat_s(\mathcal{A}_r)$ *with* $s \in \mathbf{N}$ *for* $r = 2$ *and* $s = 1$ *for* $r = 3$*.*

7. Theorem. *PDE*

$$({}_1\sigma_t + \sigma_x^2 + 2\,{}^2\sigma_y\,{}^2\sigma_x + \sigma_y^2)\mathsf{K}(x,y) = -2\mathsf{p}\mathsf{K}(x,y)[\sigma_x \mathsf{K}(x,x)] \qquad (1)$$

over the Cayley-Dickson algebra $\mathcal{A}_r$ with $2 \le r \le 3$ has a solution given by Formulas of §4 and

$$L_1 = \sigma_x + \sigma_y, \qquad (2)$$

$$L_2 = {}_1\sigma_t + \sigma_x^2 + u\sigma_y\sigma_x + \sigma_y^2, \qquad (3)$$

where $u \in \mathbf{R}$ is a real constant;

$$\mathsf{K}(x,y) = \mathsf{F}(x,y) + \mathsf{p}\ {}_\sigma\int_x^\infty \mathsf{F}(z,y)\mathsf{K}(x,z)dz; \qquad (4)$$

$$L_{2,j}\mathsf{F}(x,y) = 0 foreach j, \qquad (5)$$

$$L_{2,j}\mathsf{F}(x,y) = [{}_1\psi_j\partial_{t_j} +$$

$$u \sum_{s,p:\ i_s i_p = i_j} \psi_s\psi_p(\partial_{y_s}\partial_{x_p} + (i_p i_s)(i_s i_p)\partial_{y_p}\partial_{x_s}) + \pi_j \circ (\sigma_x^2 + \sigma_y^2)]\mathsf{F}(x,y), \qquad (6)$$

where $\pi_j : X \to X_j$ is the $\mathbf{R}$ linear projection operator (see §2.2.1);

$$L_1\mathsf{F}(x,y) = 0, i.e. \mathsf{F}(x,y) = \mathsf{F}(\frac{(\psi, x-y>}{2}), \qquad (7)$$

when the appearing integrals uniformly converge by parameters on each compact sub-domain and the operator $(I - A_x)$ is invertible and $\mathsf{F} \in Mat_s(\mathbf{R})$ and $\mathsf{K} \in Mat_s(\mathcal{A}_r)$, $s \in \mathbf{N}$ for $r = 2$, $s = 1$ for $r = 3$.

The proofs of Theorems 5-7 and further examples of solutions of PDE are written in [32].

8. Conclusion. The methods and results of this chapter can be used for integration of particular elliptic, parabolic, hyperbolic, mixed type, Sobolev type, hydrodynamical type PDE, etc. and for further investigations of PDE, their solutions and appearing classes of functions.

References

[1] M. J. Ablowitz, H. Segur. *Solitons and the inverse scattering transform* (SIAM: Philadeplhia, 1981).

[2] J.C. Baez. The octonions. *Bull. Amer. Mathem. Soc.* 39: 2, 145-205 (2002).

[3] M. Bronstein. *Symbolic integration. Transcendental functions* (Springer-Verlag: Berlin, 1997).

[4] R.V. Cadison, J.R. Ringrose. *Fundamentals of the theory of operator algebras*. V. 1,2 (Academic Press: New York, 1983).

[5] L.E. Dickson. *The collected mathematical papers*. V. 1-5 (Chelsea Publishing Co.: New York, 1975).

[6] R. Engelking. *General topology* (Mir: Moscow, 1986).

[7] G. Emch. Méchanique quantique quaternionienne et Relativitè restreinte. *Helv. Phys. Acta* 36, 739-788 (1963).

[8] G.M. Fihtengolz. *Course of differential and integral calculus*. V. 1-3 (Nauka: Moscow, 1966).

[9] I.M. Gelfand, G.E. Shilov. *Generalized functions and operations with them* (Fiz.-Mat. Lit.: Moscow, 1958).

[10] G. Grubb. *Functional calculus of pseudodifferential boundary problems* (Birkhäuser: Boston, 1996).

[11] K. Gürlebeck, K. Habetha, W. Sprössig. *Holomorphic functions in the plane and the n-dimensional space.* (Birkhäuser: Basel, 2008).

[12] K. Gürlebeck, W. Sprössig. *Quaternionic and Clifford calculus for physicists and engineers* (John Wiley and Sons: Chichester, 1997).

[13] K. Gürlebeck, W. Sprössig. *Quaternionic analysis and elliptic boundary value problem* (Birkhäuser: Basel, 1990).

[14] F. Gürsey, C.-H. Tze. *On the role of division, Jordan and related algebras in particle physics* (World Scientific Publ. Co.: Singapore, 1996).

[15] W.R. Hamilton. *Selected works. Optics. Dynamics. Quaternions* (Nauka: Moscow, 1994).

[16] L. Hörmander. *The analysis of linear partial differential operators*. V. 1-4 (Springer-Verlag: Berlin, 1985).

[17] L.I. Kamynin. Course of Mathematical Analysis (Moscow State Univ. Press: Moscow, 1995).

[18] I.L. Kantor, A.S. Solodovnikov. *Hypercomplex numbers* (Springer-Verlag: Berlin, 1989).

[19] A. Kneschke. *Differentialgleichungen und Randwertprobleme*. B. 1-3 (B.G. Teubner Verlaggesellschaft: Leipzig, 1965).

[20] A.N. Kolmogorov, S.V. Fomin. *Elements of theory of functions and functional analysis* (Nauka: Moscow, 1989).

[21] A.G. Kurosh. *Lectures on general algebra* (Nauka: Moscow, 1973).

[22] M.A. Lavrentjev, B.V. Shabat. *Methods of functions of the complex variable* (Nauka: Moscow, 1987).

[23] H.B. Lawson, M.-L. Michelson. *Spin geometry* (Princ. Univ. Press: Princeton, 1989).

[24] J. Leray. Un prologement de la transformation de Laplace qui transforme la solution unitaire d'un opérateur hyperbolique en sa solution élèmentaire. *Bull. de la Société mathém. de France.* 90, 39-156 (1962).

[25] S.V. Ludkovsky. Differentiable functions of Cayley-Dickson numbers and line integration. *J. Mathem. Sciences*, N.Y. (Springer) 141: 3, 1231-1298 (2007).

[26] S.V. Ludkovsky, F. van Oystaeyen. Differentiable functions of quaternion variables. *Bull. Sci. Math.* 127, 755-796 (2003).

[27] S.V. Ludkovsky. *Analysis over Cayley-Dickson numbers and its applications* (Lambert Academic Publishing: Saarbrücken, 2010).

[28] S.V. Ludkovsky. Differential equations over octonions. *Adv. Appl. Clifford Alg.* 21: 4 (2011), 773-797; DOI: 10.1007/s00006-011-0286-4.

[29] S.V. Ludkovsky. Feynman integration over octonions with application to quantum mechanics. *Mathematical Methods in the Appl. Sciences* 33: 9 (2010), 1148-1173.

[30] S.V. Ludkovsky, W. Sprössig. Ordered representations of normal and super-differential operators in quaternion and octonion Hilbert spaces. *Adv. Appl. Clifford Alg.*, 20: 2 (2010), 321-342.

[31] S.V. Ludkovsky. Algebras of operators in Banach spaces over the quaternion skew field and the octonion algebra. *J. Mathem. Sci.*, N.Y. (Springer) 144: 4 (2008), 4301-4366.

[32] S.V. Ludkovsky. Integration of vector hydrodynamical partial differential equations over octonions. *Complex Variables and Elliptic Equations* 58: 5 (2013), 579-609; DOI:10.1080/17476933.2011.598930.

[33] S.V. Ludkovsky. Line integration of Dirac operators over octonions and Cayley-Dickson algebras. *Comp. Meth. Funct. Theory* 12: 1 (2012), 279-306.

[34] S.V. Ludkovsky. Multidimensional Laplace transforms over quaternions, octonions and Cayley-Dickson algebras, their applications to PDE. *Advances in Pure Mathematics* 2: 2 (2012), 63-103.

[35] S.V. Ludkovsky. Multidimensional Laplace transforms over Cayley-Dickson algebras and partial differential equations. *Hypercomplex Numbers in Geometry and Physics.* 8: 1(14) (2010), 128-164 (earlier version: Los Alamos, arXiv:1004.4355 [math.CV] (2010)).

[36] S.V. Ludkovsky. The two-sided Laplace transformation over the Cayley-Dickson algebras and its applications. *J. of Mathem. Sciences*, N.Y. (Springer) 151: 5 (2008), 3372-3430.

[37] S.V. Ludkovsky. Quasi-conformal functions of quaternion and octonion variables, their integral transformations. *Far East Journal of Mathematical Sciences (FJMS)* 28: 1 (2008), 37-88.

[38] S.V. Ludkovsky. Residues of functions of octonion variables. *Far East Journal of Mathematical Sciences (FJMS)* 39: 1 (2010), 65-104.

[39] S.V. Ludkovsky. Functions of several Cayley-Dickson variables and manifolds over them. *J. Mathem. Sci.*, N.Y. (Springer) 141: 3, 1299-1330 (2007).

[40] S.V. Ludkovsky. Normal families of functions and groups of pseudoconformal diffeomorphisms of quaternion and octonion variables. *J. Mathem. Sci.*, N.Y. (Springer) 150: 4, 2224-2287 (2008).

[41] S.V. Ludkovsky. On a class of right linearly differentiable functions of Cayley-Dickson variables. *Adv. Appl. Clifford Alg.* 24: 3 (2014), 781-803; DOI: 10.1007/s00006-014-0474-0.

[42] S.V. Ludkovsky. Decompositions of PDE over Cayley-Dickson algebras. Rendic. dell'Ist. di Math. dell'Università di Trieste. *Nuova Serie* 46 (2014), 1-23.

[43] S.V. Ludkovsky. Dirac operators with generalized coefficients and their line integration. *Communic. in Mathem. and Appl.* 4: 2 (2013); 149-168.

[44] S.V. Ludkovsky. *Line integration and second order partial differential equations over Cayley-Dickson algebras*. Los Alamos, arXiv:1009.2158 [math.CV] (2010).

[45] N.M. Matveev. *Methods of integration of ordinary differential equations* (Lan: St. Petersbourg, 2003).

[46] P.W. Michor. *Manifolds of differentiable mappings* (Boston: Shiva, 1980).

[47] A.F. Nikiforov, V.B. Uvarov. *Special functions of mathematical physics* (Nauka: Moscow, 1978).

[48] I.G. Petrovsky. *Lectures on partial differential equations* (Gos. Izd. Techn. and Theoret. Liter.: Moscow, 1953).

[49] B. van der Pol, H. Bremmer. *Operational calculus based on the two-sided Laplace integral* (Cambridge Univ. Press: Cambridge, 1964).

[50] A.D. Polyanin, V.F. Zaytzev, A.I. Jurov. *Methods of solutions of non-linear equations of mathematical physics and mechanics* (Fizmatlit: Moscow, 2005).

[51] A.D. Polyanin, V.F. Zaytzev. *Reference book on non-linear equations of mathematical physics* (Fizmatlit: Moscow, 2002).

[52] I. Rubinstein, L. Rubinstein. *Partial differential equations in classical mathematical Physics* (Cambridge Univ. Press: Cambridge, 1998).

[53] R. Serôdio. On octonionic polynomials. *Adv. Appl. Clifford Alg.* 17 (2007), 245-258.

[54] R. D. Schafer. *An introduction to nonassociative algebras* (Academic Press: New York, 1966).

[55] E.H. Spanier. *Algebraic topology* (Academic Press: New York, 1966).

[56] A.G. Sveshnikov, A.B. Alshin, M.O. Korpusov, Yu.D. Pletner. *Linear and non-linear equations of Sobolev type* (Fizmatlit: Moscow, 2007).

[57] V.S. Vladimirov. *Equations of Mathematical Physics* (Nauka: Moscow, 1971).

[58] V.A. Zorich. *Mathematical Analysis* (Nauka: Moscow, 1984).

In: Partial Differential Equations
Editor: Deborah E. Richards, pp. 165-211
ISBN: 978-1-63482-643-3

Chapter 4

MIXED BOUNDARY-VALUE PROBLEM FOR DIVERGENT HYPERBOLIC PDE: EXISTENCE AND PROPERTIES OF SOLUTIONS, APPLICATIONS IN SEQUENTIAL OPTIMAL CONTROL WITH POINTWISE IN TIME STATE CONSTRAINTS*

***V. S. Gavrilov*† *and M. I. Sumin*‡**
Mechanics and Mathematics Faculty, Nizhnii Novgorod State University,
Nizhnii Novgorod, Russia

Abstract

The chapter is devoted to the study of a sequential optimization problem for a semilinear hyperbolic divergent equation with pointwise state inequality constraint and with mixed boundary conditions. The chapter consists of two main parts.

The first part is devoted to the study of the existence and uniqueness of solutions to semilinear hyperbolic divergent equations. Also in this part we study a linear hyperbolic divergent equation with a Radon measure in the right part. We study the following questions: existence, uniqueness and stability of solutions to such equations with mixed boundary condition; special integral representations of solutions to such equations; and the stability of solutions of linear hyperbolic divergent equations.

In the second part of the paper, using the results of its first part, we have studied in detail the optimal control problem with pointwise in time state constraints. The state constraint contains a functional parameter that belongs to the class of continuous functions and occurs as an additive term. We compute the first variations of functionals

*This work was supported by the Russian Foundation for Basic Research (project no. 13-02-12155-ofi_m), by the Ministry of Education and Science of the Russian Federation within the framework of project part of state tasks in the field of scientific activity in 2014-2016 (code no. 1727), and by the grant within the agreement of August 27, 2013 No. 02.B.49.21.0003 between the Ministry of Education and Science of the Russian Federation and Lobachevskii State University of Nizhnii Novgorod.

†E-mail address: vladimir.s.gavrilov@gmail.com

‡E-mail address: m.sumin@mail.ru

on the basis of a so-called two-parameter needle variation of controls. We consider necessary conditions for minimizing sequences in an optimal control problem with a pointwise in time state constraint of inequality type and with dynamics described by a semilinear hyperbolic equation in divergence form with mixed boundary condition. For the parametric optimization problem, we also consider regularity and normality conditions stipulated by the differential properties of its value function.

Keywords: sequential optimization, maximum principle for minimizing sequences, semilinear divergent hyperbolic equations, mixed boundary conditions, Radon measures, pointwise in time state constraints, boundary control, value function, sensitivity, normality, regularity, two-parameter needle variation of controls

1. Introduction

In the chapter we consider an optimal control problem with pointwise state constraints for a controlled semilinear divergent hyperbolic equation with mixed boundary conditions. Here words "mixed boundary conditions" means that the Dirichlet condition is given on one part of the boundary, and the third boundary is given condition on the other part of boundary. It should also be noted that we consider the equation with *the non-symmetric main part.*

Different aspects of the theory of divergent hyperbolic equations are considered in many works (see, e.g., [1]–[29] and the reference to them). Some of these works (see, e.g., [1, 2]) are devoted to the proof of a unique existence theorems for generalized solutions of one-dimensional wave equations belonging to L_p classes. In fact, in these studies substantiated method of characteristics. Another group of works (see, e.g., [3] and references in it) is devoted to the study of blow-up of solutions of boundary value problems and the Cauchy problem for nonlinear equations. The third group of works (see [4, 5, 6] and references in them) devoted to the study of special issues, such as arising of resonance and the asymptotic behavior of solutions. The fourth group, works [7]–[19] (see also bibliography), is devoted to the estimates of convergence rate and other issues the study of numerical methods for solution of initial–boundary value problems. The last, fifth, group of works (see [20]–[29] and references in them) is devoted to the proof of theorems of existence and uniqueness of solutions, as well as fine regularity properties of solutions.

Note that in all of these works essentially considered equations with symmetric main part. But in the study of optimal control problems whose dynamics is described by controlled hyperbolic divergent equations with initial-boundary conditions, in the case of the equation containing terms with lower order derivatives, these members go into the main part of the adjoint equations of Pontryagin maximum principle. Details can be found, for example, in [30].

In addition, as far as we know, hyperbolic divergent equations with different boundary conditions on different parts of the boundary were considered only in [25, 26]. However, works [25, 26] deals with only linear equations, and without members containing lower order derivatives.

The chapter consists of two main parts. The first part is fully devoted to the study of the existence and uniqueness of solutions to hyperbolic divergent equations, and regularity properties of these solutions. First of all, we completely enough study initial-boundary problem for semilinear divergent equations with mixed boundary conditions. We pay considerable attention to the study of regularity properties of solutions, since these properties play a central role in obtaining optimization results in the second part of the work.

In distinction to [30], in the present paper we find solutions to a initial-boundary value problems in the class of functions z which are such that: 1) for any fixed t a function z belongs to a Sobolev space, with respect to spatial variables; 2) for almost every t a function z_t is summable with square, with respect to spatial variable; 3) a function $[0,T] \ni t \mapsto z(\cdot,t)$ is continuous with respect to t, in the sense of the weak topology of the Sobolev space; 4) the inclusion $z_t \in L_\infty([0,T], L_2(\Omega))$ holds. In this paper we denote this functional class as $\mathfrak{C}^1_{2,0}(Q_T|S^0_T)$. For linear hyperbolic equation, such solutions were firstly considered in [26, Chapter 3, section 8.4]. In contrast to solutions belonging to the Sobolev space $W^1_2(Q_T)$, a consideration of solutions belonging to $\mathfrak{C}^1_{2,0}(Q_T|S^0_T)$ allowed to impose conditions that are weaker than in [30].

We also note that in obtaining the necessary optimality conditions for optimal control problems with pointwise state constraints Lagrange multiplier corresponding to the operator of pointwise state constraints is a Radon measure, and this Radon measure appears to the right of the adjoint equation corresponding to the operator of pointwise state constraints (see, e.g., [30]-[43]). In this regard, it is important to study the properties of solutions of linear hyperbolic equations in divergence form with the presence of the right side of the measure Radon. From the works in this direction, we know only [44, 45]. In contrast, in this paper we consider the more general equation with Radon measure and more general form of the boundary conditions for the equation.

Finally, in the connection with our methods of obtaining the necessary conditions, we need to examine the question of the stability of solutions of linear hyperbolic divergent equations with respect to initial conditions on a hyperplane $t = \tau$ (but not top or bottom of a cylinder $Q_T \equiv \Omega \times (0,T)$), and with respect to its position $t = \tau$. Also, we get special integral representations of solutions to hyperbolic divergent equations with Radon measure in the right-hand side part. As far as we know, no one previously studied these questions.

The second part of the chapter is devoted to an optimal control problem with pointwise state constraints for a controlled semilinear divergent hyperbolic equation. In contrast to [46], which dealt with an optimization problem for controlled initial-boundary value problem with the third boundary condition, in the present work we consider controlled initial-boundary value problem with mixed boundary conditions. The second part deals with extending the theory of Pontryagin maximum principle to parametric (i.e., parameter-dependent) problems of sequential optimization for semilinear divergent hyperbolic equations with boundary controls and with state constraints. Words "sequential optimization" mean here, that we use a concept of a sequence of admissible elements as a main concept of an optimization theory, instead of a classical concept of an optimal element. In other words, we use a sequential language of minimizing sequences instead of a classical language of optimal elements.

The continuing interest in optimal control problems for distributed systems with pointwise state constraints, which has lasted for more than four decades (see, e.g., [31]-[35], [44, 45], [47]-[49]), is explained by the diversity and complexity of these problems, as well as by their practical importance. But the majority of publications on optimal control problems with pointwise state constraints are devoted to finding necessary optimality conditions, in particular, Pontryagin's maximum principle. Other classical optimization problems related to the specified class of systems have received little attention in literature. Among these are issues related to sequential optimization problems (suboptimality conditions), regularity, normality, differential properties of the value function, stability of values of problems (sensitivity), etc. Similar issues for parametric optimization problems with pointwise state constraints were earlier considered by [38, 41] in the case of elliptic equations, by [50, 51] in the case of nonlinear hyperbolic Goursat-Darboux systems, and by [30, 42, 43] in the case of divergent hyperbolic equations.

Foremost, it should be said, that, as in [30], [42]–[45], in the present paper we consider an optimal control problem with pointwise in time state constraints. In contrast to [42, 43], here we consider an optimal control problem where a controlled hyperbolic equation is semilinear, and where corresponding initial-boundary value problems contains a boundary control. More precisely, in the paper [30] we addressed an optimization problem with linear order of growth of all source data with respect to a state variable z. In the present paper, we use an essentially narrower class of solutions (namely, the class $\mathfrak{C}^1_{2,0}(Q_T|S^0_T)$). Using of this class allows to consider source data growing superliearly, with exponents are near to bound exponents. Bound exponents are obtained from embedding theorems for Sobolev spaces.

Let us note that we do not know any works of other authors, where these questions of theory are considered for controlled nonlinear (semilinear) divergent hyperbolic equations. Difficulties in the study of optimization problems for controlled divergent hyperbolic equations are inherent to the problem class and caused by the lack of regularity properties of hyperbolic equations solutions in comparison to solutions of parabolic and elliptic equations. These regularity properties are necessary for the earlier methods of research of such problems.

Let us emphasize that we use sequential language in studying the optimization problem of the present paper. This use is an essential feature of the paper and is associated with the following important circumstances: 1) obtaining of classical optimality conditions is connected with very hard assumptions of source data of problems*, and in case of existence of an optimal control under general assumptions, classical optimality conditions are limit case of "optimality" conditions in sequential form (see [30, 38], [41]-[43], [50, 51]); 2) minimizing sequences (more precisely, minimizing approximate solutions in the sense of [53]) that we use in the paper, have regularizing properties (see, e.g., [54]), in contrast to classical optimal elements for constrained optimization problems (these elements are instable with respect to perturbations of input data (see [55, 56]).

The problem of calculating first variations of functionals has received much attention in literature. In the present paper, as in [42], to solve this problem efficiently under natural conditions on the input data of the optimization problem with pointwise state constraints, we use the so-called two-parameter (many-point) needle variation of controls [52], [57]–[59]. Such a modification for optimization problems related to hyperbolic systems with generalized solutions in Sobolev spaces is justified for the following reasons.

Firstly, it is motivated by the "instability" of classical Lebesgue points (see, e.g., [60]) of functions in the topology of the Sobolev classes they belong to.

This "instability" means the following. Let $S^n_\varepsilon(\bar{x}) \equiv \{x \in R^n : |x - \bar{x}| < \varepsilon\}$, and

*First of all it refers to optimization problems for partial differential equations.

let $\eta(x)$, $x \in D \subset R^n$, be a function that is summable over a domain D. If $\bar{x} \in D$ is a Lebesgue point of η, then, as well known,

$$\lim_{\varepsilon\to 0} \frac{1}{\operatorname{meas} S_\varepsilon^n(\bar{x})} \int\limits_{S_\varepsilon^n(\bar{x})} \eta(x)dx = \eta(\bar{x})$$

And if $\|\eta_\varepsilon - \eta\|_{\infty,D} \to 0$, $\varepsilon \to 0$, where η_ε, $\varepsilon > 0$, is a family of functions that are summable over D, then the limit relation

$$\lim_{\varepsilon\to 0} \frac{1}{\operatorname{meas} S_\varepsilon^n(\bar{x})} \int\limits_{S_\varepsilon^n(\bar{x})} \eta_\varepsilon(x)dx = \eta(\bar{x}). \tag{1.1}$$

holds. But the relation (1.1) can be failed if the convergence $\eta_\varepsilon \to \eta$ as $\varepsilon \to 0$ is not uniform, and, for example, is a convergence in some Sobolev space and if η_ε, η belongs the space. Namely, let $\eta(x) \equiv 0$ and $\bar{x} \in R^n$ be an arbitrary point. Obviously, that $\bar{x}$ is a Lebesgue point of η. Let us put

$$\eta_\varepsilon(x;\bar{x}) \equiv \begin{cases} \exp\left[\frac{|x-\bar{x}|^2}{|x-\bar{x}|^2-\varepsilon^2}\right], & |x-\bar{x}|^2 < \varepsilon^2; \\ 0, \ |x-\bar{x}|^2 \geqslant \varepsilon^2. \end{cases}$$

It is not hard to prove that

$$\lim_{\varepsilon\to 0} \|\eta_\varepsilon(\cdot;\bar{x})\|_{p,\Omega} = 0, \text{ if } p \in [1,\infty), n \geqslant 1; \ \lim_{\varepsilon\to 0} \|\eta_\varepsilon(\cdot;\bar{x})\|_{p,\Omega}^{(1)} = 0, \text{ if } p < n;$$

$$\lim_{\varepsilon\to 0} \frac{1}{\operatorname{meas} S_\varepsilon^n(\bar{x})} \int\limits_{S_\varepsilon^n(\bar{x})} \eta_\varepsilon(x;\bar{x})dx = n\int\limits_0^1 r^{n-1} e^{\frac{r^2}{r^2-1}} dr \neq 0.$$

The last limit equality means that for the point $\bar{x}$ and for the selected family $\eta_\varepsilon(\cdot;\bar{x})$ we have $\eta_\varepsilon(\cdot;\bar{x}) \to \eta$, $\varepsilon \to 0$, in the norm of W_p^1, but the limit relation (1.1) is not true.

Secondly, the classical needle variation approach may fail for divergent hyperbolic equations due to the fact that the regularity properties of solutions are much weaker than in the case of elliptic and parabolic equations. For details, see [52].

However, although two-parametrical needle variation, in contrast to the classical needle variation, and allows [30, 46] to obtain the necessary conditions, but in the case of facilities of the solution of the state equation only the Sobolev space, it is necessary to impose strict conditions on the initial data of an optimization problem (see, e.g., [30]). This raises the need for claims greater regularity of solutions, including at least continuity of a solution in time with respect to the weak topology of a Sobolev space.

The analysis of sequential optimization problems with pointwise state constraints is based on a method of [61] (as in works [30, 42, 43]) and can be divided into three main stages.

1. Approximating the original problem with pointwise state constraints by problems each of which is "equivalent" to a problem with finitely many function constraints. The original problem with pointwise state constraints is treated as a problem with infinitely many inequality-type functional constraints. This approximation permits one to use the

advantages of nonsmooth finite-dimensional analysis over its infinite-dimensional counterpart. In particular, one essential advantage from the viewpoint of obtaining results on sensitivity in the present paper is that if the subdifferential (in the sense of [62] or [63]) of a lower semicontinuous function of n variables is bounded at some point, then the function has the Lipschitz property in a neighborhood of that point (see [62, Proposition 2.9.7], [63], [64, Corollary 8.5])[†].

2. Obtaining an "approximate" maximum principle in each approximating optimization problem on the basis of a two-parameter needle variation [52] in an "ordinary" way, i.e. in analogy to the method applied in optimal control problems with finitely many inequality-type functional constraints. This approximate maximum principle is stated in terms of the adjoint functions corresponding to each constraint and satisfying the usual adjoint linear hyperbolic equations in divergence form.

3. Passing to the limit in the family of approximating maximum principles as the number of constraints tends to infinity and deriving the resulting maximum principle in the original problem with pointwise state constraints. Here, the families of adjoint equations corresponding to each inequality-type constraint in the approximating problem are glued together to form a single resulting adjoint equation corresponding to the original state constraint and containing the corresponding Radon measure in the right-hand side.

The chapter consists of the introduction and two main sections. The first section explores divergent hyperbolic equations, and consists of four subsections. In the first subsection we introduce the necessary notation. The second subsection is devoted to the existence and uniqueness of a solution of a semilinear divergent hyperbolic equation with mixed boundary conditions. Also we prove a regularity result for this solution. The third subsection deals with an unique existence theorem for a linear divergent hyperbolic equation with mixed boundary conditions and with a Radon measure in the right part of the equation. In the fourth subsections we obtain an integral representation of a solution of a linear divergent hyperbolic equation with mixed boundary conditions and with a Radon measure in input data. The second section is devoted to sequential optimization problem for a semilinear divergent hyperbolic equation with pointwise in time constraints and with mixed boundary conditions. This sections consists of five subsections. The first subsection is devoted to the problem statement. In the second subsection we formulate main optimization results. The third subsection deals with properties of solutions of state equation and adjoint equations. The fourth subsection is devoted to a calculation of first variations of functionals. Finally, in the fifth subsection we prove main optimization results.

2. Existence and Properties of Solutions of Divergent Hyperbolic Partial Differential Equations

2.1. Necessary Notation

We begin with some notation. Suppose $T > 0$ is a constant, $\Omega \subset R^n$ $(n > 1)$ is a bounded domain having a sectionally smooth boundary $S \equiv S^0 \cup S^1$, $S^0 \cap S^1 = \emptyset$, $S^i_T \equiv S^i \times (0,T)$,

[†]This is not necessarily true in the infinite-dimensional case. But for the reasonable wide class of spaces (so-called Asplund spaces) this is true (see [63, Theorem 3.52], [64, Lemma 8.5].

$i = 0, 1$, and a surface measure of S^i is not zero ($i = 0, 1$); $Q_T \equiv \Omega \times (0, T)$.

Here and in what follows the following notation is used: $\|\varphi\|_{p,\Omega}$ is a norm in the space $L_p(\Omega)$ of functions $\varphi\colon \Omega \to R$ summable to p-th power (essentially bounded for $p = \infty$); $\|\cdot\|_{2,\Omega}^{(1)}$ is a norm in the space $W_2^1(\Omega)$; $|\cdot|_X^{(0)}$ is the standard norm in the space $C(X)$ of continuous functions $\varphi\colon X \to R$ on a compact set X; $M(X)$ is the set of all Radon measures on a compact set X, $\|\mu\|$ is the total variation of a measure $\mu \in M(X)$; $L_{2,1}(Q_T)$ is a Banach space of all Lebesgue measurable functions $\varphi\colon Q_T \to R$ such that the norm $\|\varphi\|_{2,1,Q_T} \equiv \int_0^T (\int_\Omega |\varphi(x,t)|^2 dx)^{1/2} dt$ is finite; $L_{2,1}(S_T^1)$ is a Banach space of all Lebesgue measurable functions $\varphi\colon S_T^1 \to R$ such that the norm $\|\varphi\|_{2,1,S_T^1} \equiv \int_0^T (\int_{S^1} |\varphi(s,t)|^2 ds)^{1/2} dt$ is finite; $C_0^\infty(\Omega|S^0)$ is a set of functions that are infinitely differentiable and compactly supported near S^0; $\mathfrak{D}(0,T)$ is a set of functions $\varphi : (0,T) \to R$ that are infinitely differentiable and $\varphi(0) = \varphi(T) = 0$; $W_2^1(\Omega|S^0)$ is a closure of $C_0^\infty(\Omega|S^0)$ in the norm of $W_2^1(\Omega)$. By $\overset{0,1}{W}_{2,1}(S_T^1)$ denote the set of all functions $\varphi \in L_{2,1}(S_T^1)$ such that $\varphi_t \in L_{2,1}(S_T^1)$. The norm in the space $\overset{0,1}{W}_{2,1}(S_T^1)$ is defined by $\|\varphi\|_{2,1,S_T^1}^{(0,1)} \equiv \|\varphi\|_{2,1,S_T^1} + \|\varphi_t\|_{2,1,S_T^1}$. By $C^r([0,T], Y)$, where Y is a infinite-dimensional Banach space, denote the space of r times strongly continuously differentiable functions $\varphi : [0,T] \to Y$ for $r > 0$, and the space of strongly continuous functions $\varphi : [0,T] \to Y$ for $r = 0$. A norm in the space $C^r([0,T], Y)$ is defined by $|z|_Y^{(r)} \equiv \sum_{i=0}^{r} \max_{t\in[0,T]} \|z_{t^{(i)}}(t)\|_Y$. Let us put $C([0,T], Y) \equiv C^0([0,T], Y)$. By $C_s([0,T], Y)$, where Y is a infinite–dimensional Banach space, denote the space of weak continuous functions $\varphi : [0,T] \to Y$, i.e. $\lim_{t\to\tau}\langle\varphi(t), y^*\rangle = \langle\varphi(\tau), y^*\rangle$ for all $\tau \in [0,T]$, $y^* \in Y^*$. The norm in the space $C_s([0,T], Y)$ is defined by $\|\varphi\|_{C_s([0,T],Y)} \equiv \sup_{t\in[0,T]} \|\varphi(t)\|_Y$. Finally, by $\mathfrak{C}_{2,0}^1(Q_T|S_T^0)$ denote the space of functions $z : Q_T \to R$ such that $z \in C_s([0,T], W_2^1(\Omega|S^0))$, $z_t \in L_\infty([0,T], L_2(\Omega))$. A norm in the space $\mathfrak{C}_{2,0}^1(Q_T|S_T^0)$ is defined by $\|z\|_{\mathfrak{C}_2^1(Q_T)} \equiv \sup_{t\in[0,T]} \|z(\cdot,t)\|_{2,\Omega}^{(1)} + \operatorname{vraisup}_{t\in[0,T]} \|z_t(\cdot,t)\|_{2,\Omega}$.

2.2. Existence and Properties of Solutions of Semilinear Divergent Hyperbolic Partial Differential Equation

Consider the following initial–boundary value problem:

$$z_{tt} - \frac{\partial}{\partial x_i}(a_{ij} z_{x_j} + a_i z) + a(x,t,z) + b_i z_{x_i} = 0, \quad (x,t) \in Q_T, \tag{2.1}$$

$$z|_{t=0} = \varphi(x), \ z_t|_{t=0} = \psi(x), \ x \in \Omega; \ \frac{\partial z}{\partial \mathcal{N}} + \sigma(s,t) z = f(s,t), \ (s,t) \in S_T^1;$$
$$z(s,t) = 0, \quad (s,t) \in S_T^0,$$

where coefficients a_{ij}, a_i, b_i, φ, ψ, σ, w are such that

$$a_{ij} = a_{ji},\ \ \varphi \in W_2^1(\Omega|S^0),\ \ \psi \in L_2(\Omega),\ \ f \in W_{2,1}^{0,1}(S_T^1); \tag{2.2}$$
$$\nu_1|\xi|^2 \le a_{ij}(x,t)\xi_j\xi_i \le \nu_2|\xi|^2\ \forall\,(x,t) \in Q_T,\ \ \xi \in R^n\ \ (\nu_1,\ \ \nu_2 > 0);$$
$$\|a_{ij}\|_{\infty,Q_T} + \|a_{ijt}\|_{\infty,Q_T} + \|a_i\|_{\infty,Q_T} + \|a_{it}\|_{\infty,Q_T} +$$
$$+\|b_i\|_{\infty,Q_T} + \|b_{it}\|_{\infty,Q_T} + \|\sigma\|_{\infty,S_T} + \|\sigma_t\|_{\infty,S_T} \le \nu_3,\ \ i,\, j = \overline{1,\, n};$$

and the function $a\colon Q_T \times R \to R$ is measurable in the Lebesgue sense with respect to (x,t,z) and there exist functions $\mathbf{K}_0 \in L_1[0,T]$ and $\mathbf{K}_1 \in L_{2,1}(Q_T)$ such that

$$|a(x,t,z_1) - a(x,t,z_2)| \le \mathbf{K}_0(t)|z_1 - z_2| \tag{2.3}$$
$$\forall\,(x,t,z_i) \in Q_T \times R,\ \ i = 1,2;$$

$$|a(x,t,0)| \le \mathbf{K}_1(x,t)\ \ \forall\,(x,t) \in Q_T.$$

Definition 2.1. *A function $z \in \mathfrak{C}_{2,0}^1(Q_T|S_T^0)$ is said to be a solution to the initial–boundary value problem (2.1), if z satisfies the following integral identity:*

$$\int\limits_{Q_T} [-z_t\eta_t + a_{ij}z_{x_j}\eta_{x_i} + a_i z\eta_{x_i} + a(x,t,z)\eta + b_i z_{x_i}\eta]dxdt + \tag{2.4}$$
$$+\int\limits_{S_T^1} \sigma z\eta dsdt = \int\limits_{S_T^1} f\eta dsdt + \int\limits_{\Omega} \psi(x)\eta(x,0)dx$$
$$\forall \eta \in \hat{\mathfrak{C}}_{2,0}^1(Q_T|S_T^0);\ \ z(x,0) = \varphi(x),\ \ x \in \Omega.$$

Here $\hat{\mathfrak{C}}_{2,0}^1(Q_T|S_T^0) \equiv \{\eta \in \mathfrak{C}_{2,0}^1(Q_T|S_T^0) : \eta(\cdot,T) = 0\}$.

Theorem 2.1. *The problem (2.1) has a unique solution $z \in \mathfrak{C}_{2,0}^1(Q_T|S_T^0)$, and there exists a constant $B > 0$ such that*

$$\|z\|_{\mathfrak{C}_2^1(Q_T)} \le B[\|\varphi\|_{2,\Omega}^{(1)} + \|\psi\|_{2,\Omega} + \|f\|_{2,1,S_T^1}^{(1,0)} + \|a(\cdot,\cdot,0)\|_{2,1,Q_T}]. \tag{2.5}$$

The constant B depends on the dimension n, numbers T, ν_1, ν_2, $\nu_3 > 0$, a function $\mathbf{K}_0 \in L_1[0,T]$, and a domain Ω.

Proof. The proof is in three steps.
Step 1. Let us prove that a solution is unique.

Indeed, let $z_1, z_2 \in \mathfrak{C}_{2,0}^1(Q_T|S_T^0)$, and let $w \equiv z_1 - z_2$. Then $w \in \mathfrak{C}_{2,0}^1(Q_T|S_T^0)$ and w satisfies the identity

$$\int\limits_{Q_T} [-w_t\eta_t + a_{ij}w_{x_j}\eta_{x_i} + a_i w\eta_{x_i} + [a(x,t,z_2 + w) - a(x,t,z_2)]\eta + \tag{2.6}$$
$$+b_i w_{x_i}\eta]dxdt + \int\limits_{S_T^1} \sigma w\eta dsdt = 0\ \ \forall \eta \in \hat{\mathfrak{C}}_{2,0}^1(Q_T|S_T^0);\ \ w(x,0) = 0,\ \ x \in \Omega.$$

Let us introduce functions $\eta^\alpha : Q_T \to R$, $\beta_i : Q_T \to R$, $i = \overline{1,n}$ (here $\alpha \in [0,T]$ is a parameter), by relations

$$\eta^\alpha(x,t) = -\chi_{[0,\alpha]}(t) \int\limits_t^\alpha w(x,\xi)d\xi, \;\; \beta_i(x,t) = -\int\limits_0^t w_{x_i}(x,\xi)d\xi, \;\; i = \overline{1,n},$$

where χ_E is a characteristic function of a set E.

Substituting $\eta = \eta^\alpha$ into (2.6), integrating by parts, and using conditions (2.2)–(2.3), we obtain that for any $\varepsilon > 0$

$$\int_\Omega [|\eta_t^\alpha(x,\alpha)|^2 + (\nu_1 - \gamma_1\varepsilon)|\nabla_x \eta^\alpha(x,0)|^2]dx \leq$$
$$\leq \int_0^\alpha dt \int\limits_\Omega [\gamma_2 |\nabla_x \eta^\alpha|^2 + \gamma_3(t,\varepsilon)[|\eta^\alpha|^2 + |\eta_t^\alpha|^2]]dx,$$

where γ_1, $\gamma_2 >$ are some constants, $\gamma_3(t,\varepsilon) > 0$, $t \in [0,T]$, is some function. Setting $\varepsilon = \frac{\nu_1}{2\gamma_1}$ in the last inequality, we get

$$\int_\Omega [|\eta_t^\alpha(x,\alpha)|^2 + \frac{\nu_1}{2}|\nabla_x \eta^\alpha(x,0)|^2]dx \leq$$
$$\leq \int_0^\alpha dt \int_\Omega [\gamma_2 |\nabla_x \eta^\alpha|^2 + \gamma_4(t)[|\eta^\alpha|^2 + |\eta_t^\alpha|^2]]dx \;\; \forall\, \alpha \in [0,T],$$

where $\gamma_4(t) \equiv \gamma_3(t, \frac{\nu_1}{2\gamma_1})$, $t \in [0,T]$. It follows from this inequality that

$$\int\limits_\Omega \left[w^2(x,\alpha) + \frac{\nu_1}{2}\sum_{i=1}^n \beta_i^2(x,\alpha)\right]dx \leq \int\limits_0^\alpha dt \int\limits_\Omega \left[\gamma_2 \sum_{i=1}^n (\beta_i(x,\alpha) - \beta_i(x,t))^2 + \right.$$
$$\left. + \gamma_4(t)\left[\left|\int_t^\alpha w(x,\xi)d\xi\right|^2 + w^2(x,t)\right]\right]dx \leq 2\gamma_2\alpha \int_\Omega \sum_{i=1}^n \beta_i^2(x,\alpha)dx +$$
$$+2\gamma_2 \int_0^\alpha dt \int_\Omega \sum_{i=1}^n \beta_i^2 dx + \int_0^\alpha dt \int_\Omega \gamma_4(t)\left[w^2(x,t) + \right.$$
$$\left. +(\alpha - t)\int_0^\alpha w^2(x,\xi)d\xi\right]dx \leq 2\gamma_2\alpha \int_\Omega \sum_{i=1}^n \beta_i^2(x,\alpha)dx +$$
$$+2\gamma_2 \int\limits_0^\alpha dt \int\limits_\Omega \sum_{i=1}^n \beta_i^2 dx + \int\limits_0^\alpha d\xi \int\limits_\Omega \left[T\int\limits_0^\alpha \gamma_4(t)dt + \gamma_4(\xi)\right] w^2(x,\xi)dx \leq$$

$$\leq 2\gamma_2\alpha \int_\Omega \sum_{i=1}^n \beta_i^2(x,\alpha)dx +$$
$$+\int\limits_0^\alpha d\xi \int\limits_\Omega \left[T\int\limits_0^\alpha \gamma_4(t)dt + \gamma_4(\xi) + 2\gamma_2\right]\left[w^2(x,\xi) + \sum_{i=1}^n \beta_i^2(x,\xi)\right]dx.$$

Hence,

$$\int_{\Omega}[w^2(x,\alpha)+\left(\frac{\nu_1}{2}-2\gamma_2\alpha\right)\sum_{i=1}^{n}\beta_i^2(x,\alpha)]dx \le \tag{2.7}$$

$$\le \int_0^{\alpha} dt \int_{\Omega}\gamma_5(t)\left[w^2+\sum_{i=1}^{n}\beta_i^2\right]dx,$$

where $\gamma_5(t) \equiv T\int_0^T \gamma_4(\xi)d\xi + \gamma_4(t) + 2\gamma_2$.

Let $\omega_m \equiv m\theta$, $m = \overline{0,\lambda}$, where $\theta = \frac{\nu_1}{8\gamma_2}$, $\lambda = \lceil \frac{T}{\theta} \rceil$. Let us put $J_m \equiv [\omega_m, \omega_{m+1}] \cap [0,T]$, $m = \overline{0,\lambda-1}$.

Suppose that $\alpha \in J_0$ in the inequality (2.7). Then $\frac{\nu_1}{2} - 2\alpha\gamma_2 \geqslant \frac{\nu_1}{4}$, whence

$$\int_{\Omega}[w^2(x,\alpha)+\sum_{i=1}^{n}\beta_i^2(x,\alpha)]dx \le \int_0^{\alpha}dt\int_{\Omega}\gamma_6(t)[w^2(x,t)+\sum_{i=1}^{n}\beta_i^2(x,t)]dx,$$

where $\gamma_6(t) = \gamma_5(t)/\min\{1, \frac{\nu_1}{4}\}$. Applying the Gronwall lemma, we get that

$$w(x,t) \equiv 0,\ \ \beta_i(x,t) \equiv 0,\ \ (x,t) \in \Omega \times J_0,\ \ i = \overline{1,n}.$$

Arguing in the similar fashion, we obtain in the finite number of steps that

$$w(x,t) \equiv 0,\ \ \beta_i(x,t) \equiv 0,\ \ (x,t) \in \Omega \times J_m,\ \ i = \overline{1,n},\ \ m = \overline{0,\lambda-1}.$$

Thus, a difference of any two solutions of the initial–boundary value problem (2.1) is equal to zero almost everywhere in Q_T. Hence, the problem (2.1) can have no more than one solution.

Step 2. Let us prove the existence of the solution. Let a sequence $g_k \in W_2^1(\Omega|S^0)$, $k = 1, 2, \ldots$, be orthonormal in $L_2(\Omega)$, orthogonal in $W_2^1(\Omega)$ and be such that for any functions $\bar{\varphi} \in W_2^1(\Omega|S^0)$, $\bar{\psi} \in L_2(\Omega)$

$$\lim_{m\to\infty}\|\bar{\varphi}^N - \bar{\varphi}\|_{2,\Omega}^{(1)} = 0,\ \ \lim_{m\to\infty}\|\bar{\psi}^N - \bar{\psi}\|_{2,\Omega} = 0,$$

where

$$\bar{\varphi}^N(x) \equiv \sum_{m=1}^{N}\bar{\varphi}_m g_m(x),\ \bar{\psi}^N(x) \equiv \sum_{m=1}^{N}\bar{\psi}_m g_m(x),$$

$$\bar{\varphi}_j \equiv \int_{\Omega}\bar{\varphi}(y)g_j(y)dy,\ \bar{\psi}_j \equiv \int_{\Omega}\bar{\psi}(y)g_j(y)dy,\ \ j,\ N = 1, 2, \ldots$$

We will find an approximate solution z^N to the problem (2.1) in the form

$$z^N(x,t) \equiv \sum_{m=1}^{N}h_m^N(t)g_m(x),$$

where a collection of functions $h_m^N \in W_1^2[0,T]$, $m = \overline{1,N}$, is a solution to the following Cauchy problem:

$$\ddot{h}_l^N + \sum_{m=1}^{N} [\mathfrak{p}_{lm}(t)\dot{h}_m^N + \mathfrak{q}_{lm}(t)h_m^N] + \mathfrak{r}_l^N(t, h_1^N(t), \ldots, h_N^N(t)) = 0, \tag{2.8}$$

$$h_l^N(0) = \varphi_l, \ \dot{h}_l^N(0) = \psi_l, \ l = \overline{1,N},$$

where

$$\mathfrak{p}_{lm}(t) \equiv \int_\Omega c(x,t) g_l g_m dx,$$

$$\mathfrak{r}_l^N(t, h_1, \ldots, h_N) \equiv \int_\Omega a\left(x, t, \sum_{m'=1}^{N} h_{m'} g_{m'}\right) g_l dx +$$

$$+ \int_{S^1} [\sigma(s,t) \sum_{m'=1}^{N} h_{m'} g_{m'} - f(s,t)] g_l ds,$$

$$\mathfrak{q}_{lm}(t) \equiv \int_\Omega [a_{ij}(x,t) g_{mx_j} g_{lx_i} + a_i(x,t) g_m g_{lx_i} + b_i(x,t) g_{mx_i} g_l] dx,$$

Obviously, such collection exists and will be unique.

Multiplying l–th equation (2.8) by $\dot{h}_l^N(t)$, $l = \overline{1,N}$, adding all obtained equations, and integrating result over $t \in [0, \tau]$, we conclude that

$$\frac{1}{2}\int_\Omega [|z_t^N(x,\tau)|^2 + a_{ij}(x,\tau) z_{x_j}^N(x,\tau) z_{x_i}^N(x,\tau)] dx - \frac{1}{2}\int_\Omega [|\psi^N|^2 +$$

$$-a_{ij}(x,0)\varphi_{x_j}^N \varphi_{x_i}^N] dx - \int_0^\tau dt \int_\Omega \Big[\frac{1}{2} a_{ijt} z_{x_j}^N z_{x_i}^N + (a_i - b_i) z_t^N z_{x_i}^N + a_{it} z^N z_{x_i}^N -$$

$$-c|z_t^N|^2 - [a(x,t,z^N) - a(x,t,0)] z_t^N\Big] dx + \Bigg[\int_\Omega a_i z^N z_{x_i}^N dx +$$

$$+\frac{1}{2}\int_{S^1} \sigma |z^N|^2 ds - \int_{S^1} f z^N ds\Bigg]\Bigg|_{t=0}^{t=\tau} + \int_0^\tau dt \int_\Omega a(x,t,0) z_t^N dx -$$

$$-\frac{1}{2}\int_0^\alpha dt \int_{S^1} \sigma_t |z^N|^2 ds + \int_0^\alpha dt \int_{S^1} f_t z^N ds = 0.$$

Thus,

$$\frac{1}{2}\int\limits_{\Omega}[|z_t^N(x,\tau)|^2+\nu_1|\nabla_x z^N(x,\tau)|]dx\le\frac{1}{2}\int\limits_{\Omega}[|\psi^N|^2+\nu_2|\nabla\varphi^N|^2]dx+$$

$$+\int\limits_0^{\tau}dt\int\limits_{\Omega}|a(x,t,0)||z_t^N|dx-\int\limits_{\Omega}a_i z^N z_{x_i}^N dx\Bigg|_{t=0}^{t=\tau}+\int\limits_0^{\tau}dt\int\limits_{\Omega}\Big[\frac{1}{2}a_{ijt}z_{x_j}^N z_{x_i}^N+$$

$$+(a_i-b_i)z_t^N z_{x_i}^N+a_{it}z^N z_{x_i}^N-c|z_t^N|^2+k(t)|z^N||z_t^N|\Big]dx+$$

$$+\int\limits_0^{\alpha}dt\int\limits_{S^1}|f_t||z^N|ds+\Big[\int\limits_{S^1}fz^N ds-\frac{1}{2}\int\limits_{S^1}\sigma|z^N|^2ds\Big]\Bigg|_{t=0}^{t=\tau}+$$

$$+\frac{1}{2}\int\limits_0^{\alpha}dt\int\limits_{S^1}|\sigma_t||z^N|^2ds.$$

Estimating the left–hand side part of this inequality we obtain that, for any $\varepsilon>0$

$$\frac{1}{2}\int\limits_{\Omega}[|z_t^N(x,\tau)|^2+(\nu_1-\rho_1\varepsilon)|\nabla_x z^N(x,\tau)|]dx\le\rho_2(\varepsilon)\Big[\int\limits_{\Omega}[|\psi^N|^2+$$

$$+|\nabla\varphi^N|^2+|\varphi^N|^2]dx+[\|a(\cdot,\cdot,0)\|_{2,1,Q_T}+\|f\|_{2,1,S_T^1}^{(0,1)}]\max_{t\in[0,\tau]}\Big[\int\limits_{\Omega}[|z^N(x,t)|^2+$$

$$+|z_t^N(x,t)|^2+|\nabla_x z^N(x,t)|^2]dx\Big]^{\frac{1}{2}}\Big]+\int\limits_0^{\tau}dt\int\limits_{\Omega}\rho_3(t,\varepsilon)[|z^N|^2+$$

$$+|z_t^N|^2+|\nabla_x z^N|^2]dx,$$

where ρ_1 and $\rho_2(\varepsilon)$ are some positive constants, $\rho_3(t,\varepsilon)$, $t\in[0,T]$, is a nonnegative summable function.

Adding an expression $\frac{1}{2}\int\limits_{\Omega}|z^N(x,\tau)|^2dx$ to both parts of the last inequality and putting $\varepsilon\equiv\frac{\nu_1}{2\rho_3}$, we have that for any $\tau\in[0,T]$

$$y^N(\tau)\le\rho_4[y^N(0)+[\|a(\cdot,\cdot,0)\|_{2,1,Q_T}+$$
$$+\|f\|_{2,1,S_T^1}^{(0,1)}]\max_{t\in[0,\tau]}\sqrt{y^N(t)}]+\int\limits_0^{\tau}\rho_5(t)y^N(t)dt, \tag{2.9}$$

where $\rho_4\equiv 4\rho_2(\frac{\nu_1}{2\rho_1})/\min\{2,\nu_1\}$, $\rho_5(t)\equiv 4\rho_3(t,\frac{\nu_1}{2\rho_3})/\min\{2,\nu_1\}$,

$$y^N(t)\equiv\int\limits_{\Omega}[|z^N(x,t)|^2+|z_t^N(x,t)|^2+|\nabla_x z^N(x,t)|^2]dx.$$

Applying the Gronwall lemma to (2.9), we obtain

$$\max_{t\in[0,T]} \sqrt{y^N(t)} \leq \rho_6[\sqrt{y^N(0)} + \|a(\cdot,\cdot,0)\|_{2,1,Q_T} + \|f\|^{(0,1)}_{2,1,S^1_T}], \tag{2.10}$$

where $\rho_6 \equiv \rho_4 \exp\left[\int\limits_0^T \rho_5(t)dt\right]$.

It follows from (2.10) that

$$\|z^N\|_{\mathfrak{E}^1_2(Q_T)} \leq 2\rho_6[\|\varphi^N\|^{(1)}_{2,\Omega} + \|\psi^N\|_{2,\Omega}+ \\ +\|a(\cdot,\cdot,0)\|_{2,1,Q_T} + \|f\|^{(0,1)}_{2,1,S^1_T}]. \tag{2.11}$$

It is easy to see, that

$$\lim_{N\to\infty} \|\varphi^N - \varphi\|^{(1)}_{2,\Omega} = 0, \quad \lim_{N\to\infty} \|\psi^N - \psi\|_{2,\Omega} = 0. \tag{2.12}$$

Hence,

$$\lim_{N\to\infty} [\|\varphi^N\|^{(1)}_{2,\Omega} + \|\psi^N\|_{2,\Omega}] = \|\varphi\|^{(1)}_{2,\Omega} + \|\psi\|_{2,\Omega}, \tag{2.13}$$

It follows from (2.13) that there exists a constant $\rho_7 > 0$ such that

$$\|z^N\|_{\mathfrak{E}^1_2(Q_T)} \leq \rho_7 \ \forall N = 1, 2, \ldots. \tag{2.14}$$

It is not hard to see that there exist a subsequence z^{N_k}, $k = 1, 2, \ldots$, of a sequence z^N, $N = 1, 2, \ldots$, and function $z \in \mathfrak{E}^1_{2,0}(Q_T|S^0_T)$ such that

$$\lim_{k\to\infty} \max_{t\in[0,T]} \|z^{N_k}(\cdot,t) - z(\cdot,t)\|_{2,\Omega} = 0, \tag{2.15}$$

$$z^{N_k} \to z, \ k \to \infty, \text{ weakly in } W^1_2(Q_T),$$

$$z^{N_k} \to z, \ z^{N_k}_t \to z_t, \ *\text{–weakly in } L_\infty([0,T], W^1_2(\Omega|S^0)), \\ k \to \infty; \tag{2.16}$$

$$\lim_{k\to\infty} \max_{t\in[0,T]} \|z^{N_k}(\cdot,t) - z(\cdot,t)\|_{2,S^1} = 0. \tag{2.17}$$

Similarly to the argumentation provided for linear equations in [26, Chapter 3, section 8.3], we obtain that z is a solution to the problem (2.1).

Step 3. Let us obtain an a priori estimate. Indeed, in the view of (2.13), for any $\varepsilon > 0$ there exists a number $m_0(\varepsilon) \geqslant 1$ such that for all $m \geqslant m_0(\varepsilon)$

$$\|\varphi^{N_m}\|^{(1)}_{2,\Omega} + \|\psi^{N_m}\|_{2,\Omega} \leq \|\varphi\|^{(1)}_{2,\Omega} + \|\psi\|_{2,\Omega} + \varepsilon. \tag{2.18}$$

Further, it follows from (2.11) that

$$\|z^{N_m}\|_{\mathfrak{E}^1_2(Q_T)} \leq 2\rho_6[\|\varphi^{N_m}\|^{(1)}_{2,\Omega} + \|\psi^{N_m}\|_{2,\Omega} + \|a(\cdot,\cdot,0)\|_{2,1,Q_T} + \|f\|^{(0,1)}_{2,1,S^1_T}].$$

Limiting in given inequality by the numbers $m \geqslant m_0(\varepsilon)$, in the view of (2.18) we will have that for any $m \geq m_0(\varepsilon)$

$$\|z^{N_m}\|_{\mathfrak{E}_2^1(Q_T)} \leq 2\rho_6[\|\varphi\|_{2,\Omega}^{(1)} + \|\psi\|_{2,\Omega} + \|a(\cdot,\cdot,0)\|_{2,1,Q_T} + \|f\|_{2,1,S_T^1}^{(0,1)} + \varepsilon].$$

It follows from relations (2.15), (2.16) that

$$\|z\|_{\mathfrak{E}_2^1(Q_T)} \leq 2\rho_6[\|\varphi\|_{2,\Omega}^{(1)} + \|\psi\|_{2,\Omega} + \|a(\cdot,\cdot,0)\|_{2,1,Q_T} + \|f\|_{2,1,S_T^1}^{(0,1)} + \varepsilon].$$

Tending then ε to zero, we get an estimate (2.5) with $B = 2\rho_6$. Theorem is proved.

Theorem 2.2. *Let $z \in \mathfrak{E}_{2,0}^1(Q_T|S_T^0)$ be a solution to the problem (2.1). Then $z_t \in C_s([0,T], L_2(\Omega))$.*

Proof. Suppose $\eta(x,t) \equiv \zeta(t)g_l(x)$, $(x,t) \in Q_T$, ($\zeta \in C^\infty[0,T]$, $\zeta(T) = 0$) in the integral identity (2.4); then

$$\int\limits_{Q_T} [-z_t\zeta' g_l + a_{ij}z_{x_j}\zeta g_{lx_i} + a_i z\zeta g_{lx_i} + b_i z_{x_i}\zeta g_l + \tag{2.19}$$

$$+a(x,t,z(x,t))\zeta(t)g_l(x)]dxdt + \int\limits_{S_T^1} \sigma z\zeta(t)g_l(s)dsdt =$$

$$= \int\limits_{\Omega} \psi(x)\zeta(0)g_l(x)dx + \int\limits_{S_T^1} f\zeta(t)g_l(s)dsdt;$$

$$z(x,0) = \varphi(x), \quad x \in \Omega.$$

Let $V \equiv \overset{\circ}{W}{}_2^1(\Omega|S^0)$, $H \equiv L_2(\Omega)$. For any $t \in [0,T]$, let us define an operator $\mathfrak{A}(t) : V \to V^*$, and a function $\mathfrak{a}\colon (0,T) \times V \to H$ by

$$\langle \mathfrak{A}(t)v, w\rangle \equiv \int\limits_{\Omega} [a_{ij}(x,t)v_{x_j}w_{x_i} + a_i(x,t)vw_{x_i} + b_i(x,t)v_{x_i}w]dx +$$

$$+ \int\limits_{S^1} \sigma(s,t)v(s)w(s)ds - \int\limits_{S^1} f(s,t)wds, \ \forall v, \ \ w \in V;$$

$$\mathfrak{a}(t,v)(x) \equiv a(x,t,v(x)) \ \ x \in \Omega \ \forall v \in V.$$

Finally, put $\mathfrak{z}(t) \equiv z(\cdot,t)$, $t \in [0,T]$. It is not hard to see that the function $[0,T] \ni t \mapsto \mathfrak{A}(t)\mathfrak{z}(t)$ belongs to the space $L_\infty([0,T], V^*)$, and the function $[0,T] \ni t \mapsto \mathfrak{a}(t,\mathfrak{z}(t))$ belongs to $L_1([0,T], H)$.

Using the introduced notation and Fubini theorem, we can rewrite the equality (2.19) in the form

$$\int\limits_0^T \langle -\dot{\mathfrak{z}}(t), \zeta'(t)g_l\rangle dt + \int\limits_0^T \langle \mathfrak{A}(t)\mathfrak{z}(t), \zeta(t)g_l\rangle dt +$$

$$+ \int\limits_0^T \langle \mathfrak{a}(t,\mathfrak{z}(t)), \zeta(t)g_l\rangle dt = \langle \psi, \zeta(0)g_l\rangle.$$

Hence, for all $\zeta \in \mathfrak{D}(0,T)$

$$\int_0^T \langle -\dot{\mathfrak{z}}(t), \zeta'(t) g_l \rangle dt + \int_0^T \langle \mathfrak{A}(t)\mathfrak{z}(t), \zeta(t) g_l \rangle dt + \int_0^T \langle \mathfrak{a}(t, \mathfrak{z}(t)), \zeta(t) g_l \rangle dt = 0.$$

Putting $\tilde{\mathfrak{a}}(\zeta) \equiv \int_0^T \mathfrak{a}(t,\mathfrak{z}(t))\zeta dt$, $\mathsf{Z}(\zeta) \equiv \int_0^T \mathfrak{z}(t)\zeta(t)dt$, and nothing that $\mathsf{Z}'(\zeta) \equiv -\int_0^T \mathfrak{z}(t)\zeta'(t)dt$, we conclude that

$$-\mathsf{Z}'(\zeta') + \mathfrak{A}(t)\mathsf{Z}(\zeta) = \tilde{\mathfrak{a}}(\zeta).$$

In the other words,

$$\mathsf{Z}''(\zeta) + \mathfrak{A}(t)\mathsf{Z}(\zeta) = \tilde{\mathfrak{a}}(\zeta).$$

The latter means that Z'' is regular and $\mathsf{Z}'' \in L_1([0,T], V^*)$, whence $\dot{\mathfrak{z}} \in W_1^1([0,T], V^*)$.

Using [26, Lemma 8.1, p.307], we obtain that $z_t \in C_s([0,T], L_2(\Omega))$.

2.3. Existence and Properties of Solutions of Linear Divergent Hyperbolic Equations with Radon Measure in Input Data

Consider the following initial–boundary value problem:

$$\eta_{tt} - \frac{\partial}{\partial x_i}(a_{ij}\eta_{x_j} + b_i\eta) + a_i\eta_{x_i} + a\eta - f(x,t) + g(x,t)\mu(dt), \tag{2.20}$$
$$(x,t) \in Q_T;$$
$$\eta(x,T) = \varphi(x), \;\; \eta_t(x,T) = \psi(x), \;\; x \in \Omega; \;\; \left[\frac{\partial \eta}{\partial N'} + \sigma\eta\right]\Big|_{S_T^1} = \omega(s,t);$$
$$\eta(s,t) = 0, \;\; (s,t) \in S_T^0;$$

where $\frac{\partial \eta}{\partial N'} \equiv (a_{ij}z_{x_j} + b_i z)\cos\alpha_i(x,t)$, $f \in L_{2,1}(Q_T)$, $g \in C([0,T], L_2(\Omega))$, $\psi \in L_2(\Omega)$, $\varphi \in W_2^1(\Omega|S^0)$, $\omega \in \overset{\circ}{W}{}_{2,1}^{0,1}(S_T^1)$, $\mu \in M[0,T]$, and coefficients a_{ij}, a_i, b_i, a, σ, $i,\ j = \overline{1,n}$, are such that

$$a_{ij} = a_{ji}, \;\; \nu_1|\xi|^2 \le a_{ij}(x,t)\xi^i\xi^j \le \nu_2|\xi|^2 \tag{2.21}$$
$$\forall\,(x,t) \in Q_T, \;\; \xi \in R^n \;\; (\nu_1, \;\; \nu_2 > 0);$$
$$\|a_{ij}\|_{\infty,Q_T} + \|a_i\|_{\infty,Q_T} + \|b_i\|_{\infty,Q_T} + \|a\|_{\infty,Q_T} + \|a_{ijt}\|_{\infty,Q_T} +$$
$$+\|a_{it}\|_{\infty,Q_T} + \|b_{it}\|_{\infty,Q_T} + \|\sigma\|_{\infty,S_T^1} + \|\sigma_t\|_{\infty,S_T^1} \le \nu_3 \;\; i,\ j = \overline{1,n}.$$

Let us give the following

Definition 2.2. *A function $\eta \in \mathfrak{C}_{2,0}^1(Q_T|S_T^0)$ is said to be a solution to the initial–boundary value problem (2.20), if the following integral identity is fulfilled:*

$$\int_{Q_T} [-\eta_t z_t + a_{ij}\eta_{x_j} z_{x_i} + b_i \eta z_{x_i} + a_i \eta_{x_i} z + a\eta z]\, dxdt + \int_{S_T^1} \sigma\eta z ds dt +$$

$$+ \int_{\Omega} \psi(x) z(x,T) dx + \int_{Q_T} f z dx dt + \int_{[0,T]} \left[\int_{\Omega} g(x,t) z(x,t) dx \right] \mu(dt)$$

$$\forall\, z \in \mathfrak{C}^1_{2,0}(Q_T|S_T^0),\ z(\cdot,0)=0;\ \eta(x,T)=\varphi(x),\ x\in\Omega.$$

Lemma 2.1. *Under the above-mentioned conditions, there exist a unique solution* $\eta \in \mathfrak{C}^1_{2,0}(Q_T|S_T^0)$ *to the initial-boundary value problem (2.20), and a constant* $\bar{c}_0 > 0$ *such that*

$$\|\eta\|_{\mathfrak{C}^1_2(Q_T)} \le \bar{c}_0[\|\varphi\|^{(1)}_{2,\Omega} + \|\psi\|_{2,\Omega} + \|f\|_{2,1,Q_T} +$$
$$+ \max_{t\in[0,T]} \|g(\cdot,t)\|_{2,\Omega}\|\mu\| + \|\omega\|^{(0,1)}_{2,1,S_T^1}].$$

The constant $\bar{c}_0 > 0$ *depends only on* T, ν_1, ν_2, $\nu_3 > 0$, *on the domain* Ω *and on the dimension* n.

Proof. Let us describe the scheme of the proof. Firstly, we approximate (in the $*$–weak sense) the Radon measure μ by a sequence of Radon measures μ^k, $k = 1, 2, \ldots$ Each of approximating measures is absolutely continuous with respect to Lebesgue measure. Then, for each $k = 1, 2, \ldots$, we write out initial–boundary value problem (2.20), where μ is replaced by μ^k. Using the result of Theorem 2.1. and taking then the limit for $k \to \infty$, we obtain the required results.

2.4. Integral Representations of Solutions of Linear Divergent Hyperbolic Equations with Radon Measure in Input Data

In this section, we get a special integral representation of the solution to a homogeneous third initial–boundary value problem for a linear divergent hyperbolic partial differential equation with Radon measure in the right–hand side part. To formulate this section result, let us introduce the following notation.

Suppose $t_1 < t_2$, $t_1, t_2 \in [0, T]$. By definition, put $Q_{(t_1,t_2)} \equiv \Omega \times (t_1, t_2)$, $Q_{[t_1,t_2]} \equiv \Omega \times [t_1, t_2]$, $S^i_{(t_1,t_2)} \equiv S^i \times (t_1, t_2)$, $i = 0, 1$. Suppose functions a_{ij}, a_i, b_i, $i, j = \overline{1, n}$, a, σ satisfy (2.21). Suppose $g \in C([0, T], L_2(\Omega))$, $\mu \in M[0, T]$. By definition, put $\chi(t,\tau) \equiv \{1,\ 0 \le t \le \tau \le T;\ 0,\ 0 \le \tau < t \le T\}$. Consider the following initial–boundary value problem:

$$\mathfrak{f}_{tt} - \frac{\partial}{\partial x_i}(a_{ij}\mathfrak{f}_{x_j} + b_i\mathfrak{f}) + a_i\mathfrak{f}_{x_i} + a\mathfrak{f} = g(x,t)\mu(dt),\ (x,t)\in Q_T; \tag{2.22}$$

$$\mathfrak{f}(x,T) = \mathfrak{f}_t(x,T) = 0,\ x\in\Omega;\ \frac{\partial\mathfrak{f}}{\partial\mathcal{N}'} + \sigma(s,t)\mathfrak{f} = 0,\ (s,t)\in S_T^1;$$
$$\mathfrak{f}(s,t) = 0,\ (s,t)\in S_T^0.$$

By $\mathfrak{f}[a, g, \mu]$ denote the solution of this problem.

Let us put $\mathfrak{p}[a, g](x, t, \tau) \equiv \mathfrak{f}[a, g, \delta_\tau](x, t)$, where δ_τ is a Radon δ–measure which concentrated at the point $t = \tau$.

Let us define a function $\mathfrak{x}[a,g](x,t,\tau)$, $(x,t) \in Q_T$, $\tau \in [0,\,T]$, for $(x,t) \in Q_{[0,\,\tau]}$ as a solution to the initial–boundary value problem

$$\mathfrak{x}_{tt} - \frac{\partial}{\partial x_i}(a_{i\,j}\mathfrak{x}_{x_j} + b_i\mathfrak{x}) + a_i\mathfrak{x}_{x_i} + a\mathfrak{x} = 0, \ \ (x,t) \in Q_{[0,\,\tau]}, \tag{2.23}$$

$$\mathfrak{x}|_{t=\tau} = 0, \ \ \mathfrak{x}_t|_{t=\tau} = -g(x,\tau), \ \ x \in \Omega, \ \ \frac{\partial \mathfrak{x}}{\partial \mathcal{N}'} + \sigma(s,t)\mathfrak{x} = 0, \ \ (s,t) \in S^1_{(0,\,\tau)},$$
$$\mathfrak{x}(s,t,\tau) = 0, \ \ (s,t) \in S^0_{(0,\,\tau)},$$

and as a solution to the problem

$$\mathfrak{x}_{tt} - \frac{\partial}{\partial x_i}(a_{i\,j}\mathfrak{x}_{x_j} + b_i\mathfrak{x}) + a_i\mathfrak{x}_{x_i} + a\mathfrak{x} = 0, \ \ (x,t) \in Q_{[\tau,\,T]}, \tag{2.24}$$

$$\mathfrak{x}|_{t=\tau} = 0, \ \ \mathfrak{x}_t|_{t=\tau} = -g(x,\tau), \ \ x \in \Omega, \ \ \frac{\partial \mathfrak{x}}{\partial \mathcal{N}'} + \sigma(s,t)\mathfrak{x} = 0, \ \ (s,t) \in S^1_{(\tau,\,T)},$$
$$\mathfrak{x}(s,t,\tau) = 0, \ \ (s,t) \in S^0_{(\tau,\,T)},$$

for $(x,t) \in Q_{[\tau,\,T]}$

The function $\mathfrak{x}[a,g]$ can be interpreted as a solution to the initial–boundary value problem

$$\mathfrak{x}_{tt} - \frac{\partial}{\partial x_i}(a_{i\,j}\mathfrak{x}_{x_j} + b_i\mathfrak{x}) + a_i\mathfrak{x}_{x_i} + a\mathfrak{x} = 0, \ \ (x,t) \in Q_T, \tag{2.25}$$

$$\mathfrak{x}|_{t=\tau} = 0, \ \ \mathfrak{x}_t|_{t=\tau} = -g(x,\tau), \ \ x \in \Omega, \ \ \frac{\partial \mathfrak{x}}{\partial \mathcal{N}'} + \sigma(s,t)\mathfrak{x} = 0, \ \ (s,t) \in S^1_T,$$
$$\mathfrak{x}(s,t,\tau) = 0, \ \ (s,t) \in S^0_T, \tag{2.26}$$

where initial conditions are given on a section of the cylinder Q_T by a hyperplane $t = \tau$.

In the connection with the selected research method of the considered optimal control problem, the question arises about an integral representation of the solution $\mathfrak{f}[a,g,\mu]$ to the initial–boundary value problem (2.22). Let us remember the scheme of this method. Firstly, we approximate the source problem with pointwise state constraints by problems with finite number of functional constraints. Secondly, applying "standard" methods, we obtain "approximating" maximum principle in each "approximating" problem. Thirdly, we go to the limit in the family of approximating maximum principles as a number of functional inequality constraints tends to infinity. To this aim, we must "glue" family of adjoint equations (each of adjoint equations corresponds to some functional inequality constraint) into one adjoint equation. The "glued" adjoint equation corresponds to the source state constraint, and involves a Radon measure in the right–hand side part. To provide this "gluing", we prove result on a representation of a solution to the problem (2.22) in the form of an integral of Green $\mathfrak{p}[a,g]$ function by measure μ with respect to τ. To prove this representation result, we need prove that the solution $\mathfrak{x}[a,g]$ of the problem (2.25) depends continuously on a position of the initial conditions hyperplane. Additionally, we need connection of Green function $\mathfrak{p}[a,g]$ of problem (2.22) with $\mathfrak{x}[a,g]$.

Theorem 2.3. *The inclusion* $\mathfrak{x}[a,g] \in C([0,\,T], \mathfrak{E}^1_2(Q_T|S^0_T))$ *holds, i.e., for any* $\tau \in [0,\,T]$ *there exists a trace* $\mathfrak{x}[a,g](\cdot,\cdot,\tau) \in \mathfrak{E}^1_{2,0}(Q_T|S^0_T)$. *This trace depends continuously on*

$\tau \in [0, T]$ *in the norm of* $\mathfrak{C}^1_{2,0}(Q_T|S^0_T)$. *In addition,*

$$\mathfrak{p}[a,g](x,t,\tau) \equiv \mathfrak{r}[a,g](x,t,\tau)\chi(t,\tau), \quad (x,t) \in Q_T, \quad \tau \in [0, T]; \tag{2.27}$$

$$\mathfrak{f}[a,g,\mu](x,t) = \int\limits_{[0,T]} \mathfrak{p}[a,g](x,t,\tau)\mu(d\tau), \text{ for a.e. } (x,t) \in Q_T, \tag{2.28}$$

and the following a priori estimate holds:

$$\max_{\tau \in [0, T]} \|\mathfrak{r}[a,g](\cdot,\cdot,\tau)\|_{\mathfrak{C}^1_2(Q_T)} \leq \tilde{c} \max_{t \in [0, T]} \|g(\cdot,t)\|_{2,\Omega}. \tag{2.29}$$

Here the constant $\tilde{c} > 0$ *depends only on numbers* T, ν_1, ν_2, $\nu_3 > 0$, *on the domain* Ω, *and on a dimension* n.

To prove this theorem, we need two lemmas. First of them is the consequence of [24, Ch. 1, Theorem 6.1] and [65, Theorem 2.2 on page 157].

Lemma 2.2. *Suppose* $\Omega \subset R^n$ *is a bounded domain with the sectionally smooth boundary; then there exists an orthonormal (in the space* $L_2(\Omega)$*) sequence* $h_k \in W^1_2(\Omega|S^0)$, $k = 1, 2, \ldots$, *such that for any* $\varphi \in W^1_2(\Omega|S^0)$, $\psi \in L_2(\Omega)$, *the following equalities are fulfilled:*

$$\lim_{N\to\infty} \|\varphi^N - \varphi\|^{(1)}_{2,\Omega} = 0, \quad \lim_{N\to\infty} \|\psi^N - \psi\|_{2,\Omega} = 0.$$

Here $\varphi^N(x) \equiv \sum\limits_{k=1}^{N} \varphi_k h_k(x)$, $\psi^N(x) \equiv \sum\limits_{k=1}^{N} \psi_k h_k(x)$, $\varphi_k \equiv \int\limits_{\Omega} \varphi h_k \, dx$, $\psi_k \equiv \int\limits_{\Omega} \psi h_k \, dx$, k, $N = 1, 2, \ldots$.

Lemma 2.3. *Suppose* $h_k \in W^1_2(\Omega)$, $k = 1, 2, \ldots$, *be an orthonormal (in the space* $L_2(\Omega)$*) sequence from the previous lemma; then*

$$\lim_{N\to\infty} |g_0^N - g_0|^{(0)}_{L_2(\Omega)} = 0, \quad \lim_{N\to\infty} |g_1^N - g_1|^{(r)}_{W^1_2(\Omega)} = 0,$$

if $g_0 \in C^r([0, T], L_2(\Omega))$, $g_1 \in C^r([0, T], W^1_2(\Omega|S^0))$ *(*$r \geqslant 0$ *is a fixed integer), where*

$$g_0^N(x,t) \equiv \sum_{k=1}^{N} g_{0k}(t)h_k(x), \quad {g_1}^N(x,t) \equiv \sum_{k=1}^{N} g_{1k}(t)h_k(x),$$

$$g_{0k}(t) \equiv \int\limits_{\Omega} g_0(x,t)h_k(x)dx, \quad g_{1k}(t) \equiv \int\limits_{\Omega} g_1(x,t)h_k(x)dx,$$

$$k, \ N \geq 1, \quad t \in [0, T].$$

Moreover, the set $C^2([0,T], W^1_2(\Omega))$ *is dense in the space* $C([0,T], L_2(\Omega))$.

Proof. The proof of the lemma's first assertion is analogous to the proof of [65, Chapter IV, §4, Lemma 4.1]. By this reason, the proof of the first assertion is omitted. The second assertion follows immediately from the first assertion and the classic Weierstrass theorem about uniform approximation of continuous functions by algebraic polynoms. Lemma is proved.

The proof of Theorem 2.3. The proof is in three steps.
Step 1. Let us show that the equality (2.27) holds. First of all, according to Theorem 2.1., the function $\mathfrak{r}[a,g]$ is uniquely determined in the cylinder Q_T, for any $\tau \in [0, T]$. Besides, in view of Definition 2.2., the identities hold

$$\int_{Q_T} [-\mathfrak{p}_t[a,g](x,t,\tau)z_t + a_{ij}\mathfrak{p}_{x_j}[a,g](x,t,\tau)z_{x_i} + \tag{2.30}$$

$$+b_i\mathfrak{p}[a,g](x,t,\tau)z_{x_i} + a_i\mathfrak{p}_{x_j}[g](x,t,\tau)z + a\mathfrak{p}[g](x,t,\tau)z]dxdt+$$

$$+\int_{S_T^1} \sigma\mathfrak{p}[a,g](x,t,\tau)zdsdt = \int_\Omega g(x,\tau)z(x,\tau)dx$$

$$\forall\, z \in \mathfrak{C}^1_{2,0}(Q_T|S_T^0),\ z(\cdot,0)\equiv 0;\ \ \mathfrak{p}[a,g](x,T,\tau)\equiv 0,\ \ x\in\Omega,\ \ \tau\in[0,\, T].$$

In these identities, let us consider three cases.

Case A. Suppose $\tau = 0$; then the function $\mathfrak{p}[a,g](x,t,0)$, $(x,t) \in Q_T$, is the solution to the homogeneous initial–boundary value problem

$$\mathfrak{p}_{tt} - \frac{\partial}{\partial x_i}(a_{ij}\mathfrak{p}_{x_j} + b_i\mathfrak{p}) + a_i\mathfrak{p}_{x_i} + a\mathfrak{p} = 0,\ \ (x,t)\in Q_T;$$

$$\mathfrak{p}(x,T) = \mathfrak{p}_t(x,T) = 0,\ x\in\Omega;\ \frac{\partial\mathfrak{p}}{\partial N'} + \sigma(s,t)\mathfrak{p} = 0,\ (x,t)\in S_T^1,$$

$$\mathfrak{p}(x,t) = 0,\ (x,t)\in S_T^0.$$

According to Theorem 2.1., this problem has only trivial solution. Hence, in the case $\tau = 0$, the equality (2.27) holds.
Case B. Suppose $\tau = T$; then the function $\mathfrak{p}[a,g](\cdot,\cdot,T)$ coincides identically with the function $\mathfrak{r}[a,g](\cdot,\cdot,T)$ whence the equality (2.27) holds, if $\tau = T$.
Case C. Suppose $\tau \in (0, T)$. In the identities (2.30), let z be a linear combination of the form $\sum_{k=1}^{N} c_k(t)h_k(x)$, where N is some natural number, and functions c_k, $k = \overline{1, N}$, are piecewise differentiable on $[0, T]$, $c_k|_{[0,\tau]} \equiv 0$, $k = \overline{1, N}$. Here $h_k \in W_2^1(\Omega|S^0)$, $k = 1, 2, \ldots$, is the sequence from Lemma 2.2.. Because the set of restrictions of all such z is everywhere dense in the set of all functions belonging to $W_2^1(Q_{(\tau,T)}|S^0_{(\tau,T)})$ and such that $z|_{t=\tau} = 0$, we conclude that

$$\int_{Q_{(\tau,\,T)}} [-\mathfrak{p}_t[a,g](x,t,\tau)z_t + a_{ij}\mathfrak{p}_{x_j}[a,g](x,t,\tau)z_{x_i} + b_i\mathfrak{p}[a,g](x,t,\tau)z_{x_i} +$$

$$+a_i\mathfrak{p}_{x_j}[a,g](x,t,\tau)z + a\mathfrak{p}[a,g](x,t,\tau)z]\, dxdt+$$

$$+\int_{S^1_{(\tau,\,T)}} \sigma\mathfrak{p}[a,g](s,t,\tau)zdsdt = 0,\ \ \forall\, z\in\mathfrak{C}^1_{2,0}(Q_{(\tau,T)}|S^0_{(\tau,T)}),\ \ z(\cdot,\tau)\equiv 0;$$

$$\mathfrak{p}[a,g](x,T,\tau)\equiv 0,\ \ x\in\Omega.$$

Therefore, if $(x,t) \in Q_{(\tau,T)}$, then the function $\mathfrak{p}[a,g](\cdot,\cdot,\tau)$ is the solution to the homogeneous initial–boundary value problem

$$\mathfrak{p}_{tt} - \frac{\partial}{\partial x_i}(a_{ij}\mathfrak{p}_{x_j} + b_i\mathfrak{p}) + a_i\mathfrak{p}_{x_i} + a\mathfrak{p} = 0, (x,t) \in Q_{(\tau,T)},$$

$$\mathfrak{p}(x,T) = \mathfrak{p}_t(x,T) = 0,\ x \in \Omega,\ \left[\frac{\partial \mathfrak{p}}{\partial \mathcal{N}'} + \sigma\mathfrak{p}\right]_{S^1_{(\tau,T)}} = 0,\ \mathfrak{p}|_{S^0_{(\tau,T)}} = 0.$$

According to Theorem 2.1., this problem has only trivial solution whence

$$\mathfrak{p}[a,g](x,t,\tau) \equiv 0,\quad (x,t) \in Q_{(\tau,T)}.$$

Hence, the relations (2.30) can be rewritten in the form

$$\int\limits_{Q_{(0,\tau)}} [-\mathfrak{p}_t[a,g](x,t,\tau)z_t + a_{ij}\mathfrak{p}_{x_j}[a,g](x,t,\tau)z_{x_i} + b_i\mathfrak{p}[a,g](x,t,\tau)z_{x_i} +$$

$$+a_i\mathfrak{p}_{x_j}[g](x,t,\tau)z + a\mathfrak{p}[g](x,t,\tau)z]dxdt + \int\limits_{S^1_{(0,\tau)}} \sigma\mathfrak{p}[a,g](s,t,\tau)zdsdt =$$

$$= \int\limits_{\Omega} g(x,\tau)z(x,\tau)\,dx\ \ \forall\, z \in \mathfrak{C}^1_{2,0}(Q_T|S^0_T),\ \ z(\cdot,0) \equiv 0\ \ \tau \in [0,\, T];$$

$$\mathfrak{p}[a,g](x,\tau,\tau) \equiv 0,\ \ x \in \Omega,\ \ \tau \in [0,\, T].$$

In view of the last equalities, if $(x,t) \in Q_{(0,\tau)}$, then the function $\mathfrak{p}[a,g](\cdot,\cdot,\tau)$ is the solution to the initial–boundary value problem

$$\mathfrak{p}_{tt} - \frac{\partial}{\partial x_i}(a_{ij}\mathfrak{p}_{x_j} + b_i\mathfrak{p}) + a_i\mathfrak{p}_{x_i} + a\mathfrak{p} = 0,\ \ (x,t) \in Q_{(0,\tau)},$$

$$\mathfrak{p}(x,\tau) = 0,\ \ \mathfrak{p}_t(x,\tau) = -g(x,\tau),\ \ x \in \Omega,\ \ \left[\frac{\partial \mathfrak{p}}{\partial \mathcal{N}'} + \sigma\mathfrak{p}\right]\Bigg|_{S^1_{(0,\tau)}} = 0,$$

$$\mathfrak{p}|_{S^0_{(0,\tau)}} = 0.$$

Thus, in the case $\tau \in (0,\, T)$, the equality (2.27) holds. Hence, (2.27) is completely proved.

Step 2. Using the Galerkin method, let us show an inclusion

$$\mathfrak{x}[a,g] \in C([0,T], \mathfrak{C}^1_{2,0}(Q_T|S^0_T)).$$

Suppose $g \in C^2([0,T], W^1_2(\Omega|S^0))$. Let $h_k \in W^1_2(\Omega)$, $k = 1,2,\dots$, be the sequence from Lemma 2.2., and let $g^N \equiv \sum\limits_{m=1}^{N} g_m(\tau)h_m(x)$, $g_m(\tau) \equiv \int_\Omega g(x,\tau)h_m(x)dx$, $m = \overline{1,\, N}$, $N = 1,2,\dots$, $\tau \in [0,\, T]$. Then, by virtue of Lemma 2.2.,

$$\lim_{N\to\infty} |g^N - g|^{(2)}_{W^1_2(\Omega)} = 0. \tag{2.31}$$

We will find an approximation $\mathfrak{x}^N[a, g]$ to the function $\mathfrak{x}[a, g]$ in the form

$$\mathfrak{x}^N[a, g](x, t, \tau) \equiv \sum_{k=1}^{N} e_k^N(t, \tau) h_k(x),$$

where a collection of functions $e_k^N \in C^1(\Delta)$, $\Delta \equiv [0, T]^2$, $k = \overline{1, N}$, is a unique continuous solution to the integral equation

$$e^N(t, \tau) + \int_t^T (y - t)\mathcal{A}^N(y)e^N(y, \tau)\, d\tau = \bar{g}^N(\tau)(\tau - t), \quad (t, \tau) \in \Delta. \tag{2.32}$$

Here the matrix–valued function $\mathcal{A}^N(t) = [\alpha_{km}(t)]_{k,\, m=\overline{1, N}}$, $t \in [0, T]$, and vector–valued functions $e^N(t, \tau) \equiv [e_1^N(t, \tau), \ldots, e_N^N(t, \tau)]^*$, $\bar{g}^N(\tau) \equiv [g_1(\tau), \ldots, g_N(\tau)]^* \in R^N$ are such that

$$\alpha_{km}(t) \equiv \int_\Omega [a_{ij}(x, t)h_{kx_j}h_{mx_i} + b_i(x, t)h_k h_{mx_i} + a_i(x, t)h_{kx_i}h_m +$$

$$+a(x, t)h_k h_m]dx + \int_{S^1} \sigma(s, t)h_k(s)h_m(s)ds, \;\; k,\, m = 1, 2, \ldots, \;\; t \in [0, T].$$

Differentiating (2.32) with respect to t twice, we obtain that (2.32) is equivalent to the Cauchy problem

$$e_{tt}^N(t, \tau) + \mathcal{A}^N(t)e^N(t, \tau) = 0, \;\; e^N(t, \tau)|_{t=\tau} = 0, \;\; e_t^N(t, \tau)|_{t=\tau} = -\bar{g}^N(\tau),$$

Differentiating (2.32) with respect to τ once, and, then, with respect to t twice, we have

$$(e_\tau^N)_{tt} + \mathcal{A}^N(t)e_\tau^N = 0, \;\; e_\tau^N|_{t=\tau} = \bar{g}^N(\tau), \;\; (e_\tau^N)_t|_{t=\tau} = -\bar{g}^{N\prime}(\tau).$$

Applying reasoning similar to that in the proof of an energetic inequality from [24, Chapter IV, §3] and in the proof of an a priori estimate from Theorem 2.1., we conclude that

$$\|\mathfrak{x}^N[a, g](\cdot, \cdot, \tau_1) - \mathfrak{x}^N[a, g]](\cdot, \cdot, \tau_2)\|_{\mathfrak{C}_2^1(Q_T)} \leq \tag{2.33}$$
$$\leq T\sqrt{C}|\tau_1 - \tau_2|^{1/2}|g^N|^{(1)}_{W_2^1(\Omega)},$$
$$\|\mathfrak{x}^N[a, g](\cdot, \cdot, \tau)\|_{\mathfrak{C}_2^1(Q_T)} \leq \sqrt{CT}|g^N|^{(0)}_{L_2(\Omega)} \;\; \forall \tau, \;\; \tau_1, \;\; \tau_2 \in [0, T].$$

Let $\tau_1, \tau_2 \in [0, T]$ be fixed.

Furthermore, in analogy to the existence proof for solutions to an initial–boundary value problem in Theorem 2.1., we obtain that there exist a subsequence N_m, $m = 1, 2, \ldots,$ of the sequence $N = 1, 2, \ldots,$ and functions $\mathfrak{x}[a, g](\cdot, \cdot, \tau_i)$, $i = 1,\, 2$, such that

$$\mathfrak{x}^{N_m}[a, g](\cdot, \cdot, \tau_i) \to \mathfrak{x}[a, g](\cdot, \cdot, \tau_i) \;\text{ weak in } W_2^1(Q_T), \tag{2.34}$$
$$\max_{t \in [0, T]} \|\mathfrak{x}^{N_m}[a, g](\cdot, t, \tau_i) - \mathfrak{x}[a, g](\cdot, t, \tau_i)\|_{2,\Omega} \to 0,$$
$$\|\mathfrak{x}^{N_m}[a, g](\cdot, \cdot, \tau_i) - \mathfrak{x}[a, g](\cdot, \cdot, \tau_i)\|_{2,S_T^1} \to 0, \;\; i = 1,\, 2, \;\; m \to \infty.$$

Moreover, each of functions $\mathfrak{x}[a,g](\cdot,\cdot,\tau_i)$, $i = 1, 2$, is a solution to initial–boundary value problem (2.23) for $(x,t) \in Q_{[0,\,\tau_i]}$, and is a solution to initial–boundary value problem (2.24) for $(x,t) \in Q_{[\tau_i,\,T]}$, where $\tau = \tau_i$, $i = 1, 2$. Using the limit relations (2.31) and (2.34) together with the weak compactness of a closed ball of Hilbert space, we get

$$\|\mathfrak{x}[a,g](\cdot,\cdot,\tau_1) - \mathfrak{x}[a,g](\cdot,\cdot,\tau_2)\|_{\mathfrak{C}_2^1(Q_T)} \le T\sqrt{C}|\tau_1 - \tau_2|^{1/2}|g|^{(2)}_{W_2^1(\Omega)}; \tag{2.35}$$

$$\|\mathfrak{x}[a,g](\cdot,\cdot,\tau)\|_{\mathfrak{C}_2^1(Q_T)} \le \sqrt{CT}|g|^{(0)}_{L_2(\Omega)}. \tag{2.36}$$

Thus, if $g \in C^2([0,T], W_2^1(\Omega|S^0))$, then $\mathfrak{x}[a,g] \in C([0,T], \mathfrak{C}^1_{2,0}(Q_T|S^0_T))$, and, additionally, $\mathfrak{x}[a,g]$ depends linearly on $g \in C^2([0,T], W_2^1(\Omega|S^0))$. Using the density of the space $C^2([0,T], W_2^1(\Omega|S^0))$ in the space $C([0,\,T],\,L_2(\Omega))$, and linearly depending $\mathfrak{x}[a,g]$ on $g \in C^2([0,T], W_2^1(\Omega|S^0))$, we obtain that $\mathfrak{x}[a,g] \in C([0,T], \mathfrak{C}^1_{2,0}(Q_T|S^0_T))$ for all functions $g \in C([0,T], L_2(\Omega))$.

Step 3. Let us prove other assertions of the Lemma. From the inequality (2.36) and density $C^2([0,\,T], W_2^1(\Omega|S^0))$ in $C([0,\,T],\,L_2(\Omega))$ it follows that the estimation (2.29) holds. Since, by the second step of the proof, $\mathfrak{x}[a,g] \in C([0,T], \mathfrak{C}^1_{2,0}(Q_T|S^0_T))$ for all $g \in C([0,T], L_2(\Omega))$, then functions

$$\int\limits_{[0,T]} \mathfrak{p}[a,g](x,t,\tau)\mu(d\tau), \quad \int\limits_{[0,T]} \mathfrak{p}_t[a,g](x,t,\tau)\mu(d\tau),$$

$$\int\limits_{[0,T]} \mathfrak{p}_{x_i}[a,g](x,t,\tau)\mu(d\tau), \ i = \overline{1,n}$$

are square summable over Q_T for any Radon measure $\mu \in M[0,\,T]$. Besides, using a definition of generalized derivatives in the sense of Sobolev, we conclude that

$$\begin{aligned} \frac{\partial}{\partial t}\int_{[0,T]} \mathfrak{p}[a,g](x,t,\tau)\mu(d\tau) &= \int_{[0,T]} \mathfrak{p}_t[a,g](x,t,\tau)\mu(d\tau), \\ \frac{\partial}{\partial x_i}\int_{[0,T]} \mathfrak{p}[a,g](x,t,\tau)\mu(d\tau) &= \int_{[0,T]} \mathfrak{p}_{x_i}[a,g](x,t,\tau)\mu(d\tau), \\ i = \overline{1,\,n},\ (x,t) &\in Q_T. \end{aligned} \tag{2.37}$$

Let us show that (2.28) take place. Indeed, because $\mathfrak{p}[a,g] \equiv \mathfrak{f}[a,g,\delta_\tau]$, then, writing out the corresponding integral identity, integrating this identity in $\tau \in [0,T]$ by a measure μ, and taking into account (2.37), we get that the function $\zeta(x,t) \equiv \int\limits_0^T \mathfrak{p}[a,g](x,t,\tau)\mu(d\tau)$, $(x,t) \in Q_T$, is a solution to the problem (2.22). Because the problem (2.22) has a unique solution, then the equality (2.28) holds. Theorem 2.3. is completely proved.

Finally, let us prove the following result.

Lemma 2.4. *Let us consider the initial–boundary value problem*

$$z_{tt} - \frac{\partial}{\partial x_i}(a_{ij}z_{x_j} + a_i z) + b_i z_{x_i} + az = f(x,t), \ (x,t) \in Q_T, \tag{2.38}$$
$$z(x,0) = 0, \ z_t(x,0) = \psi(x), \ x \in \Omega, \ \left[\frac{\partial z}{\partial \mathcal{N}} + \sigma z\right]\Big|_{S_T^1} = \omega,$$
$$z(s,t) = 0, \ (s,t) \in S_T^0,$$

where $f \in L_{2,1}(Q_T)$, $\psi \in L_2(\Omega)$, $\omega \in W_{2,1}^{0,1}(S_T^1)$, and coefficients a_{ij}, a_i, b_i, a, σ satisfy the conditions (2.21).

Suppose function $z \in \mathfrak{C}_{2,0}^1(Q_T|S_T^0)$ is a solution to the problem (2.38), $\tau \in [0,T]$. Suppose a function $g : \Omega \times [0,T] \to R$ has the trace $g(\cdot,\tau) \in L_2(\Omega)$ for any $\tau \in [0,T]$, and this trace depends continuously on $\tau \in [0,T]$ in the norm of the space $L_2(\Omega)$. Then

$$\int_\Omega g(x,\tau)z(x,\tau)dx = \int_{Q_T} f\mathfrak{p}[a,g](x,t,\tau)dxdt+ \tag{2.39}$$
$$+\int_\Omega \psi\mathfrak{p}[a,g](x,0,\tau)dx + \int_{S_T^1} \mathfrak{p}[a,g](s,t,\tau)\omega dsdt.$$

Proof. Because $z \in \mathfrak{C}_{2,0}^1(Q_T|S_T^0)$ is a unique solution to the problem (2.38), then the restriction $z|_{Q_\tau}$ is a unique solution to the problem

$$z_{tt} - \frac{\partial}{\partial x_i}(a_{ij}z_{x_j} + a_i z) + b_i z_{x_i} + az = f(x,t), \ (x,t) \in Q_\tau, \tag{2.40}$$
$$z(x,0) = 0, \ z_t(x,0) = \psi(x), \ x \in \Omega, \ \left[\frac{\partial z}{\partial \mathcal{N}} + \sigma z\right]\Big|_{S_\tau^1} = \omega,$$
$$z(s,t) = 0, \ (s,t) \in S_\tau^0.$$

In view of (2.40) we can write out the following identity which holds for all functions $\eta \in \hat{\mathfrak{C}}_{2,0}^1(Q_T|S_T^0)$:

$$\int_{Q_\tau}[-z_t\eta_t + a_{ij}z_{x_j}\eta_{x_i} + a_i z\eta_{x_i} + b_i z_{x_i}\eta + az\eta]dxdt + \int_{S_\tau^1}\sigma z\eta dsdt = \tag{2.41}$$
$$= \int_{S_\tau^1}\omega\eta dsdt + \int_{Q_\tau} f\eta dxdt + \int_\Omega \psi(x)\eta(x,0)dx; \ z(x,0) = 0, \ x \in \Omega.$$

However, in view of the lemma's conditions, all conditions of the Theorem 2.1. of a solution $\mathfrak{x} \in \mathfrak{C}_{2,0}^1(Q_T|S_T^0)$ to the adjoint problem (2.23) are fulfilled, whence the following identity holds for all $z \in \mathfrak{C}_{2,0}^1(Q_T|S_T^0)$, $z(\cdot,0) = 0$:

$$\int_{Q_\tau}[-z_t\mathfrak{x}_t + a_{ij}z_{x_j}\mathfrak{x}_{x_i} + a_i z\mathfrak{x}_{x_i} + b_i z_{x_i}\mathfrak{x} + az\mathfrak{x}]dxdt + \int_{S_\tau^1}\sigma z\mathfrak{x} dsdt = \tag{2.42}$$

$$= \int\limits_{\Omega} g(x,\tau)z(x,\tau)dx; \ \ \mathfrak{r}(x,\tau) = 0, \ \ x \in \Omega.$$

Substituting $\mathfrak{r}[a,g]$ for η in (2.41), and substituting the solution to the problem (2.40) for z in (2.42) and taking into account that left–hand side parts of (2.41) and (2.42) coincide, we conclude that

$$\int\limits_{\Omega} g(x,\tau)z(x,\tau)dx = \int\limits_{Q_\tau} f(x,t)\mathfrak{r}[a,g](x,t,\tau)dxdt +$$
$$+ \int\limits_{\Omega} \psi(x)\mathfrak{r}[a,g](x,0,\tau)dx + \int\limits_{S^1_\tau} \mathfrak{r}[a,g](s,t,\tau)\omega(s,t)dsdt.$$

The last equality can be rewritten in the form

$$\int\limits_{\Omega} g(x,\tau)z(x,\tau)dx = \int\limits_{Q_T} f(x,t)\mathfrak{r}[a,g](x,t,\tau)\chi(t,\tau)dxdt + \tag{2.43}$$
$$+ \int\limits_{\Omega} \psi(x)\mathfrak{r}[a,g](x,0,\tau)\chi(0,\tau)dx + \int\limits_{S^1_T} \mathfrak{r}[a,g](s,t,\tau)\chi(t,\tau)\omega(s,t)dsdt.$$

According to equality (2.3.) from Lemma 2.27, we have that

$$\mathfrak{r}[a,g](x,t,\tau)\chi(t,\tau) = \mathfrak{p}[a,g](x,t,\tau).$$

Therefore, (2.43) can be rewritten in the form (2.39). The lemma is completely proved.

3. Sequential Optimization for Semilinear Divergent Hyperbolic Equation with Pointwise in Time State Constraints and with Mixed Boundary Conditions

3.1. Optimization Problem Statement

We begin with some notation. Suppose $U \subset R^m$ is a compact set, $V \subset R$ is a segment, $T > 0$ is a constant, $\Omega \subset R^n$ $(n > 1)$ is a bounded domain having a sectionally smooth boundary $S \equiv S^0 \cup S^1$, $S^0 \cap S^1 = \emptyset$, $S^i_T \equiv S^i \times (0,T)$, $i = 0,1$, and a surface measure of S^i is not zero $(= 0,1)$; $Q_T \equiv \Omega \times (0,T)$, $\mathcal{D} \equiv \{\pi \equiv (u,v,w) \ : \ \pi \in \mathcal{D}_1 \times \mathcal{D}_2 \times \mathcal{D}_3\}$, $\mathcal{D}_1 \equiv \{u \in L^m_\infty(Q_T) \ : \ u(x,t) \in U$ for a.e. $(x,t) \in Q_T\}$, $\mathcal{D}_2 \equiv \{v \in L_\infty(\Omega) : v(x) \in V$ for a.e. $x \in \Omega\}$, $\mathcal{D}_3 \equiv \{w \in W^{0,1}_{2,1}(S^1_T) \ : \ w \in \mathbf{W}\}$, where $\mathbf{W}$ is a convex closed bounded subset of $W^{0,1}_{2,1}(S^1_T)$.

Consider the following parametric optimization problem:

$$I_0(\pi) \to \inf, \ \ \pi \in \mathcal{D}, \ \ I_1(\pi) \in \mathcal{M} + q, \ \ q \in C(X) \text{ is a parameter,} \tag{P_q}$$

where $\mathcal{M}$ is a set of all continuous nonpositive functions on the compact set $X \subseteq [0, T]$, the functional $I_0 \colon \mathcal{D} \to R$ and the operator $I_1 : \mathcal{D} \to C(X)$ are defined by

$$I_0(\pi) \equiv \int_\Omega G(x, z[\pi](x,T), v(x))dx,$$

$$I_1(\pi)(\tau) \equiv \int_\Omega \Phi(x, \tau, z[\pi](x,\tau), v(x))dx, \quad \tau \in [0, T],$$

$z[\pi] \in \mathfrak{C}^1_{2,0}(Q_T|S^0_T)$ is a unique generalized solution (see Theorem 2.1.) to a initial–boundary value problem

$$z_{tt} - \frac{\partial}{\partial x_i}(a_{ij}z_{x_j} + a_i z) + a(x,t,z,u) + b_i z_{x_i} = 0, \quad (x,t) \in Q_T, \tag{3.1}$$

$$z|_{t=0} = \varphi(x), \; z_t|_{t=0} = v(x), \; x \in \Omega; \; \frac{\partial z}{\partial \mathcal{N}} + \sigma(s,t)z = w(s,t), \; (s,t) \in S^1_T;$$

$$z(s,t) = 0, \quad (s,t) \in S^0_T;$$

corresponding to a triple $\pi \equiv (u, v, w) \in \mathcal{D}$. Here $\frac{\partial z}{\partial \mathcal{N}} \equiv (a_{ij}z_{x_j} + a_i z)\cos\alpha_i$, and $\alpha_i(x,t)$ is an angle between an outwards normal to S_T and Ox_i–axis.

Assume that

a) the functions a_{ij}, a_{ijt}, a_i, a_{it}, b_i, b_{it}, i, $j = \overline{1, n}$, are Lebesgue measurable on Q_T;

b) the functions σ and σ_t are Lebesgue measurable on S^1_T;

c) the function $a \colon Q_T \times R \times U \to R$, together with $\nabla_z a$, is measurable in the Lebesgue sense with respect to (x, t, z, u) and continuous with respect to (z, u) for a.e. $(x, t) \in Q_T$;

d) the function $G \colon \Omega \times R \times V \to R$, together with $\nabla_z G$, $\nabla_v G$, is measurable in the Lebesgue sense with respect to (x, z, v) and continuous with respect to (z, v) for a.e. $x \in \Omega$;

e) the function $\Phi \colon \Omega \times [0,T] \times R \times R \to R$, together with gradients $\nabla_z\Phi$, $\nabla_v\Phi$, is measurable in the Lebesgue sense with respect to $(x, t, z, v) \in \Omega \times [0, T] \times R \times V$ and continuous with respect to $(t, z, v) \in [0, T] \times R \times V$ for a.e. $x \in \Omega$;

f) the functions $a \colon Q_T \times R \times U \to R$ and $\nabla_z a \colon Q_T \times R \times U \to R$ are Lebesgue measurable with respect to $(x, t, z, u) \in Q_T \times R \times U$ and continuous with respect to $(z, u) \in \mathbb{R} \times U$ for a.e. $(x, t) \in Q_T$; there exists a function $K_0 \in L_1[0, T]$ such that

$$|\nabla_z a(x,t,z,u)| \leq K_0(t) \;\; \forall\, (x,t,z,u) \in Q_T \times \mathbb{R} \times U;$$

moreover, there exists $K_1 \in L_{2,1}(Q_T)$ such that

$$|a(x,t,0,u)| \leq K_1(x,t) \;\; \forall u \in U \text{ for a.e. } (x,t) \in Q_T;$$

g) the following conditions and estimates are fulfilled:

$$a_{ij} = a_{ji},\ \varphi \in W_2^1(\Omega|S^0),\ \nu_1|\xi|^2 \le a_{ij}(x,t)\xi_j\xi_i \le \nu_2|\xi|^2$$
$$\forall\,(x,t) \in Q_T,\ \xi \in R^n\ (\nu_1,\ \nu_2 > 0);$$
$$\|a_{ij}\|_{\infty,1,Q_T} + \|a_{ijt}\|_{\infty,Q_T} + \|a_i\|_{\infty,1,Q_T} + \|a_{it}\|_{\infty,Q_T} +$$
$$+\|b_i\|_{\infty,1,Q_T} + \|b_{it}\|_{\infty,Q_T} + \|\sigma\|_{\infty,S_T^1} + \|\sigma_t\|_{\infty,S_T^1} \le \nu_3,\ \ i,\, j = \overline{1,\, n};$$

h) the following conditions are fulfilled:

$$|G(x,z,v)| + |\nabla_v G(x,z,v)| \le K_2[1+|z|^{\gamma_1}],$$
$$|\nabla_z G(x,z,v)| \le K_2[1+|z|^{\gamma_2}]\ \forall\,(x,z,v) \in \Omega \times R \times V;$$
$$|\Phi(x,\tau,z,v)| + |\nabla_v \Phi(x,\tau,z,v)| \le K_2[1+|z|^{\gamma_1}],$$
$$|\nabla_z \Phi(x,\tau,z,v)| \le K_2[1+|z|^{\gamma_2}]\ \forall\,(x,\tau,z,v) \in \Omega \times [0,T] \times R \times V;$$
$$|\nabla_z G(x,z_1,v_1) - \nabla_z G(x,z_2,v_2)| \le K_2|z_1 - z_2|^{\gamma_2} + K_3(|v_1 - v_2|),$$
$$|\nabla_v G(x,z_1,v_1) - \nabla_v G(x,z_2,v_2)| \le K_2|z_1 - z_2|^{\gamma_1} + K_3(|v_1 - v_2|)$$

$$\forall\,(x,z_i,v_i) \in \Omega \times R \times V,\ \ i = 1,\,2;$$
$$|\nabla_z \Phi(x,\tau,z',v') - \nabla_z \Phi(x,\tau,z'',v'')| \le K_2|z' - z''|^{\gamma_2} + K_3(|v' - v''|),$$
$$|\nabla_v \Phi(x,\tau,z',v') - \nabla_v \Phi(x,\tau,z'',v'')| \le K_2|z' - z''|^{\gamma_1} + K_3(|v' - v''|)$$
$$\forall\,(x,\tau,z',v'),\ \ (x,\tau,z'',v'') \in \Omega \times [0,T] \times R \times V;$$

where $\gamma_1,\ \gamma_2 \in [1,+\infty)$, if $n = 2$; $\gamma_1 \in \left[1, \frac{2n}{n-2}\right)$, $\gamma_2 \in \left[1, \frac{n}{n-2}\right)$, if $n > 2$; and $K_3\colon [0, \operatorname{meas} V] \to [0,+\infty)$ is a nonnegative nondecreasing function such that $\lim\limits_{\xi\to+0} K_3(\xi) = K_3(0) = 0$;

i) the following estimates are fulfilled:

$$|\Phi(x,t_1,z,v) - \Phi(x,t_2,z,v)| + |\nabla_z \Phi(x,t_1,z,v) - \nabla_z \Phi(x,t_2,z,v)| +$$
$$+|\nabla_v \Phi(x,t_1,z,v) - \nabla_v \Phi(x,t_2,z,v)| \le K_4(|t_1 - t_2|)$$
$$\forall\,(x,t_i,z,v) \in \Omega \times [0,T] \times R \times V,\ \ i = 1,2;$$

where a function $K_4\colon [0,T] \to [0,+\infty)$ is such that $\lim\limits_{\tau\to+0} K_4(\tau) = K_4(0) = 0$;

j) there exists a function $K_5 \in L_1[0,T]$ such that

$$|\nabla_z a(x,t,z_1,u) - \nabla_z a(x,t,z_2,u)| \le K_5(t)|z_1 - z_2|^{\gamma_2}$$
$$\forall\,(x,t,z_i,u) \in Q_T \times R \times U,\ \ i = 1,\,2.$$

By definition, put $\mathcal{D}_q^\varepsilon \equiv \{\pi \in \mathcal{D} : I_1(\pi)(\tau) - q(\tau) \le \varepsilon, \tau \in X\}, \varepsilon \geqslant 0$, $\beta(q) \equiv \beta_{+0}(q) \equiv \lim\limits_{\varepsilon\to+0} \beta_\varepsilon(q)$, where

$$\beta_\varepsilon(q) \equiv \{\inf_{\pi\in\mathcal{D}_q^\varepsilon} I_0(\pi),\ \text{if}\ \mathcal{D}_q^\varepsilon \neq \emptyset; +\infty,\ \text{if}\ \mathcal{D}_q^\varepsilon = \emptyset\}, \varepsilon \geqslant 0.$$

The function $\beta\colon C(X) \to R \cup \{+\infty\}$ is called the value function of the problem (P_q). It is obvious that $\beta(q) \le \beta_0(q)\ \forall q \in C(X)$, where $\beta_0\colon C(X) \to R$ is a classic value function. Suppose $\beta(q) < +\infty$. According to Warga [53], a minimizing approximate solution (m.a.s.) in the problem (P_q) is a sequence of triples $\pi^i \in \mathcal{D}$, $i = 1, 2, \dots$, such that

$$I_0(\pi^i) \le \beta(q) + \delta^i, \quad \pi^i \in \mathcal{D}_q^{\varepsilon^i}, \quad i = 1, 2, \dots, \tag{3.2}$$

where δ^i, ε^i, $i = 1, 2, \dots$, $\delta^i, \varepsilon^i \to 0$, $i \to \infty$, are sequences of nonnegative numbers.

3.2. Main Optimization Results

The main results of the present paper deals with necessary conditions for elements of minimizing sequences, that is, the Pontryagin maximum principle for m.a.s. and the properties of normality, regularity, and sensitivity. In the sequel, we use the following standard notation: $H(x, t, z, u, \eta) \equiv -\eta a(x, t, z, u)$.

The following theorem gives us the necessary conditions for a m.a.s., which is referred to as the maximum principle for the m.a.s..

Theorem 3.1. *Let $\pi^i \equiv (u^i, v^i, w^i) \in \mathcal{D}$, $i = 1, 2, \dots$, be a m.a.s. to the problem (P_q); then there exist a sequence of numbers $\gamma^i \geqslant 0$, $i = 1, 2, \dots$, $\gamma^i \to 0$, $i \to \infty$, a sequence of nonnegative numbers λ^i, $i = 1, 2, \dots$, and a sequence of nonnegative Radon measures $\mu^i \in M(X)$, $i = 1, 2, \dots$, where μ^i is concentrated on the set*

$$X_i \equiv \{\tau \in X : |I_1(\pi^i)(\tau) - q(\tau)| \le \gamma^i\}, \tag{3.3}$$

such that the following conditions are fulfilled:

a) a nontriviality condition of Lagrange multipliers:

$$\pi^i \in \mathcal{D}_q^{\gamma^i}, \quad \lambda^i + \|\mu^i\| = 1; \tag{3.4}$$

b) a maximum condition with respect to u:

$$\int_{Q_T} [\max_{u' \in U} H(x, t, z[\pi], u', \eta[\pi, \lambda, \mu](x, t)) - \\ -H(x, t, z[\pi], u(x, t), \eta[\pi, \lambda, \mu](x, t))] dx dt \le \gamma, \tag{3.5}$$

c) a transversality condition with respect to v:

$$\max_{v' \in \mathcal{D}_2} \left\{ \int_{\Omega} [\eta[\pi, \lambda, \mu](x, 0) + \lambda \nabla_v \Phi(x, z[\pi](x, T), v))](v - v') dx + \right. \\ \left. + \int_X \mu(d\tau) \int_{\Omega} \nabla_v \Phi(x, \tau, z[\pi](x, \tau), v(x))(v(x) - v'(x)) dx \right\} \le \gamma, \tag{3.6}$$

d) a transversality condition with respect to w:

$$\max_{w' \in \mathcal{D}_3} \int_{S_T^1} \eta[\pi, \lambda, \mu](s, t)(w(s, t) - w'(s, t)) ds dt \le \gamma, \tag{3.7}$$

for $\pi = \pi^i$, $\lambda = \lambda^i$, $\mu = \mu^i$, $\gamma = \gamma^i$, $i = 1, 2, \ldots$.

Here $\eta[\pi, \lambda, \mu]$ *is a solution to the adjoint initial–boundary value problem*

$$\eta_{tt} - \frac{\partial}{\partial x_i}(a_{ij}\eta_{x_j} + b_i\eta) + a_i\eta_{x_i} + \nabla_z a(x, t, z[\pi], u)\eta = \tag{3.8}$$

$$= \nabla_z \Phi(x, t, z[\pi], v)\mu(dt),$$

$$\eta|_{t=T} = 0, \;\; \eta_t|_{t=T} = -\lambda \nabla_z G(x, z[\pi](x, T), v(x)), \;\; x \in \Omega,$$

$$\left[\frac{\partial \eta}{\partial N'} + \sigma\eta\right]\Bigg|_{S_T^1} = 0, \;\; \eta|_{S_T^0} = 0,$$

where $\frac{\partial \eta}{\partial N'} \equiv (a_{ij}(x, t)\eta_{x_j} + b_i(x, t)\eta) \cos \alpha_i(x, t)$.

It follows from Theorem 3.1. that, under the condition $I_0(\pi^0) = \beta(q)$, the control $\pi^0 \in \mathcal{D}_q^0$ satisfies the ordinary maximum principle for $\pi^i \equiv \pi^0$, $\gamma^i \equiv 0$, $(\lambda^i, \mu^i) \equiv (\lambda, \mu)$, $i = 1, 2, \ldots$

Next, following, say, [37, 39, 40], we introduce some natural definitions.

Definition 3.1. *A sequence* $\pi^i \in \mathcal{D}$, $i = 1, 2, \ldots$, *is said to be stationary in problem* (P_q) *if there exists a sequence of nonnegative numbers* $\gamma^i \to 0$, $i \to \infty$, *and a bounded sequence of pairs* (λ^i, μ^i), *where* $\lambda^i \geqslant 0$, *and* $\mu^i \in M[0, T]$ *is a nonnegative Radon measure concentrated on the set* X_i, *such that* $\pi^i \in \mathcal{D}_q^{\gamma^i}$, $i = 1, 2, \ldots$, *inequalities (3.5)–(3.7) are satisfied, and all limit points of the sequence of pairs* (λ^i, μ^i), $i = 1, 2, \ldots$, *(in the* $*$*-weak sense for the second component), are nonzero.*

Definition 3.2. *A stationary sequence* $\pi^i \in \mathcal{D}$, $i = 1, 2, \ldots$, *in problem* (P_q) *is said to be normal if the first components of all limit points of each corresponding sequence* (λ^i, μ^i), $i = 1, 2, \ldots$, *are nonzero. Problem* (P_q) *is said to be normal if all of its stationary sequences are normal. A stationary sequence* $\pi^i \in \mathcal{D}$, $i = 1, 2, \ldots$, *in problem* (P_q) *is said to be regular if the first components of all limit points of some corresponding sequence* (λ^i, μ^i), $i = 1, 2, \ldots$, *are nonzero. Problem* (P_q) *is said to be regular if in it there exist regular stationary sequences.*

Next, let us state regularity and normality conditions for problem (P_q) (e.g., see [39, 40]). The consideration of such conditions is possible owing to the presence of a parameter in the original optimization problem. First, let us state a theorem on the stability of the optimal value of problem (P_q) in the case of its normality.

Theorem 3.2. *If problem* (P_q) *is normal, then its value function* β *satisfies the Lipschitz condition in a neighborhood of the point* $q \in C(X)$.

It turns out that the converse assertion holds in a sense as well.

Theorem 3.3. *Let the value function* β *of problem* (P_q) *satisfy the Lipschitz condition in a neighborhood of the point* q. *Then problem* $(P_{q'})$ *has regular m.a.s. for all* q' *in that neighborhood.*

In the general case where the value function $\beta(q)$ does not necessarily possess the Lipchitz property, we have the following general result.

Theorem 3.4. *For any point* $q \in \mathrm{dom}\beta \equiv \{q' \in C(X) : \beta(q') < +\infty\}$ *and any continuous positive function* $\xi \in C(X)$ *all m.a.s. are regular in problem* $(P_{q'})$ *for almost all points* q' *on the ray* $\{q + t\xi \,:\, t \geqslant 0\}$, *i.e., the property that any m.a.s. in problem* $(P_{q+t\xi})$ *with these* q *and* ξ *is regular is a property of general position for* $t \geqslant 0$.

3.3. Properties of Solutions of the State and Adjoint Equations

First of all, we need results concerning the main initial–boundary value problem (3.1). In order to formulate these results, let us introduce some notation. For any triples $\pi^i \equiv (u^i, v^i, w^i) \in \mathcal{D}$, $i = 1, 2$, by definition, put $\mathcal{R}(u^1, u^2) \equiv \{(x,t) \in Q_T \,:\, u^1(x,t) \neq u^2(x,t)\}$, $\mathcal{R}_t(u^1, u^2) \equiv \{x \in \Omega \,:\, (x,t) \in \mathcal{R}(u^1,u^2)\}$, $d(\pi^1, \pi^2) \equiv \|v^1 - v^2\|_{\infty,\Omega} + \mathrm{meas}\,\mathcal{R}(u^1,u^2) + \|w^1 - w^2\|^{(0,1)}_{2,1,S_T}$. Equip the set $\mathcal{D}$ with the metric $d(\cdot,\cdot)$. Then $\mathcal{D}$ is a complete metric space (see, e.g., [66]).

The following result follows from Theorem 2.1. and assumptions on input data of problem (P_q).

Lemma 3.1. *For any triple* $\pi \equiv (u, v, w) \in \mathcal{D}$ *there exists a unique solution* $z[\pi] \in \mathfrak{C}^1_{2,0}(Q_T|S^0_T)$ *to the initial–boundary value problem (3.1), and there exists a constant* $c_0 > 0$ *such that*

$$\|z[\pi]\|_{\mathfrak{C}^1_2(Q_T)} \leq c_0 \;\; \forall \pi \in \mathcal{D}. \tag{3.9}$$

This constant is independent of a triple $\pi \in \mathcal{D}$. *Moreover, for any two triples* $\pi^i \equiv (u^i, v^i, w^i) \in \mathcal{D}$, $i = 1, 2$, *the following inequality is fulfilled*

$$\|z[\pi^1] - z[\pi^2]\|_{\mathfrak{C}^1_2(Q_T)} \leq c_1\Big[\|v^1 - v^2\|_{\infty,\Omega} + \|w^1 - w^2\|^{(0,1)}_{2,1,S^1_T} + \tag{3.10}$$
$$+ \int_0^T K_0(t)\sqrt{\mathrm{meas}\,\mathcal{R}_t(u^1,u^2)}dt + \int_0^T \|K_1(\cdot,t)\|_{2,\mathcal{R}_t(u^1,u^2)}dt\Big],$$

where $c_1 > 0$ *is a constant that is independent of triples* π^1, $\pi^2 \in \mathcal{D}$.

The following result follows from lemma 3.1. and from embedding theorem for the space $W^1_2(\Omega)$.

Lemma 3.2. *Suppose sequences of triples* $\pi^{i,k} \equiv (u^{i,k}, v^{i,k}, w^{i,k}) \in \mathcal{D}$, $k = 1, 2$, $i = 1, 2, \ldots$, *are such that*

$$\lim_{i \to \infty} d(\pi^{i,1}, \pi^{i,2}) = 0;$$

then

$$\lim_{i \to \infty} \|z[\pi^{i,1}] - z[\pi^{i,2}]\|_{\mathfrak{C}^1_2(Q_T)} = 0,$$
$$\lim_{i \to \infty} \max_{t \in [0,T]} \|z[\pi^{i,1}](\cdot,t) - z[\pi^{i,2}](\cdot,t)\|_{2,S} = 0.$$

Finally, from lemmas 3.1.–3.2. the following result is concluded.

Lemma 3.3. *The functional* $I_0 : \mathcal{D} \to R$ *and the operator* $I_1 : \mathcal{D} \to C(X)$ *are uniformly continuous and uniformly bounded on the complete metric space* $\mathcal{D}$. *Moreover, the set* $\{I_1(\pi) : \pi \in \mathcal{D}\}$ *is a precompact set in the space* $C(X)$.

Let us formulate facts concerning the stability of solutions of maximum principle's adjoint equations under perturbations of controls. These facts will be used to prove main results. First of all, after investigation of special questions of the theory of linear hyperbolic equations with Radon measures in right–hand side parts, let us define Pontryagin maximum principle's adjoint functions of $\mathfrak{p}_0[\pi^1,\pi^2]$ and $\mathfrak{p}_1[\pi^1,\pi^2]$, $\pi^i \equiv (u^i,v^i,w^i) \in \mathcal{D}$, $i = 1,2$, by

$$\mathfrak{p}_0[\pi^1,\pi^2](x,t) \equiv \mathfrak{p}[\mathfrak{a}[\pi^1,\pi^2],\mathfrak{g}_0[\pi^1,\pi^2]](x,t,T),$$
$$\mathfrak{p}_1[\pi^1,\pi^2](x,t,\tau) \equiv \mathfrak{p}[\mathfrak{a}[\pi^1,\pi^2],\mathfrak{g}_1[\pi^1,\pi^2]](x,t,\tau),\quad (x,t)\in Q_T,\quad \tau\in[0,T],$$

where

$$\mathfrak{a}[\pi^1,\pi^2](x,t) \equiv \int_0^1 \nabla_z a(x,t,z_2(x,t)+\gamma\Delta z(x,t),u^1(x,t))d\gamma,$$
$$\mathfrak{g}_0[\pi^1,\pi^2](x) \equiv -\int_0^1 \nabla_z G(x,z_2(x,T)+\gamma\Delta z(x,T))d\gamma,$$
$$\mathfrak{g}_1[\pi^1,\pi^2](x,\tau) \equiv -\int_0^1 \nabla_z \Phi(x,z_2(x,\tau)+\gamma\Delta z(x,\tau),v^2(x)+\gamma\Delta v(x))d\gamma.$$

Here $z_i \equiv z[\pi^i]$, $i=1,2$, $\Delta z \equiv z_1 - z_2$, $\Delta v \equiv v^1 - v^2$. For brevity, set $\mathfrak{g}_0[\pi,\pi] \equiv \mathfrak{g}_0[\pi]$, $\mathfrak{g}_1[\pi,\pi] \equiv \mathfrak{g}_1[\pi]$, $\mathfrak{p}_0[\pi,\pi] \equiv \mathfrak{p}_0[\pi]$, $\mathfrak{p}_1[\pi,\pi] \equiv \mathfrak{p}_1[\pi]$, $\mathfrak{a}[\pi,\pi] \equiv \mathfrak{a}[\pi]$. Let us remember that $\mathfrak{p}[a,g](x,t,\tau) \equiv \mathfrak{f}[a,g,\delta_\tau](x,t)$, $(x,t)\in Q_T$, $\tau\in[0,T]$, where $\mathfrak{f}[a,g,\delta_\tau]$ is a solution to the problem (2.22) for $\mu = \delta_\tau$.

From assumptions on input data of problem (P_q), and from Theorem 2.3., it follows that the following lemma holds.

Lemma 3.4. *If sequences* $\pi^{i,k} \in \mathcal{D}$, $k = \overline{1,4}$, $i = 1,2,\ldots$, *are such that*

$$\lim_{i\to\infty} d(\pi^{i,1},\pi^{i,3}) = 0,\quad \lim_{i\to\infty} d(\pi^{i,2},\pi^{i,4}) = 0,$$

then

$$\lim_{i\to\infty} \|\mathfrak{p}_0[\pi^{i,1},\pi^{i,2}] - \mathfrak{p}_0[\pi^{i,3},\pi^{i,4}]\|_{\mathfrak{C}_2^1(Q_T)} = 0,$$
$$\lim_{i\to\infty}\sup_{\tau\in[0,T]} \|\mathfrak{p}_1[\pi^{i,1},\pi^{i,2}](\cdot,\cdot,\tau) - \mathfrak{p}_1[\pi^{i,3},\pi^{i,4}](\cdot,\cdot,\tau)\|_{\mathfrak{C}_2^1(Q_T)} = 0.$$

3.4. A Calculation of First Variations of Functionals

In the sequel, in a definition and a calculation of first variations of functionals, a notion of an iterative limit will play the main role. Besides, corresponding generalizations of a classic notion of Lebesgue point will also play the important role.

Definition 3.3. *[57, 58, 59] Let* $G \subset R^n$ *be an open set. A point* $x \in G$ *is said to be an* (l,m)*–Lebesgue point of a summable function* $f : G \to R^1$, $1 \le l \le m \le n$, *if* $f(x) \ne \infty$ *and*

$$\lim_{h\to 0}\frac{1}{(2h)^{m-l+1}}\int_{x_l-h}^{x_l+h}\cdots\int_{x_m-h}^{x_m+h}|f(x_1,\ldots,x_{l-1},y_1,\ldots,y_{m-l+1},x_{m+1},\ldots,x_n)-$$
$$-f(x)|dy_1\ldots dy_{m-l+1} = 0.$$

It is easy to see that an $(1, n)$–Lebesgue point is a Lebesgue point in an usual sense [60].

Lemma 3.5. *[52], [57]-[59] For any fixed l, m, $1 \le l \le m \le n$, almost all points of an open set G are (l, m)–Lebesgue points of a summable function $f\colon G \to R^1$.*

On the base of these notions and results, we calculate the first variations of functionals $I_0(\cdot)$ and $I_1(\cdot)(\tau)$ $\forall \tau \in [0, T]$. Let $\pi \equiv (u, v, w) \in \mathcal{D}$ be arbitrary. Let us construct a collection of variation parameters $\mathfrak{m} \equiv (\{(x^i, t^i),\ \gamma^{i,r},\ u^{i,r},$ $i = \overline{1, i_1},\ r = \overline{1, r_0(i)}\}, \tilde{v}, \tilde{w})$, where $(x^i, t^i) \in Q_T$, $\gamma^{i,r} \ge 0$, $i = \overline{1, i_1}$, $r = \overline{1, r_0(i)}$, $\sum_{i=1}^{i_1} \sum_{r=1}^{r_0(i)} \gamma^{i,r} \le 1$, $\tilde{v} \in \mathcal{D}_2$, $\tilde{w} \in \mathcal{D}_3$, $u^{i,r} \in U^*$, $i = \overline{1, i_1}$, $r = \overline{1, r_0(i)}$; $U^* \subseteq U$ is a countable set, U^* is everywhere dense in U . Let us denote the set of all such collections $\mathfrak{m}$ by $\mathfrak{M}$.

A triple $\pi_\varepsilon \equiv (u_\varepsilon, v_\varepsilon, w_\varepsilon) \in \mathcal{D}$ is called a variation of triple $\pi \equiv (u, v, w) \in \mathcal{D}$, where $\varepsilon \equiv (\varepsilon_1, \varepsilon_2)$, $\varepsilon_1, \varepsilon_2 \geqslant 0$, $0 \le \varepsilon_1, \varepsilon_2 \le \varepsilon_0 < 1$, if

$$u_\varepsilon(x, t) \equiv \begin{cases} u^{i,r}, & (x, t) \in Q^\varepsilon_{i,r},\ i = \overline{1, i_1},\ r = \overline{1, r_0(i)}; \\ u(x, t), & (x, t) \in Q_T \setminus \bigcup\limits_{i=1}^{i_1} \bigcup\limits_{r=1}^{r_0(i)} Q^\varepsilon_{i,r}; \end{cases}$$

$$v_\varepsilon(x) \equiv v(x) + \varepsilon_1^n \varepsilon_2 (\tilde{v}(x) - v(x)) \equiv v(x) + \varepsilon_1^n \varepsilon_2 \delta v(x), \quad x \in \Omega;$$
$$w_\varepsilon(s, t) \equiv w(s, t) + \varepsilon_1^n \varepsilon_2 (\tilde{w}(s, t) - w(s, t)) \equiv w(s, t) + \varepsilon_1^n \varepsilon_2 \delta w(s, t),$$
$$(s, t) \in S^1_T;$$

where $Q^\varepsilon_{i,r} \equiv Q^{\varepsilon_1,\varepsilon_2}_{i,r} \equiv Q^{\varepsilon_1}_{i,r,1} \times Q^{\varepsilon_2}_{i,r,2}$, $Q^{\varepsilon_1}_{i,r,1} \equiv \prod_{\alpha=1}^{n} (x^i_\alpha - \varepsilon_1, x^i_\alpha - \varepsilon_1(r-1)]$, $Q^{\varepsilon_2}_{i,r,2} \equiv (t^i - \varepsilon_2 \sum_{\alpha=1}^{r} \gamma^{i,\alpha}, t^i - \varepsilon_2 \sum_{\alpha=1}^{r-1} \gamma^{i,\alpha}]$. Here $\varepsilon_0 > 0$ is a small enough number depending on $\gamma^{i,r}$ and (x^i, t^i), $i = \overline{1, i_1}$, $r = \overline{1, r_0(i)}$, such that sets $Q^{\varepsilon_0}_i \equiv Q^{\varepsilon_0}_{i,1} \times Q^{\varepsilon_0}_{i,2} \equiv \prod_{\alpha=1}^{n} [x^i_\alpha - \varepsilon_0 r_0(i), x^i_\alpha] \times [t^i - \varepsilon_0 \sum_{\alpha=1}^{r_0(i)} \gamma^{i,\alpha}, t^i]$, $i = \overline{1, i_0(i)}$, do not pairwise intersect.

Let us formulate the following obvious result.

Lemma 3.6. *There exists a constant* $L \equiv 1 + \operatorname{meas} V + \operatorname{diam} \mathbf{W}$ *such that for any* $\pi \equiv (u, v, w) \in \mathcal{D}$

$$d(\pi, \pi_\varepsilon) \le L \varepsilon_1^n \varepsilon_2.$$

Let us prove that the following lemma holds.

Lemma 3.7. *1) For any triple $\pi \equiv (u, v, w) \in \mathcal{D}$ there exists a subset $Q_0[\pi] \subseteq Q_T$ such that* $\operatorname{meas} Q_0[\pi] = \operatorname{meas} Q_T$ *and for all $\mathfrak{m} \equiv (\{(x^i, t^i),\ \gamma^{i,r},\ u^{i,r},\ i = \overline{1, i_1},\ r = \overline{1, r_0(i)}\}, \tilde{v}, \tilde{w}) \in \mathfrak{M}$, $(x^i, t^i) \in Q_0[\pi]$, $i = \overline{1, i_1}$, there exists a variation*

$$\delta I_0(\pi; \mathfrak{m}) \equiv \lim_{\varepsilon_1 \to +0} \frac{1}{\varepsilon_1^n} \lim_{\varepsilon_2 \to +0} \frac{I_0(\pi_{\varepsilon_1,\varepsilon_2}) - I_0(\pi)}{\varepsilon_2}.$$

In addition, the following representation holds

$$\delta I_0(\pi;\mathfrak{m}) \equiv \int\limits_{\Omega} [\mathfrak{p}_0[\pi](x,0) + \nabla_v G(x, z[\pi](x,T), v)]\delta v dx + \int\limits_{S_T^1} \mathfrak{p}_0[\pi]\delta w ds dt -$$

$$-\sum_{i=1}^{i_1}\sum_{r=1}^{r_0(i)} \gamma^{i,r}\mathfrak{p}_0[\pi](x^i,t^i)\Delta_u a(x^i,t^i,z[\pi](x^i,t^i);u^{i,r},u(x^i,t^i)).$$

2) For any point $\tau \in [0,T]$ and any triple $\pi \equiv (u,v,w) \in \mathcal{D}$ there exists a subset $Q_1[\pi,\tau] \subseteq Q_T$ such that $\mathrm{meas}\, Q_1[\pi,\tau] = \mathrm{meas}\, Q_T$ *and for all collections* $\mathfrak{m} \equiv (\{(x^i,t^i),\ \gamma^{i,r},\ u^{i,r},\ i=\overline{1,i_1},\ r=\overline{1,r_0(i)}\},\tilde v,\tilde w) \in \mathfrak{M}$, $(x^i,t^i) \in Q_1[\pi,\tau]$, $i=\overline{1,i_1}$, *there exists a variation*

$$\delta I_1(\pi,\tau;\mathfrak{m}) \equiv \lim_{\varepsilon_1\to+0}\frac{1}{\varepsilon_1^n}\lim_{\varepsilon_2\to+0}\frac{I_1(\pi_{\varepsilon_1,\varepsilon_2})(\tau)-I_1(\pi)(\tau)}{\varepsilon_2}.$$

This variation can be represented in the form

$$\delta I_1(\pi,\tau;\mathfrak{m}) \equiv \int\limits_{\Omega} [\mathfrak{p}_1[\pi](x,0,\tau) + \nabla_v \Phi(x,\tau,z[\pi](x,\tau),v)]\delta v dx +$$

$$+\int\limits_{S_T^1} \mathfrak{p}_1[\pi](s,t,\tau)\delta w ds dt - \sum_{i=1}^{i_1}\sum_{r=1}^{r_0(i)} \gamma^{i,r}\mathfrak{p}_1[\pi](x^i,t^i,\tau)\Delta_u a(x^i,t^i,$$

$$z[\pi](x^i,t^i);u^{i,r},u(x^i,t^i)).$$

Proof. Let us prove only second assertion of the lemma, because a calculation of the iterative limit $\delta I_0(\pi;\mathfrak{m})$ is completely analogous.

Let a triple $\pi \equiv (u,v,w) \in \mathcal{D}$ be fixed. Linearizing the equation (3.1), and setting $z_\varepsilon \equiv z[\pi_\varepsilon]$, $z \equiv z[\pi]$, $\Delta_\varepsilon z \equiv z_\varepsilon - z$, $\Delta_\varepsilon v \equiv v_\varepsilon - v$, $\Delta_\varepsilon w \equiv w_\varepsilon - w$, we obtain that

$$\Delta_\varepsilon z_{tt} - \frac{\partial}{\partial x_i}(a_{ij}\Delta_\varepsilon z_{x_j} + a_i \Delta_\varepsilon z) + b_i \Delta_\varepsilon z_{x_j} +$$
$$+\mathfrak{a}[\pi_\varepsilon,\pi]\Delta_\varepsilon z = -\Delta_u a(x,t,z;u_\varepsilon,u);$$
$$\Delta_\varepsilon z|_{t=0} = 0,\ \ \Delta_\varepsilon z_t|_{t=0} = \Delta_\varepsilon v, x\in\Omega;\ \ \left[\frac{\partial(\Delta_\varepsilon z)}{\partial \mathcal{N}} + \sigma \Delta_\varepsilon z\right]\Bigg|_{S_T^1} = \Delta_\varepsilon w,$$
$$\Delta_\varepsilon z|_{S_T^0} = 0.$$

Linearizing the increment $\Delta_\varepsilon I_1(\tau) \equiv I_1(\pi_\varepsilon)(\tau) - I_1(\pi)(\tau)$ of the functional $I_1(\cdot)(\tau)$, we have

$$\Delta_\varepsilon I_1(\tau) = \int\limits_{\Omega}\left[\int\limits_0^1 \nabla_v \Phi(x,\tau,z(x,\tau)+y\Delta_\varepsilon z(x,\tau), v + y\Delta_\varepsilon v)dy\right]\Delta_\varepsilon v dx -$$

$$-\int\limits_{\Omega} \mathfrak{g}_1[\pi_\varepsilon,\pi](x,\tau)\Delta_\varepsilon z(x,\tau)dx.$$

Applying lemma 2.4. to the last equality, we obtain

$$\Delta_\varepsilon I_1(\tau) = \left\{ \int\limits_\Omega \left[\mathfrak{p}_1[\pi_\varepsilon,\pi](x,0,\tau) + \int\limits_0^1 \nabla_v \Phi(x,\tau,z(x,\tau) + y\Delta_\varepsilon z(x,\tau), v + \right.\right.$$

$$\left. \left. + y\Delta_\varepsilon v)dy \right] \Delta_\varepsilon v dx + \int\limits_{S_T^1} \mathfrak{p}_1[\pi_\varepsilon,\pi](s,t,\tau)\Delta_\varepsilon w ds dt \right\} +$$

$$+ \left\{ - \int\limits_{Q_T} \Delta_\varepsilon \mathfrak{p}_1(x,t,\tau)\Delta_u a(x,t,z;u_\varepsilon,u) dx dt \right\} +$$

$$+ \left\{ - \int\limits_{Q_T} \mathfrak{p}_1[\pi](x,t,\tau)\Delta_u a(x,t,z;u_\varepsilon,u) dx dt \right\} \equiv$$

$$\equiv \{\Delta_\varepsilon I_1^{(1)}(\tau)\} + \{\Delta_\varepsilon I_1^{(2)}(\tau)\} + \{\Delta_\varepsilon I_1^{(3)}(\tau)\}.$$

Here $\Delta_\varepsilon \mathfrak{p}_1 \equiv \mathfrak{p}_1[\pi_\varepsilon,\pi] - \mathfrak{p}_1[\pi]$.

Using assumptions on an integrand Φ, lemma 3.2., lemma 3.4., and the definition of a variation π_ε of controls, we conclude

$$\lim_{\varepsilon_1\to+0} \frac{1}{\varepsilon_1^n} \lim_{\varepsilon_2\to+0} \frac{\Delta_\varepsilon I_1^{(1)}(\tau)}{\varepsilon_2} = \int\limits_{S_T^1} \mathfrak{p}_1[\pi](s,t,\tau)\delta w ds dt + \tag{3.11}$$

$$+ \int\limits_\Omega [\nabla_v \Phi(x,\tau,z(x,\tau),v) + \mathfrak{p}_1[\pi](x,0,\tau)]\delta v dx.$$

Let

$$\mathbf{b}_{i,r}(x,t) \equiv \begin{cases} -\Delta_u a(x,t,z(x,t);u^{i,r},u(x,t)),\ (x,t)\in Q_{i,r}^{\varepsilon_0}; \\ 0,\ \ (x,t)\in \mathbb{R}^{n+1}\setminus Q_{i,r}^{\varepsilon_0}; \end{cases}$$

$$\mathbf{B}_{i,r}(t) \equiv \|\mathbf{b}_{i,r}(\cdot,t)\|_{2,\mathbb{R}^n}.$$

It is not hard to see that $\mathbf{B}_{i,r} \in L_1(\mathbb{R})$. Hence,

$$\mathbf{b}_{i,r}(\cdot,t) \in L_2(\mathbb{R}^n) \text{ for a.e. } t\in[0,T].$$

Let us introduce a maximal function

$$(M\mathbf{B}_{i,r})(t) \equiv \sup_{\delta>0} \frac{1}{2\delta} \int\limits_{t-\delta}^{t+\delta} |\mathbf{B}_{i,r}(\kappa)| d\kappa.$$

By virtue of the classic maximal function theorem (see, e.g., proposition a) of Theorem 1 on p.15 in [60]),

$$(M\mathbf{B}_{i,r})(t) \ \text{ is finite for a.e. } t\in\mathbb{R}^1,\ \ i=\overline{1,i_1},\ \ r=\overline{1,r_0(i)}. \tag{3.12}$$

Let us require that the following assumptions on points (x^i, t^i) are fulfilled:

1) points (x^i, t^i), $i = \overline{1, i_1}$, are $(1, n+1)$, $(1, n)$–Lebesgue points ($t \equiv x_{n+1}$) (see Definition 3.3.) of all functions

$$-\mathfrak{p}_1[\pi](\cdot,\cdot,\tau)\Delta_u a(\cdot,\cdot,z(\cdot,\cdot);u',u(\cdot,\cdot)),\ \ u' \in U^*; \tag{3.13}$$

2) almost all points of sections $Q_T^{t^i} \equiv \{(x,t) \in Q_T : (x,t) = (x,t^i)\}$, $i = \overline{1, i_1}$, are $(n+1, n+1)$–Lebesgue points of all functions (3.13);

3) the following conditions holds:

$$(M\mathbf{B}_{i,r})(t^i) \text{ is finite}, \ \ i = \overline{1, i_1}, \ \ r = \overline{1, r_0(i)}.$$

Due to Lemma 3.5. and conditions (3.12), such selection of points (x^i, t^i) is possible, and, additionally, the Lebesgue measure of all such points set $Q_1[\pi, \tau]$ coincides with the Lebesgue measure of the cylinder Q_T.

Let us calculate similar limit for the summand $\Delta_\varepsilon I_1^{(2)}(\tau)$:

$$\left|\frac{\Delta_\varepsilon^{(2)} I(\tau)}{\varepsilon_2}\right| \le \sum_{i=1}^{i_1}\sum_{r=1}^{r_0(i)} \int\limits_{Q_{i,r,1}^{\varepsilon_1}} \left[\frac{1}{\varepsilon_2}\int\limits_{Q_{i,r,2}^{\varepsilon_2}} |\Delta_\varepsilon \mathfrak{p}_1(x,t,\tau)||\mathbf{b}_{i,r}(x,t)|dt\right]dx =$$

$$= \frac{1}{\varepsilon_2}\sum_{i=1}^{i_1}\sum_{r=1}^{r_0(i)} \int\limits_{Q_{i,r,2}^{\varepsilon_2}} dt \int\limits_{Q_{i,r,1}^{\varepsilon_1}} |\Delta_\varepsilon \mathfrak{p}_1(x,t,\tau)||\mathbf{b}_{i,r}(x,t)|dx \le$$

$$\le \frac{1}{\varepsilon_2}\sum_{i=1}^{i_1}\sum_{r=1}^{r_0(i)} \int\limits_{Q_{i,r,2}^{\varepsilon_2}} \|\Delta_\varepsilon \mathfrak{p}_1(\cdot,t,\tau)\|_{2,Q_{i,1}^{\varepsilon_0}} \|\mathbf{b}_{i,r}(\cdot,t)\|_{2,Q_{i,1}^{\varepsilon_0}} dt \le$$

$$\le \sup_{\xi\in[0,T]} \|\Delta_\varepsilon \mathfrak{p}_1(\cdot,\cdot,\xi)\|_{\mathfrak{C}_2^1(Q_T)} \sum_{i=1}^{i_1}\sum_{r=1}^{r_0(i)} \frac{1}{\varepsilon_2}\int\limits_{Q_{i,r,2}^{\varepsilon_2}} \|\mathbf{b}_{i,r}(\cdot,t)\|_{2,Q_{i,1}^{\varepsilon_0}} dt \le$$

$$\le \sup_{\xi\in[0,T]} \|\Delta_\varepsilon \mathfrak{p}_1(\cdot,\cdot,\xi)\|_{\mathfrak{C}_2^1(Q_T)} \sum_{i=1}^{i_1}\sum_{r=1}^{r_0(i)} \frac{2\delta^{i,r}}{2\varepsilon_2\delta^{i,r}} \int\limits_{t^i-\varepsilon_2\delta^{i,r}}^{t^i+\varepsilon_2\delta^{i,r}} \|\mathbf{b}_{i,r}(\cdot,t)\|_{2,Q_{i,1}^{\varepsilon_0}} dt \le$$

$$\le \sup_{\xi\in[0,T]} \|\Delta_\varepsilon \mathfrak{p}_1(\cdot,\cdot,\xi)\|_{\mathfrak{C}_2^1(Q_T)} \sum_{i=1}^{i_1}\sum_{r=1}^{r_0(i)} 2\delta^{i,r}(M\mathbf{B}_{i,r})(t^i),$$

where $\delta^{i,r} = \sum\limits_{l=1}^{r} \gamma^{i,l}$.

Hence, in view of Lemma 3.4., the definition of a variation π_ε, and conditions on points (x^i, t^i),

$$\lim_{\varepsilon_1\to+0}\frac{1}{\varepsilon_1^n}\lim_{\varepsilon_2\to+0}\frac{\Delta_\varepsilon I_1^{(2)}(\tau)}{\varepsilon_2} = 0. \tag{3.14}$$

Using the notation introduced before, let us rewrite an expression for $\Delta_\varepsilon I_1^{(3)}(\tau)$:

$$\frac{1}{\varepsilon_2}\Delta_\varepsilon I_2^{(3)}(\tau) \equiv \sum_{i=1}^{i_1}\sum_{r=1}^{r_0(i)} \int_{Q_{i,r,1}^{\varepsilon_1}} \left[\frac{1}{\varepsilon_2}\int_{Q_{i,r,2}^{\varepsilon_2}} \mathbf{b}_{i,r}(x,t)\mathfrak{p}_1[\pi](x,t,\tau)dt\right] dx.$$

Using a maximal function definition, the classic Lebesgue dominated convergence theorem, and the (l,m)–Lebesgue points definition, we obtain that

$$\lim_{\varepsilon_1\to+0}\frac{1}{\varepsilon_1^n}\lim_{\varepsilon_2\to+0}\frac{1}{\varepsilon_2}\Delta_\varepsilon I_1^{(3)}(\tau) = -\sum_{i=1}^{i_1}\sum_{r=1}^{r_0(i)}\gamma^{i,r}\mathfrak{p}_1[\pi](x^i,t^i,\tau)\Delta_u a(x^i,t^i,$$
$$z[\pi](x^i,t^i);u^{i,r},u(x^i,t^i)).$$

Combining this equality, (3.11) and (3.14), we get second assertion of this lemma. This completes the proof of the lemma.

3.5. Proof of the Main Optimization Results

In this section, we prove the optimization main results essentially related to the presence of a parameter in problem (P_q).

Proof of Theorem 3.1. Suppose $\pi^k \equiv (u^k, v^k, w^k)$, $k = 1,2,\ldots$, is a m.a.s. in problem (P_q). Consider the problem

$$J(\pi) \to \inf,\quad \pi \in \mathcal{D}, \tag{3.15}$$

where $J(\pi) \equiv \max\{I_0(\pi) - \beta(q),\ I_1(\pi)(\tau) - q(\tau),\ \tau \in X\}$. Obviously, the sequence π^k, $k = 1,2,\ldots$, is also a minimizing sequence in problem (3.15), and the value of problem (3.15) is equal to zero. Let $\hat{X}^k \equiv \{\tau^{k,j} : j = 1,\ldots,l_k\} \subset X$ be a $1/k$-net in X, $\hat{X}^k \subseteq \hat{X}^{k+1}$, $k = 1,2,\ldots$. Consider the family of auxiliary problems $J_k(\pi) \to \inf$, $\pi \in \mathcal{D}$, where $J_k(\pi) \equiv \max\{I_0(\pi) - \beta(q);\ I_1(\pi)(\tau) - q(\tau),\ \tau \in \hat{X}^k\}$. By virtue of Lemma 3.3., a functional $J_k(\cdot)$ is continuous and bounded on $\mathcal{D}$. Using a precompactness of the family $\{I_1(\pi) : \pi \in \mathcal{D}\} \subset C(X)$ in the space $C(X)$ (see Lemma 3.3.), it is not hard to show that $\lim\limits_{k\to\infty}\inf\limits_{\pi\in\mathcal{D}} J_k(\pi) = \inf\limits_{\pi\in\mathcal{D}} J(\pi) = \lim\limits_{k\to\infty} J(\pi^k) = 0$. It follows that there exists a sequence $\varkappa^k \geq 0$, $k = 1,2,\ldots$, $\varkappa^k \to 0$, $k \to \infty$, such that $J_k(\pi^k) \leq \inf\limits_{\pi\in\mathcal{D}} J_k(\pi) + \varkappa^k$. Hence, according to the Ekeland variational principle [66], there exists a sequence $\bar{\pi}^k \equiv (\bar{u}^k, \bar{v}^k, \bar{w}^k) \in \mathcal{D}$, $k = 1,2,\ldots$, such that for any $k = 1,2,\ldots$, $\bar{\pi}^k$ is a solution to the problem

$$J_k(\pi) + \sqrt{\varkappa^k}d(\bar{\pi}^k,\pi) \to \min,\quad \pi \in \mathcal{D}, \tag{3.16}$$

and

$$d(\pi^k,\bar{\pi}^k) \leq \sqrt{\varkappa^k},\quad J_k(\bar{\pi}^k) \leq J_k(\pi^k). \tag{3.17}$$

Suppose $\pi \in \mathcal{D}$, $I \in R$. By definition, put

$$\hat{I}_j(\pi) = \begin{cases} I_0(\pi),\ j = 0; \\ I_1(\pi)(\tau^{k,j}),\ j = \overline{1,l_k}; \end{cases} \qquad F_j(I) = \begin{cases} I - \beta(q),\ j = 0; \\ I - q(\tau^{k,j}),\ j = \overline{1,l_k}. \end{cases}$$

Then, obviously, $J_k(\pi) \equiv \max_{s=\overline{0,l_k}} F_j(\hat{I}_j(\pi))$. Let $\Gamma_k \equiv \{j = \overline{0,l_k} : J_k(\bar{\pi}^k) = F_j(\hat{I}_j(\bar{\pi}^k))\}$. Then

$$J_k(\bar{\pi}^k) = F_j(\hat{I}_j(\bar{\pi}^k)),\ \ j \in \Gamma_k;\ \ J_k(\bar{\pi}^k) > F_j(\hat{I}_j(\bar{\pi}^k)),\ \ j \notin \Gamma_k.$$

Let $\bar{k}$ be a number of elements in Γ_k. Introduce first variations vector

$$\delta\hat{I}(\bar{\pi}^k;\mathfrak{m}) \equiv (\delta\hat{I}_0(\bar{\pi}^k;\mathfrak{m}), \delta\hat{I}_1(\bar{\pi}^k;\mathfrak{m}), \ldots, \delta\hat{I}_{l_k}(\bar{\pi}^k;\mathfrak{m})) \in R^{l_k+1}.$$

By $\mathcal{K}(\bar{\pi}^k)$ denote the set of all first variations vectors. Using standard methods of optimal control, it can be shown that $\mathcal{K}(\bar{\pi}^k)$ is convex. Let us project $\mathcal{K}(\bar{\pi}^k)$ on the subspace of R^{l_k+1} spanned vectors e_j, $j \in \Gamma_k$ (e_j, $j = \overline{0,l_k}$, is the standard basis of R^{l_k+1}), and let us denote the projection by $\mathcal{K}_{\bar{k}}(\bar{\pi}^k)$. Consider the set $\mathcal{K}_{\bar{k}}^- \equiv \{\sum_{j\in\Gamma_k} x_j e_j : x_j \le -2L\sqrt{\varkappa^k},\ j \in \Gamma_k\}$, where the constant $L \equiv 1 + \operatorname{meas} V + 2A$ is defined in Lemma 3.6.. Let us show that the following lemma holds.

Lemma 3.8. $\mathcal{K}_{\bar{k}}^- \cap \mathcal{K}_{\bar{k}}(\bar{\pi}^k) = \emptyset$.

Proof Assume the converse. Then there exists $\mathfrak{m} \in \mathfrak{M}$ such that $\delta\hat{I}_j(\bar{\pi}^k;\mathfrak{m}) \le -2L\sqrt{\varkappa^k}$, $j \in \Gamma_k$. In view of first variations' form of functionals in Lemma 3.7., we get $F_j(\hat{I}_j(\bar{\pi}^k_\varepsilon)) - F_j(\hat{I}_j(\bar{\pi}^k)) = \hat{I}_j(\bar{\pi}^k_\varepsilon) - \hat{I}_j(\bar{\pi}^k) = \varepsilon_1^n\varepsilon_2\delta\hat{I}_j(\bar{\pi}^k;\mathfrak{m}) + \varepsilon_1^n\varepsilon_2\omega_1(\varepsilon_1) + \varepsilon_2\omega_2(\varepsilon_1,\varepsilon_2)$, where ω_1 and ω_2 are such that $\lim_{\varepsilon_2\to 0}\omega_2(\varepsilon_1,\varepsilon_2) = \lim_{\varepsilon_1\to 0}\omega_1(\varepsilon_1) = 0$. Hence,

$$\begin{aligned} F_j(\hat{I}_j(\bar{\pi}^k_\varepsilon)) - F_j(\hat{I}_j(\bar{\pi}^k)) &= \hat{I}_j(\bar{\pi}^k_\varepsilon) - \hat{I}_j(\bar{\pi}^k) = \varepsilon_1^n\varepsilon_2\delta\hat{I}_j(\bar{\pi}^k;\mathfrak{m}) + \\ +\varepsilon_1^n\varepsilon_2\omega_1(\varepsilon_1) + \varepsilon_2\omega_2(\varepsilon_1,\varepsilon_2) &\le -\frac{19}{10}L\varepsilon_1^n\varepsilon_2\sqrt{\varkappa^k} + \varepsilon_1^n\varepsilon_2\omega_1(\varepsilon_1) + \\ +\left[-\frac{19}{10}L\varepsilon_1^n\sqrt{\varkappa^k} + \omega_2(\varepsilon_1,\varepsilon_2)\right]\varepsilon_2 &< -\frac{19}{10}L\varepsilon_1^n\varepsilon_2\sqrt{\varkappa^k} + \varepsilon_1\varepsilon_2^n\omega_1(\varepsilon_1) = \\ = -\frac{9}{5}L\varepsilon_1^n\varepsilon_2\sqrt{\varkappa^k} + \left[-\frac{1}{10}L\sqrt{\varkappa^k} + \omega_1(\varepsilon_1)\right]\varepsilon_1^n\varepsilon_2 &< -\frac{9}{5}L\varepsilon_1^n\varepsilon_2\sqrt{\varkappa^k} \end{aligned}$$

for all enough small ε. Therefore, $F_j(\hat{I}_j(\bar{\pi}^k_\varepsilon)) \le F_j(\hat{I}_j(\bar{\pi}^k)) - \frac{9}{5}L\varepsilon_1^n\varepsilon_2\sqrt{\varkappa^k}$, $j \in \Gamma_k$. According to Lemma 3.6., it follows that

$$\begin{aligned} F_j(\hat{I}_j(\bar{\pi}^k_\varepsilon)) + \sqrt{\varkappa^k}d(\bar{\pi}^k_\varepsilon, \pi^k) &\le F_j(\hat{I}_j(\bar{\pi}^k)) - \frac{9}{5}L\varepsilon_1^n\varepsilon_2\sqrt{\varkappa^k} + L\varepsilon_1^n\varepsilon_2\sqrt{\varkappa^k} + \\ +\sqrt{\varkappa^k}d(\bar{\pi}^k, \pi^k) &= J_k(\bar{\pi}^k) + \sqrt{\varkappa^k}d(\bar{\pi}^k, \pi^k) - \frac{4}{5}L\varepsilon_1^n\varepsilon_2\sqrt{\varkappa^k}. \end{aligned}$$

Thus, $J_k(\bar{\pi}^k_\varepsilon) + \sqrt{\varkappa^k}d(\bar{\pi}^k_\varepsilon, \pi^k) < J_k(\bar{\pi}^k) + \sqrt{\varkappa^k}d(\bar{\pi}^k, \pi^k)$. But this contradicts to an optimality of $\bar{\pi}^k$ in problem (3.16). The lemma is proved.

Since $\mathcal{K}_{\bar{k}}^- \cap \mathcal{K}_{\bar{k}}(\bar{\pi}^k) = \emptyset$, these sets are separable; i.e., there exists a vector $\lambda^{\bar{k}} \in R^{\bar{k}}$, $\lambda_j^{\bar{k}} \geqslant 0$, $j \in \Gamma_k$, $\sum_{j\in\Gamma_k}\lambda_j^{\bar{k}} = 1$, such that

$$\sum_{j\in\Gamma_k}\lambda_j^{\bar{k}}\delta\hat{I}_j(\bar{\pi}^k;\mathfrak{m}) \geqslant \sum_{j\in\Gamma_k}\lambda_j^{\bar{k}}x_j \ \ \forall\,\mathfrak{m} \in \mathfrak{M}\ \forall\, x = \sum_{j\in\Gamma_k} x_j e_j \in \mathcal{K}_{\bar{k}}^-.$$

Putting $x_j = -2L\sqrt{\varkappa^k}$, $j \in \Gamma_k$, in the last inequality, and completing vectors $\lambda^{\bar{k}} \in R^{\bar{k}}$ to vectors $\lambda^k \in R^{l_k+1}$ by zeros, we conclude that

$$\lambda_j^k \geqslant 0,\ j = 0, \dots, l_k;\ \lambda_0^k + \sum_{j=1}^{l_k} \lambda_j^k = 1; \tag{3.18}$$

$$\lambda_j^k (J_k(\bar{\pi}^k) - (I_1(\bar{\pi}^k)(\tau^{k,j}) - q(\tau^{k,j}))) = 0,\ j = 1, \dots, l_k; \tag{3.19}$$

$$\lambda_0^k \delta I_0(\bar{\pi}^k; \mathfrak{m}) + \sum_{j=1}^{l_k} \delta I_1(\bar{\pi}^k, \tau^{k,j}; \mathfrak{m}) \geqslant -2L\sqrt{\varkappa^k},\quad \forall \mathfrak{m} \in \mathfrak{M}. \tag{3.20}$$

In view of first variations expressions (see Lemma 3.7.), from the last inequality it follows that

$$H(x,t,z[\bar{\pi}^k](x,t),u,\lambda_0^k \mathfrak{p}_0[\bar{\pi}^k](x,t) + \sum_{j=1}^{l_k} \lambda_j^k \mathfrak{p}_1[\bar{\pi}^k](x,t,\tau^{k,j})) - \tag{3.21}$$

$$-H(x,t,z[\bar{\pi}^k](x,t),\bar{u}^k(x,t),\lambda_0^k \mathfrak{p}_0[\bar{\pi}^k](x,t) + \sum_{j=1}^{l_k} \lambda_j^k \mathfrak{p}_1[\bar{\pi}^k](x,t,\tau^{k,j})) \leq$$

$$\leq 2L\sqrt{\varkappa^k}\ \ \forall u \in U \text{ for a.e. } (x,\, t) \in Q_T;$$

$$\int_\Omega \left[\lambda_0^k \mathfrak{p}_0[\bar{\pi}^k](x,0) + \sum_{j=1}^{l_k} \lambda_j^k \mathfrak{p}_1[\bar{\pi}^k](x,0,\tau^{k,j})\right] (\bar{v}^k(x) - \tilde{v}(x))\, dx + \tag{3.22}$$

$$+\sum_{j=1}^{l_k} \lambda_j^k \int_\Omega \nabla_v \Phi(x,\tau^{k,j}, z[\pi^k](x,\tau^{k,j}), v^k)(v^k - \tilde{v})dx \leq 2\sqrt{\varkappa^k}\ \forall \tilde{v} \in \mathcal{D}_2;$$

$$\int_{S_T^1} [\lambda_0^k \mathfrak{p}_0[\bar{\pi}^k](s,t) + \sum_{j=1}^{l_k} \lambda_j^k \mathfrak{p}_1[\bar{\pi}^k](s,t,\tau^{k,j})](\bar{w}^k - \tilde{w})\, dsdt \leq 2\sqrt{\varkappa^k} \tag{3.23}$$

$$\forall \tilde{w} \in \mathcal{D}_3.$$

By definition, put $\mu^k \equiv \sum_{j=1}^{l_k} \lambda_j^k \delta_{\tau^{k,j}}$, where δ_τ is a Radon δ–measure concentrated in the point τ. Then $\sum_{j=1}^{l_k} \lambda_j^k \mathfrak{p}_1[\bar{\pi}^k](x,t,\tau^{k,j}) = \int_X \mathfrak{p}_1[\bar{\pi}^k](x,t,\tau)\mu^k(d\tau)$. According to Theorem 2.3.,

$$\int_X \mathfrak{p}_1[\bar{\pi}^k](x,t,\tau)\mu^k(d\tau) \equiv \mathfrak{f}[\mathfrak{a}[\bar{\pi}^k], \mathfrak{g}_1[\bar{\pi}^k], \mu^k](x,t).$$

Hence, relations (3.18), (3.19), (3.21)–(3.23) can be rewritten in the form

$$\lambda_0^k \geqslant 0,\ \ \lambda_0^k + \|\mu^k\| = 1; \tag{3.24}$$

$$\|\mu^k\| J_k(\bar{\pi}^k) - \int_X [I_1(\bar{\pi}^k)(\tau) - q(\tau)]\mu^k(d\tau) = 0; \tag{3.25}$$

$$H(x,t,z[\bar{\pi}^k](x,\,t),u,\eta[\bar{\pi}^k,\lambda_0^k,\mu^k](x,t)) - H(x,t,z[\bar{\pi}^k](x,\,t), \tag{3.26}$$
$$\bar{u}^k(x,t),\eta[\bar{\pi}^k,\lambda_0^k,\mu^k](x,t)) \le 2L\sqrt{\varkappa^k}\ \ \forall\, u \in U \text{ for a.e. } (x,\,t) \in Q_T;$$

$$\int_X \mu^k(d\tau) \int_\Omega \nabla_v \Phi(x,\tau,z[\bar{\pi}^k](x,\tau),\bar{v}^k(x))(\bar{v}^k(x) - \tilde{v}(x))dx + \tag{3.27}$$
$$+ \int_\Omega \eta[\bar{\pi}^k,\lambda_0^k,\mu^k](x,0)(\bar{v}^k(x) - \tilde{v}(x))dx \le 2L\sqrt{\varkappa^k}\ \ \forall\,\tilde{v} \in \mathcal{D}_2;$$

$$\int_{S_T^1} \eta[\bar{\pi}^k,\lambda_0^k,\mu^k](s,t)(\bar{w}^k - \tilde{w})\,dsdt \le 2L\sqrt{\varkappa^k}\ \ \forall\,\tilde{w} \in \mathcal{D}_3. \tag{3.28}$$

Using assumptions on source data of problem (P_q), Lemma 3.2., Lemma 3.4., the first inequality in (3.17), from relations (3.26), (3.27), and (3.28) we obtain relations (3.5), (3.6), and (3.7) respectively. From the second inequality in (3.17), relation (3.25), and an uniform continuity of I_1 on $\mathcal{D}$ (see Lemma 3.3.), it follows that the nondegenerate condition holds, and the measure μ^k is concentrated on the set X_k. This completes the proof of Theorem 3.1..

Let us approximate the original problem with pointwise state constraints by problems with finitely many functional constraints. We denote a $1/k$-net in X by $\hat{X}^k \equiv \{\tau^{k,j} : j = 1,\dots,l_k\} \subset X$, $\hat{X}^k \subseteq \hat{X}^{k+1}$, $k = 1,2,\dots$ Consider a sequence of families of optimization problems depending on the vector parameter $q^k \equiv (q_1^k,\dots,q_{l_k}^k) \in R^{l_k}$ and approximating the original family (P_q):

$$I_0(\pi) \to \inf,\ \ I^k(\pi) \in \mathcal{M}^k + q^k,\ \ \pi \in \mathcal{D}, \qquad q^k \in R^{l_k} \text{ is a parameter, } (P^k_{q^k})$$

where $\mathcal{M}^k \equiv \{y \in R^{l_k} : y_i \le 0,\ i = 1,\dots,l_k\}$, $I^k(\pi) \equiv (I_1^k(\pi),\dots,I_{l_k}^k(\pi))$, $I_i^k(\pi) \equiv I_1(\pi)(\tau^{k,i})$, and $\tau^{k,\,i} \in \hat{X}^k$. Just as in problem (P_q), in problem $(P^k_{q^k})$ the value function $\beta_k : R^{l_k} \to R \cup \{+\infty\}$ is defined by

$$\beta_k(q^k) \equiv \lim_{\varepsilon \to +0} \beta_{k,\,\varepsilon}(q^k),$$
$$\beta_{k,\,\varepsilon}(q^k) \equiv \{\inf_{\pi \in \mathcal{D}^{k,\,\varepsilon}_{q^k}} I_0(\pi),\ \mathcal{D}^{k,\,\varepsilon}_{q^k} \ne \emptyset;\ +\infty,\ \mathcal{D}^{k,\,\varepsilon}_{q^k} = \emptyset\},$$

where $\mathcal{D}^{k,\,\varepsilon}_{q^k} \equiv \{\pi \in \mathcal{D} : I_j^k(\pi) \le q_j^k + \varepsilon,\ j = 1,\dots,l_k\}$. We have the following approximation lemma, which is similar to the corresponding lemmas in [40, 50, 51] and follows from the precompactness of the image of the operator I_1 in $C(X)$ (see Lemma 1.4 in [42].

Lemma 3.9. *Let $\beta(q) < +\infty$ and $q \in C(X)$. Then the sequence of vectors $\bar{q}^k \in R^{l_k}$, $\bar{q}^k \equiv (\bar{q}_1^k,\dots,\bar{q}_{l_k}^k)$, $\bar{q}_i^k = q(\tau^{k,\,i})$, $i = 1,\dots,l_k$, $k = 1,2,\dots$, satisfies the limit relation $\beta_k(\bar{q}^k) \to \beta(q)$, $k \to \infty$.*

In the sequel, we need the following definition and results in [63, 67, 69, 70] (see also [64]). Suppose that $A \subset R^s$ is a nonempty closed set, $\varepsilon \geqslant 0$, and $x \in A$. The nonempty set

$$\hat{N}(x;\, A) \equiv \left\{ x^* \in R^s \,:\, \limsup_{u \xrightarrow{A} x} \frac{\langle x^*, u - x\rangle}{|u - x|} \leq 0 \right\}$$

is called the cone of Fréchet normals to the set A at the point x. We define the Mordukhovich basic/limiting normal cone at a point $\hat{x} \in A$ by the formula $N(\hat{x};\, A) \equiv \limsup_{x \xrightarrow{A} \hat{x}} \hat{N}(x;\, A)$ (see [63, 67, 69, 70]).

For a lower semicontinuous function $f \,:\, R^s \to R \cup \{+\infty\}$ and for $\bar{x} \in \mathrm{dom} f$, the Fréchet subdifferential $\hat{\partial} f(\bar{x})$ of the function f at the point $\bar{x} \in \mathrm{dom}\, f$ is given by the relation

$$\hat{\partial} f(\bar{x}) \equiv \left\{ x^* \in R^s \,:\, \liminf_{x \to \bar{x}} \frac{f(x) - f(\bar{x}) - \langle x^*,\, x - \bar{x}\rangle}{|x - \bar{x}|} \geqslant 0 \right\},$$

or, which is equivalent, by the relation

$$\hat{\partial} f(\bar{x}) \equiv \left\{ x^* \in R^s \,:\, (x^*, -1) \in \hat{N}((\bar{x},\, f(\bar{x})); \mathrm{epi} f) \right\}.$$

For any $\bar{x} \in \mathrm{dom} f$, the sets

$$\partial f(\bar{x}) \equiv \{x^* \in R^s \,:\, (x^*, -1) \in N((\bar{x},\, f(\bar{x})); \mathrm{epi} f)\},$$
$$\partial^\infty f(\bar{x}) \equiv \{x^* \in R^s \,:\, (x^*, 0) \in N((\bar{x},\, f(\bar{x})); \mathrm{epi} f)\}$$

are referred to as the subdifferential and the singular subdifferential, respectively, of the function f at the point $\bar{x}$ in the sense of [67, 68]. If f is a lower semicontinuous function, then

$$\partial f(\bar{x}) = \limsup_{x \xrightarrow{f} \bar{x}} \hat{\partial} f(x), \quad \partial^\infty f(\bar{x}) = \limsup_{x \xrightarrow{f} \bar{x};\, \bar{\varepsilon} \downarrow 0} \bar{\varepsilon} \hat{\partial} f(x), \tag{3.29}$$

where $x \xrightarrow{f} \bar{x}$ means that $x \to \bar{x}$, $f(x) \to f(\bar{x})$. We have $\partial^\infty f(\bar{x}) = \{0\}$ if the function f satisfies the Lipschitz condition in a neighborhood of the point x. The following assertion holds ([63, 70], see also [64]).

Lemma 3.10. *Let $A \subset R^s$ be a nonempty closed set. Then the set $\{x \in A \,:\, \hat{N}(x; A) \neq \{0\}\}$ that is, the set of all boundary points of A at which there exists a nonzero Fréchet normal is everywhere dense in the set of all boundary points of A. In addition, for any lower semicontinuous function $f \,:\, R^s \to R \cup \{\pm\infty\}$ the set $\{x \in \mathrm{dom} f \,:\, \hat{\partial} f(x) \neq \emptyset\}$ is everywhere dense in* $\mathrm{dom} f$.

The definition of a Fréchet subdifferential of a function f at a point x directly implies the following assertion.

Lemma 3.11. *Let $f \,:\, R^s \to R \cup \{\pm\infty\}$ be a lower semicontinuous function, and let $x \in \mathrm{dom} f$. If $(x^*, -\eta) \in \hat{N}((x, f(x)); epi\, f)$, $\eta > 0$, then for each $\varepsilon > 0$ there exists a neighborhood S_ε of x such that $\eta f(x') - \eta f(x) - \langle x^*, x' - x\rangle + \varepsilon |x' - x| \geqslant 0 \;\forall x' \in S_\varepsilon$.*

The following lemma on the relationship of Lagrange multipliers and Fréchet normals is hold.

Lemma 3.12. *Let $\beta_k(q^k) < \infty$ and let $(\zeta^k, -\varkappa^k) \in N((q^k, \beta_k(q^k)); \mathrm{epi}\beta_k)$ be an arbitrary vector. Then there exist a sequence of nonnegative numbers $\gamma^i \to 0$, $i \to \infty$, a sequence of controls $\pi^i \in \mathcal{D}^{k,\gamma^i}_{q^k}$, $i = 1, 2, \ldots$, and a bounded sequence of Lagrange multipliers $\lambda^i \equiv (\lambda^i_0, \lambda^i_1, \ldots, \lambda^i_{l_k}) \in R^{l_k}$, $i = 1, 2, \ldots$, such that*

$$|\lambda^i| \neq 0, \ \ \lambda^i_j \geqslant 0, \ \ j = \overline{0, l_k}; \ \ \lambda^i_j(I^k_j(\pi^i) - q^k_j) \geqslant -\gamma^i, \ \ j = \overline{1, l_k}, \tag{3.30}$$

inequalities (3.5)–(3.7) hold for $\pi = \pi^i$, $\alpha = \lambda^i_0$, $\mu = \mu^i \equiv \sum_{j=1}^{l_k} \lambda^i_j \delta_{\tau^{k,j}}$, $\gamma = \gamma^i$, and the relation

$$\zeta^k + \sum_{j=1}^{l_k} \lambda_j e^j = 0, \tag{3.31}$$

holds, where $\lambda \equiv (\varkappa^k, \lambda_1, \ldots, \lambda_{l_k}) \neq 0$ is some limit point of the sequence of vectors λ^i, $i = 1, 2, \ldots$; δ_τ is the Radon δ-measure concentrated at a point $\tau \in X$; and

$$e^j \equiv (\underbrace{0, \ldots, 0}_{j-1}, 1, 0, \ldots, 0) \in R^{l_k}.$$

Proof. The proof is similar to that of lemma 6.4 in [43], so it is omitted.

Let us give the following natural definitions.

Definition 3.4. *A sequence of controls $\pi^i \in \mathcal{D}$, $i = 1, 2, \ldots$, is said to be **stationary** in problem $(P^k_{q^k})$ if there exists a bounded sequence of vectors $\lambda^i \equiv (\lambda^i_0, \lambda^i_1, \ldots, \lambda^i_{l_k}) \in R^{l_k+1}$, $i = 1, 2, \ldots$, and a sequence of nonnegative numbers γ^i, $\gamma^i \to 0$, such that $\pi^i \in \mathcal{D}^{k,\gamma^i}_{q^k}$, relations (3.30) are satisfied, inequalities (3.5)–(3.7) hold for $\pi = \pi^i$, $\alpha = \lambda^i_0$, $\mu = \mu^i \equiv \sum_{j=1}^{l_k} \lambda^i_j \delta_{\tau^{k,j}}$, and $\gamma = \gamma^i$, and all limit points of the sequence λ^i, $i = 1, 2, \ldots$, are nonzero.*

Definition 3.5. *A stationary sequence $\pi^i \in \mathcal{D}$, $i = 1, 2, \ldots$, in problem $(P^k_{q^k})$ is said to be **normal** if the first components of all limit points of each corresponding sequence of vectors λ^i, $i = 1, 2, \ldots$, are nonzero. Problem $(P^k_{q^k})$ is said to be **normal** if all of its stationary sequences are normal.*

*A stationary sequence $\pi^i \in \mathcal{D}$, $i = 1, 2, \ldots$, in problem $(P^k_{q^k})$ is said to be **regular** if the first components of all limit points of some corresponding sequence of vectors λ^i, $i = 1, 2, \ldots$, are nonzero. Problem $(P^k_{q^k})$ is said to be **regular** if there exist regular stationary sequences in it.*

We introduce sets of multipliers $L^{k,\nu}_{q^k} \equiv \{-\sum_{j=1}^{l_k} \lambda_j e^j \in R^{l_k} : \lambda = (\lambda_0, \lambda_1, \ldots, \lambda_{l_k}) \in R^{l_k+1}, \ \lambda \neq 0, \ \lambda_0 = \nu$, in problem $(P^k_{q^k})$, there exists a stationary sequence for which the corresponding, in accordance with the definition of a stationary sequence, sequence of vectors λ^i, $i = 1, 2, \ldots$, has the limit point being the vector $\lambda\}$, $\nu = 0, 1$; $M^{k,0}_{q^k} \equiv L^{k,0}_{q^k} \cup \{0\}$, and $M^{k,1}_{q^k} \equiv L^{k,1}_{q^k}$.

The following assertion is a straightforward consequence of Lemma 3.12. and the definition of generalized subdifferentials [63, 64, 70].

Theorem 3.5. *Let $\beta_k(q^k) < +\infty$. Then $\partial\beta_k(q^k) \subseteq M^{k,1}_{q^k}$ and $\partial^\infty\beta_k(q^k) \subseteq M^{k,0}_{q^k}$, where $\partial\beta_k$ and $\partial^\infty\beta_k$ are the ordinary and singular generalized subdifferentials, respectively, in the sense of [67, 68].*

From Theorem 3.52 in [63] (see also Corollary 8.5 in [64]) and Theorem 3.5., we obtain the following important result in which the finite dimension of the space R^{l_k} plays an important role.

Lemma 3.13. *If in some neighborhood O_{q^k} of the point q^k all problems $(P^k_{y^k})$, $y^k \in O_{q^k}$, are normal, i.e., $M^{k,0}_{y^k} = \{0\}$, $y^k \in O_{q^k}$, and moreover, the sets $M^{k,1}_{y^k}$ are uniformly bounded by a constant K in some norm $\|\cdot\|$ (for example, the Euclidean norm $|\cdot|$), then the value function β_k satisfies the Lipschitz condition in the norm dual to $\|\cdot\|$ on O_{q^k} with the same constant K.*

In addition, the lower semicontinuity of the value function of the approximating problem, its monotonicity with respect to each of l_k arguments, the results in [71], and Lemma 3.12. imply the following assertion.

Theorem 3.6. *If the condition $\hat{\partial}\beta_k(q^k) \neq \emptyset$ is satisfied in problem $(P^k_{q^k})$, then problem $(P^k_{q^k})$ is regular. The set of all such points $q^k \in R^{l_k}$ has full measure in* $\mathrm{dom}\beta_k$.

Proof of Theorem 3.2. The proof of given theorem similarly to proof of theorem 5.2 in [43], so it is omitted.

Proof of Theorem 3.3. Let us state the main idea of the proof. We consider the family of problems $(\bar{P}_\rho) \equiv (P_{q+\rho\tilde{q}})$ depending on a real parameter ρ, where $\tilde{q} \equiv 1$. Since the function β satisfies the Lipschitz condition in a neighborhood of the point q, it follows that the function $\bar{\beta}(\rho) \equiv \beta(q + \rho\tilde{q})$ of one variable satisfies the Lipschitz condition in a neighborhood of zero. Therefore, by virtue of the first formula in (3.29), there exist sequences ρ^i, ζ^i, and $\varkappa^i$, $i = 1, 2, \ldots$, of real numbers such that

$$\rho^i \to 0, \ \ \bar{\beta}(\rho^i) \to \bar{\beta}(0), \ \ (\zeta^i, -\varkappa^i) \to (\zeta, -\varkappa) \neq 0, \ \ i \to \infty, \ \ \varkappa > 0,$$
$$(\zeta^i, -\varkappa^i) \in N((\rho^i, \bar{\beta}(\rho^i)); \mathrm{epi}\bar{\beta}).$$

By using the same considerations as in the statement of the auxiliary problem (6.5) in [43], we find that any m.a.s. $\pi^{i,k} \in \mathcal{D}$, $k = 1, 2, \ldots$, in the sense of (3.2) in problem $(\bar{P}_{\rho^i})$ is a m.a.s. (together with ρ^i) in the problem

$$\varkappa^i I_0(\pi) - \zeta^i\rho' + \omega|\rho' - \rho^i| \to \inf, \ \ I_1(\pi) \in \mathcal{M} + q + \rho'\tilde{q}, \qquad (3.32)$$
$$\rho' \in (-\omega, \omega), \ \ \pi \in \mathcal{D},$$

as well, where $\omega > 0$ is large enough to ensure that $\rho^i \in (-\omega, \omega)$, $i = 1, 2, \ldots$. Then, by writing out necessary conditions for m.a.s. in problem (3.32) and by passing to the limit as $i \to \infty$, we obtain the m.a.s. $\pi^s \in \mathcal{D}$, $s = 1, 2, \ldots$, in (P_q) satisfying all relations of Theorem 3.1.; moreover, $\lambda_0^s \to \varkappa > 0$, $s \to \infty$, which implies the regularity of problem (P_q).

Proof of Theorem 3.4. Indeed, one can readily see that the function $\tilde{\beta}(t) \equiv \beta(q + t\xi)$ is monotone nonincreasing on the ray $t \geqslant 0$. Consequently, by the classical result of the theory of functions of a real variable, $\tilde{\beta}$ is differentiable in the classical sense almost everywhere for $t \geqslant 0$. Therefore, $\hat{\partial}\tilde{\beta}(t) \neq \emptyset$ for almost all $t \geqslant 0$, which implies that, for almost all $t \geqslant 0$, there exists a nonzero Fréchet normal $(\zeta, -\varkappa) \in N((t, \tilde{\beta}(t)); \mathrm{epi}\tilde{\beta})$, $\varkappa > 0$. By the definition of a basic normal cone, there exist sequences t^i, ζ^i, and $\varkappa^i$, $i = 1, 2, \ldots$, of real numbers such that

$$t^i \to t,\ \ \tilde{\beta}(t^i) \to \tilde{\beta}(t),\ \ (\zeta^i, -\varkappa^i) \to (\zeta, -\varkappa) \neq 0,\ \ i \to \infty,\ \ \varkappa > 0,$$
$$(\zeta^i, -\varkappa^i) \in N((t^i, \tilde{\beta}(t^i)); \mathrm{epi}\bar{\beta}).$$

Just as in the proof of Theorem 3.3., hence we find that any m.a.s. $\pi^{i,k} \in \mathcal{D}$, $k = 1, 2, \ldots$, in the sense of inequality 3.2 in problem $(P_{q+t^i\xi})$ is a m.a.s. (together with t^i) in the problem

$$\varkappa^i I_0(\pi) - \zeta^i t' + \omega|t' - t^i| \to \inf,\ \ I_1(\pi) \in \mathcal{M} + q + \rho'\tilde{q},\qquad (3.33)$$
$$t' \in (-\omega + t, \omega + t),\ \ \pi \in \mathcal{D},$$

as well, where $\omega > 0$ is large enough to ensure that $t^i \in (-\omega + t, \omega + t)$, $i = 1, 2, \ldots$. Next, by writing out necessary conditions for a m.a.s. in problem (3.33) and by passing to the limit as $i \to \infty$, we obtain the m.a.s. $\pi^s \in \mathcal{D}$, $s = 1, 2, \ldots$, in problem $(P_{q+t\xi})$ satisfying all relations of Theorem 3.1.; moreover, $\lambda_0^s \to \varkappa > 0$, $s \to \infty$, which implies the regularity of problem $(P_{q+t\xi})$ for almost all $t \geqslant 0$.

References

[1] Il'in V.A., Kuleshov A.A. On some properties of generalized solutions of the wave equation of the classes L_p and W_p^1 with $p \geq 1$, *Diff. Uravn.;* 2012, vol.48, No.11, pp.1493-1500 [in Russian].

[2] Il'in V.A., Kuleshov A.A. A necessary and sufficient condition for membership in the class L_p for $p \geq 1$ generalized solution of the mixed problem for the wave equation, *Diff. Uravn.;* 2012, vol.48, No.12, pp.1607-1611 [in Russian].

[3] Del Santo D., Mitidieri E. Destruction of solutions to a hyperbolic system: the critical case, *Diff. Uravn.;* 1998, vol.34, No.9, pp.1155-1161 [in Russian].

[4] Karachalios N., Stavrakakis N. Asymptotic behavior of solutions of some nonlinearly damped equations on $\mathbb{R}^N$, *Topological Methods in Nonlinear Analysis*; 2001, vol.18, pp.73-87.

[5] Ramm A.G. On the behavior of solutions of the boundary value problem for a hyperbolic equation for $t \to \infty$, *Izv. vuzov. Matematika;* 1966, No.1, pp.124-138 [in Russian].

[6] Kolesov A.Yu., Rozov N.H. Multifrequency parametric resonance in nonlinear wave equations, *Math. Russian Academy of Sciences. Ser. mathematical;* 2002, vol.66, No.6, pp.49-64 [in Russian].

[7] Bales L., Lasiecka I. Negative norm estimates for fully discrete finite element approximations to the wave equation with nonhomogeneous L_2 Dirichlet boundary data, *Mathematics of computation;* 1995, vol.64, No.209, pp.89-115.

[8] Lasiecka I., Sokolowski J. Regularity and strong convergence of a variational approximation to a nonhomogeneous Dirichlet hyperbolic boundary problem, *SIAM J. Math. Anal.;* 1988, vol.19, pp.528-540.

[9] Bukesova N.N., Zhelezovsky S.E. On the rate of convergence of the Galerkin method for a class of quasilinear operator differential equations, *Comp. Math. Math. Phys.;* 1999, vol.39, No.9, pp.1519-1531 [in Russian].

[10] Vorovich I.I. On some direct methods in the nonlinear theory of oscillations of shallow shells, *Izvestiya. USSR Academy of Sciences. Ser. Mathematical;* 1957, vol.21, pp.747-784 [in Russian].

[11] Zhelezovsky S.E. Bubnov-Galerkin method for abstract quasilinear problem of stationary action, *Diff. Uravn.;* 1995, vol.31, No.7, pp.1222-1231 [in Russian].

[12] Zhelezovsky S.E. Estimates of the rate of convergence of the Galerkin method for abstract hyperbolic equation, *Mat. notes;* 2001, vol.69, Issue 2, pp.223-234 [in Russian].

[13] Zhelezovsky S.E. Estimates of the rate of convergence of the projection-difference method for hyperbolic equations, *Izv. vuzov. Matematika;* 2002, No.1, pp.21-30 [in Russian].

[14] Zhelezovsky S.E. On error estimates in the Galerkin method for hyperbolic equations, *Siberian Math. Zh.;* 2005, vol.46, No.2, pp.374-389 [in Russian].

[15] Zhelezovsky S.E. On the justification of the Galerkin method for hyperbolic equations, *Diff. Uravn.;* 2007, vol.43, No.3, pp.402-410 [in Russian].

[16] Zhelezovsky S.E. Study of convergence of projection-difference method for hyperbolic equations, *Siberian Math. Zh.;* 2007, vol.48, No.1, pp.93-102 [in Russian].

[17] Zhelezovsky S.E., Bukesova N.N. Error estimates for the projection method for an abstract quasilinear hyperbolic equations, *Izv. vuzov. Matematika;* 1999, No.5, pp.94-96 [in Russian].

[18] Zhelezovsky S.E., Lyashko A.D. Error Estimates for the Galerkin method for quasilinear hyperbolic equations, *Diff. Uravn.;* 2001, vol.37, No7, pp.941-949 [in Russian].

[19] Ishmukhametov A.Z. On the approximation of hyperbolic operator-differential equations of the second order, *Comp. Math. Math. Phys.;* 1987, vol.27, No.8, pp.1154-1165 [in Russian].

[20] Lasiecka I., Lions J.-L., Triggiani R. Nonhomogeneous boundary value problems for second-order hyperbolic operators, *J. Mat. Pures Appl.;* 1986, vol.65, No.2, pp.149-192.

[21] Lasiecka I., Triggiani R. Sharp regularity theory for second order hyperbolic equations of Neumann type, I: L_2 nonhomogeneous data, *Ann. Mat. Pura Appl.;* 1990, vol.157, pp.285-367.

[22] Lasiecka I., Triggiani R. Regularity theory of hyperbolic equations with non-homogeneous Neumann boundary conditions, II: General boundary data, *J. Diff. Eq.;* 1991, vol.94, pp.112-164.

[23] Ladyzhenskaya O.A. On the solution of non-stationary operator equations, *Mat. Sat.;* 1956, vol.39, No.4, pp.491-524 [in Russian].

[24] Ladyzhenskaya O.A. *Boundary–Value Problems of Mathematical Physics*; Nauka, Moscow, 1973. [in Russian]

[25] Lions J.-L. *Quelques méthodes de résolution des problèmes aux limites non linéaire*; Dunod, Gauthier-Villars, Paris, 1969.

[26] Lions J.-L., Magenes E. *Problemes aux limites non homogènes et applications*; volume 1, Dunod, Paris, 1968.

[27] Lomovtsev F.E. Hyperbolic second order differential equations with discontinuous coefficients, *Diff. Uravn.;* 1997, vol.33, No.10, pp.1394-1403 [in Russian].

[28] Nikitin A.A. On a mixed problem for the wave equation with the third and the first boundary conditions, *Diff. Uravn.;* 2007, vol.43, No.12, pp.1692-1699 [in Russian].

[29] Yakubov S.Ja. Uniformly Cauchy problem for abstract hyperbolic equations, *Izv. vuzov. Matematika;* 1970, No.12, pp.108-113 [in Russian].

[30] Gavrilov V.S., Sumin M.I. Perturbation Method in the Theory of Pontryagin Maximum Principle for Optimal Control of Divergent Semilinear Hyperbolic Equations with Pointwise State Constraints, *In book "Control Theory and its Applications"*, New-York: Nova Science Publishers Inc.,2011, Chapter 4, 83-144.

[31] Mackenroth U. Convex Parabolic Boundary Control Problems with Pointwise State Constraints, *J. Math. Anal. Appl.;* 1982, vol.87, pp.256-277.

[32] Mackenroth U. On Some Elliptic Optimal Control Problems with State Constraints, *Optimization;* 1986, vol.17, pp.595-607.

[33] Bergounioux M. A Penalization Method for Optimal Control of Elliptic Problems with State Constraints, *SIAM J. Control Optim.;* 1992, vol.30, No.2, pp.305-323.

[34] Casas E. Boundary Control of Semilinear Elliptic Equations with Pointwise State Constraints, *SIAM J. Control Optim.;* 1993, vol.31, No.4, pp.993-1006.

[35] Bonnans J.F., Casas E. An Extension of Pontryagin's Principle for State-Constrained Optimal Control of Semilinear Elliptic Equations and Variational Inequalities, *SIAM J. Control Optim.;* 1995, vol.33, No.1, pp.274-298.

[36] Raymond J.-P., Zidani H. Pontryagin's Principle for State-Constrained Control Problems Governed by Parabolic Equations with Unbounded Controls, *SIAM J. Control Optim.;* 1998, vol.36, No.6, pp.1853-1879.

[37] Sumin M.I. Suboptimal Control of Distributed Parameters Systems: Minimizing Sequences and the Value Function, *Computational Mathematics and Mathematical Physics*, 1997, vol.37, No.1, pp.21-39.

[38] Sumin M.I. Optimal Control of Semilinear Elliptic Equation with State Constraint: Maximum Principle for Minimizing Sequence, Regularity, Normality, Sensitivity, *Control and Cybernetics;* 2000, vol.29, No.2, pp.449-472.

[39] Sumin M.I. Suboptimal Control of Semilinear Elliptic Equations with Phase Constraints, I: The Maximum Principle for Minimizing Sequences and Normality, *Russian Math. (Izv. vuzov);* 2000, vol.44, No.6, pp.31-42.

[40] Sumin M.I. Suboptimal Control of Semilinear Elliptic Equations with Phase Constraints, II: Sensitivity, Genericity of the Regular Maximum Principle, *Russian Math. (Izv. vuzov);* 2000, vol.44, No.8, pp.50-60.

[41] Sumin M.I. Suboptimal Control of a Semilinear Elliptic Equation with a Phase Constraint and a Boundary Control, *Differential Equations;* 2001, vol.37, No.2, pp.281-300.

[42] Gavrilov V.S., Sumin M.I. Parametric Optimization for a Hyperbolic Equation in Divergence Form with a Pointwise State Constraint: I, *Differential Equations;* 2011, vol.47, No.4, pp.547-559.

[43] Gavrilov V.S., Sumin M.I. Parametric Optimization for a Hyperbolic Equation in Divergence Form with a Pointwise State Constraint: II, *Differential Equations;* 2011, vol.47, 5, 726-737.

[44] Mordukhovich B.S., Raymond J.-P. Dirichlet Boundary Control of Hyperbolic Equations in the Presence of State Constraints, *Appl. Math. Optim.;* 2004, vol.49, pp.145-157.

[45] Mordukhovich B.S., Raymond J.-P. Neumann Boundary Control of Hyperbolic Equations with Pointwise State Constraints, *SIAM J. Control Optim.;*, 2005, vol.43, No.4, pp.1354-1372.

[46] Gavrilov V.S., Sumin M.I. Sequential optimization for semilinear divergent hyperbolic equation with a boundary control and state inequality constraint, *Control and Cybernetics;* 2014, vol.43, No.2. pp.183-226.

[47] Novozhenov M.M., Plotnikov V.I. Generalized Lagrange Multiplyers Rule for Distributed Systems with State Constraints, *Diff. Uravn.*, 1982, vol.18, No.4. pp.584-692 [in Russian].

[48] Li X., Yong J. *Optimal Control Theory for Infinite Dimensional Systems;* Birkhäuser Verlag, Basel, 1995.

[49] Casas E., Raymond J.-P., Zidani H. Pontryagin's Principle for Local Solutions of Control Problems with Mixed Control-State Constraints, *SIAM J. Control Optim.;* 2000, vol.39, No.4, pp.1182-1208.

[50] Gavrilov V.S., Sumin M.I. Parametric Optimization of Nonlinear Goursat-Darboux Systems with State Constraints, *Computational Mathematics and Mathematical Physics;* 2004, vol.44, No.6, pp.949-968.

[51] Gavrilov V.S., Sumin M.I. A Parametric Problem of the Suboptimal Control of the Goursat–Darboux System with a Pointwise Phase Constraint, *Russian Math. (Izv. vuzov);* 2005, vol.49, No.6, pp.37-48.

[52] Sumin M.I. The First Variation and Pontryagin's Maximum Principle in Optimal Control for Partial Differential Equations, *Computational Mathematics and Mathematical Physics;* 2009, vol.49, No.6, pp.958-978.

[53] Warga J. *Optimal Control of Differential and Functional Equations;* Academic Press, New York, 1972.

[54] Sumin M.I., Trushina, E.V. Minimizing sequences in optimal control with approximately given input data and the regularizing properties of the Pontryagin maximum principle, *Computational Mathematics and Mathematical Physics;* 2008, vol.48, No.2, pp.209-224.

[55] Sumin M.I. Regularized parametric Kuhn–Tucker theorem in a Hilbert space, *Computational Mathematics and Mathematical Physics;* 2011, vol.51, No.9, pp.1489-1509.

[56] Sumin M.I. On the Stable Sequential Kuhn–Tucker Theorem and its Applications, *Applied Mathematics;* 2012, vol.3, Issue 10A (Special issue "Optimization"), pp.1334-1350.

[57] Sumin M.I. Optimal Control Problems for Lumped and Distributed Systems with Differentiable and Nondifferentiable Functionals and Functions Defining Systems, Candidate's Dissertation in Mathematics and Physics. Gor'kov. Gos. Univ., Gor'kii, 1983. [in Russian]

[58] Sumin M.I. On First Variation in Optimal Control Theory for Distributed Parameters Systems, *Diff. Uravn.;* 1991, vol.27, No.12. pp.2179-2181 [in Russian].

[59] Sumin M.I. Mathematical Theory of Suboptimal Control for Distributed Systems, Doctoral Dissertation in Mathematics and Physics, Nizhegorodskii Gos. Univ., Nizhnii Novgorod, 2000. [in Russian].

[60] Stane E.M. *Singular Integrals and Differentiability Properties of Functions;* Princeton Univ. Press, Princeton, 1970.

[61] Sumin M.I. On Minimizing Sequences in Optimal Control Problems with Constrained State Coordinates, *Diff. Uravn.;* 1986, vol.22, No.10, pp.1719-1731 [in Russian].

[62] Clarke F. *Optimization and Nonsmooth Analysis;* Wiley, New York, 1983.

[63] Mordukhovich B.S. *Variational Analysis and Generalized Differentiation, I: Basic Theory;* Springer, Berlin., 2006

[64] Mordukhovich B.S., Shao Y. Nonsmooth sequential analysis in Asplund spaces, *Trans. Amer. Math. Soc.;* 1996, vol.346, No.4, pp.1235-1280.

[65] Osipov Yu.S., Vasil'ev F.P., Potapov M.M. *Fundamentals of the Dynamic Regularization Method;*, Mosk. Gos. Univ., Moscow, 1999 [in Russian].

[66] Ekeland I. On the Variational Principle, *J. Math. Anal. Appl.*, 1974, vol.47, pp.324–353.

[67] Mordukhovich B.S. Maximum principle in problems of time optimal control with state constraints, *J. Applied Math. Mech.;* 1976, vol.40, pp.960-969.

[68] Mordukhovich B.S. Metric approximation and necessary optimality conditions for general classes of extremal problems, *Soviet. Math. Dokl.;* 1980, vol.22, pp.526-530.

[69] Mordukhovich B.S. *Approximation Methods in Problems of Optimization and Control;* Nauka, Moscow, 1988 [in Russian].

[70] Mordukhovich B.S. *Variational Analysis and Generalized Differentiation, II: Applications.* Springer, Berlin, 2006.

[71] Ward A.L. Differentiability of Vector Monotone Functions, *Proc. London Math. Soc.;* 1935, vol.32, No.2, pp.339-362.

INDEX

A

B

C

D

E

F

G

H

I

L

M

N

O

P

Q

R

S

T

U

V

W

Z